Accuracy

in

Molecular Processes

ITS CONTROL AND RELEVANCE TO LIVING SYSTEMS

Accuracy

in

Molecular Processes

ITS CONTROL AND RELEVANCE TO LIVING SYSTEMS

Edited by

T. B. L. KIRKWOOD AND R. F. ROSENBERGER

National Institute for Medical Research
London, UK

and

D. J. GALAS

University of Southern California
Los Angeles, USA

London   New York
CHAPMAN AND HALL

First published in 1986 by
Chapman and Hall Ltd
11 New Fetter Lane, London EC4P 4EE
Published in the USA by
Chapman and Hall
29 West 35th Street, New York, NY 10001

© 1986 Chapman and Hall

Printed in Great Britain at the University Press, Cambridge

ISBN 0 412 26940 6

British Library Cataloguing in Publication Data

Accuracy in molecular processes : its control and
relevance to living systems.
1. Molecular biology
I. Kirkwood, T. B. L.   II. Rosenberger, R. F.
III. Galas, D. J.
574.8'8   QH506

ISBN 0-412-26940-6

Library of Congress Cataloging in Publication Data

Accuracy in molecular processes.
Includes bibliographies and index.
1. Molecular biology – Miscellanea.   2. Molecular
biology. I. Kirkwood, T. B. L. II. Rosenberger, R. F.
III. Galas, D. J.
QH506.A23 1986   574.87'328   85-25505
ISBN 0-412-26940-6

Contents

Contributors

Dr R. P. Anderson
Department of Biology
University of Iowa
Iowa City
Iowa 52242
USA

Dr C. Blomberg
Department of Theoretical Physics
Royal Institute of Technology
S-100 44
Stockholm
Sweden

Dr E. W. Branscomb
Lawrence Livermore National
  Laboratory
Biomedical Sciences Division
University of California
PO Box 5507 L-452
Livermore
California 94550
USA

Dr R. H. Buckingham
Institut de Biologie Physico-chimique
13, rue Pierre et Marie Curie
75005 Paris
France

Dr M. Ehrenberg
Department of Molecular Biology
Uppsala University
Biomedical Centre, Box 590
S-751 24 Uppsala
Sweden

Professor A. R. Fersht
Department of Chemistry
Imperial College of Science and
  Technology
London SW7 2AY
UK

Professor D. J. Galas
Molecular Biology, ACBR 126
University of Southern California
Los Angeles
California 90007
USA

Professor J. A. Gallant
Department of Genetics SK-50
University of Washington
Seattle
Washington 98195
USA

Dr M. F. Goodman
University of Southern California
Department of Biological Sciences
Molecular Biology Section
Los Angeles
California 90089–1481
USA

Dr H. Grosjean
Associated Professor
University of Brussels
Co-director
Biological Chemistry Laboratory
Department of Molecular Biology
67, rue des Chevaux
B-1640 Rhode-St-Genese
Belgium

Dr R. Holliday
Genetics Division
National Institute for Medical
   Research
The Ridgeway
Mill Hill
London NW7 1AA
UK

Dr T. B. L. Kirkwood
Computing Laboratory
National Institute for Medical
   Research
The Ridgeway
Mill Hill
London NW7 1AA
UK

Professor C. G. Kurland
Department of Molecular Biology
Uppsala University
Biomedical Centre, Box 590
S-751 24 Uppsala
Sweden

Professor J. R. Menninger
Department of Biology
University of Iowa
Iowa City
Iowa 52242
USA

Dr J. Ninio
Institut Jacques Monod
Tour 43, 2 Place Jussieu
75251 Paris cedex 05
France

Professor M. I. Page
Department of Chemical Sciences
The Polytechnic
Huddersfield HD1 3DH
UK

Dr R. F. Rosenberger
Genetics Division
National Institute for Medical
   Research
The Ridgeway
Mill Hill
London NW7 1AA
UK

Dr S. G. Sedgwick
Genetics Division
National Institute for Medical
   Research
The Ridgeway
Mill Hill
London NW7 1AA
UK

Preface

Molecular biology proceeds at unremitting pace to unfold new secrets of the living world. Biology, long regarded as an inexact companion to physics and chemistry, has undergone transformation. Now, chemical and physical principles are tools in understanding highly complex biomolecular processes, whose origin lies in a history of chance, constraint and natural selection. The accuracy of these processes, often remarkably high, is crucial to their self-perpetuation, both individually and collectively, as ingredients of the organism as a whole.

In this book are presented thirteen chapters which deal with various facets of the accuracy problem. Subjects covered include: the specificity of enzymes; the fidelity of synthesis of proteins; the replication and repair of DNA: general schemes for the enhancement of biological accuracy; selection for an optimal balance between the costs and benefits of accuracy; and the possible relevance of molecular mistakes to the process of ageing. The viewpoints are distinct, yet complementary, and the book as a whole offers to researchers and students the first comprehensive account of this growing field.

The idea of a book on accuracy in molecular processes was inspired first by a workshop organized in 1978 by Jacques Ninio (a contributor to this volume) with the sponsorship of the European Molecular Biology Organization. So successful was this meeting that two further workshops on similar lines were held in 1981 and 1985. Many of the contributors to this book participated in these workshops, and the book has benefited substantially from the sustained, informal exchange of views which the workshops have helped to bring about. The book is entirely independent, however, of these conferences.

We are grateful to all contributors for the care and patience with which they have written and, where necessary, revised their chapters, and to numerous of our colleagues for helpful comments and suggestions. We thank, in particular, Dr Alan Crowden of Chapman and Hall for his support and encouragement.

T. B. L. Kirkwood
R. F. Rosenberger
D. J. Galas

Acknowledgements

The following acknowledgements are made by authors with reference to their individual chapters

R. H. Buckingham and H. Grosjean
(Chapter 5)

The authors thank Drs C. G. Kurland, E. Murgola and J. Ninio for stimulating discussions during the writing of Chapter 5. We acknowledge Dr M. Grunberg-Manago and Professor H. Chantrenne for their interest and encouragement and the following organizations for financial support: in France, the Centre National de la Recherche Scientifique (Groupe de Recherche no. 18, A.T.P. 'Biologie Moléculaire du Gène', 'Microbiologie 1979' et 'Internationale 1980'), the Délégation Générale à la Recherche Scientifique et Technique (Convention 80.E. 0872), the Ligue Nationale contre le Cancer (Comité de la Seine) and the Commissariat à l'Energie Atomique; in Belgium, 'Actions concertées', Fonds National de la Recherche Scientifique Collective (contrat nî 2.4520.81) and Patrimoine de l'Université de Liège.

M. F. Goodman and E. W. Branscomb
(Chapter 8)

We wish to express appreciation to Dr John Petruska and Dr John J. Hopfield for the generous amounts of time spent instructing us in the subtleties and complexities of nucleotide base-pairings. We are grateful to Dr Christopher K. Mathews, Dr David Korn, Dr Martin Kamen, Dr Robert Ratliff, Dr Bruce Alberts, and Ms Randi Hopkins for their interest and help in numerous aspects of the work. Special thanks are due Dr Maurice J. Bessman and Dr Bernard C. Abbott. We also wish to thank Mrs Leilani Corell for expert and patient preparation of the 'diverging' series of drafts of this manuscript. The work was supported in part by grants GM21422 and CA17358 from the National Institutes of Health and grant 595 from the American Heart Association, Los Angeles Affiliate. The work was also performed in part under the auspices of the United States Department of Energy by the Lawrence Livermore National Laboratory under contract number W-7405-ENG-48 and Interagency Agreement 81-D-X0533 with the Environmental Protection Agency.

J. Ninio
(Chapters 10 and 13)

I am indebted to Leslie Orgel for his teachings in fundamental chemical kinetics. My thanks also to Manolo Gouy for checking all equations, to Pierre Claverie for discussions in chemical physics, to Eduardo Mizraji for contributing to my enzymological culture, to David Galas for suggestions concerning the manuscript and to Charles Kurland for his constant support within and outside the ribosome club.

    I also wish to thank David Galas and Vic Norris for style suggestions for Chapter 13.

1 An introduction to the problem of accuracy

D. J. GALAS, T. B. L. KIRKWOOD
and R. F. ROSENBERGER

1.1 Setting the scene

The primary concern of this book is about how cells copy and maintain the information which is stored as base sequences in their DNA and how they use this information to specify the structure of proteins. It is generally accepted that these processes of information transfer are the most essential and basic functions any living organism has to perform.

Two of the outstanding features of the information transfer processes are the accuracy and the speed with which they operate. All of them involve selecting the correct monomer from a pool of quite similar molecular species, for example the right nucleic acid out of the four available alternatives, or the right amino acid out of the twenty present in the cytoplasm. In the most accurate of the operations, the replication of DNA, the process has the astonishingly low error rate of about one mistake in 10^8. Further, selection occurs at a speed which allows the polymerization of many monomers per second. The mechanisms used to effect polymer synthesis and the problems living cells encounter in doing this are discussed in depth in the following chapters. In the present introductory chapter, we will attempt to bring these systems into a general focus.

Historically, the first serious biological encounter with the concept of random change came with the intellectual ferment that brought forth the Darwinian theory of evolution (see Eisley, 1958). While Darwin discussed the importance of the random variations of characteristics of organisms and commented on the fundamental role that chance must therefore play in evolution, it awaited the rise of genetics and, finally, molecular biology for biologists to come to grips with the mechanistic reality of Darwin's essential variations. One of the earliest attempts to probe the accuracy of replication of the hereditary molecule was made, in fact, by a theoretical physicist (Schrödinger, 1944), who brought to bear the heady optimism of the new quantum theory of physics on the nature of the genetic mechanism. These bold speculations helped to kindle the intellectual spirit that led in a few short years to the identification of Watson–Crick base-pairing as the chemical rule for replication of genetic information and the elucidation of the genetic code as the set of rules for translating genes into proteins.

At first, the focus of attention on these discoveries was, quite naturally, on the remarkable properties of base-pairing in providing deterministic chemical rules for the replication and translation of genetic information. The possibility of error was recognized, however, since without error there could be no mutation, and without mutation there would be no evolution. In relation to protein synthesis, Pauling (1957) pointed out that the molecular difference between the amino acids valine and isoleucine (see Chapter 3) was so small that it should be very difficult for the protein synthetic apparatus to discriminate sharply between them, as it apparently does. Pauling thereby posed in concrete terms the important question that most of the chapters deal with in some form, namely, what details of the mechanisms for information transfer are responsible for their extraordinary accuracy.

Shortly after Pauling's challenge, Loftfield (1963) showed that discrimination against valine was indeed much stronger than simple chemical differences would predict. This was followed by the discovery that the first steps in the charging of isoleucine tRNA were actually not very strongly discriminatory against valine, but that a subsequent step destroyed the activated valine (Baldwin and Berg, 1966). Subsequently, Yarus (1972a,b) and Eldred and Schimmel (1972) discovered that aminoacyl-tRNA ligases (synthetases) could actually deacylate their own tRNAs that are mischarged with an incorrect amino acid. Thus, the accuracy of tRNA charging was seen to be actively guarded by the charging enzymes (see Chapter 4). An analogous sort of active monitoring or 'proofreading' was proposed by Brutlag and Kornberg (1972) in the replication of DNA (see Chapter 8).

The accuracy of protein synthesis is, of course, not only due to the accuracy with which tRNAs are charged. Over a period of time parallel with the above studies on tRNA charging, firstly the ribosome was characterized and shown to be central to protein synthesis and its accuracy, secondly the phenomenon of 'informational suppression' was discovered (for review see Steege and Söll, 1979), and thirdly the misreading of codons induced by aminoglycoside antibiotics, like streptomycin, was discovered. These phenomena are discussed in Chapters 5 and 6.

Some of the first ideas on the role of the ribosome in determining the accuracy of protein synthesis were provided by the inventive and catalytic work of Luigi Gorini. In a series of highly original studies he examined the effects of aminoglycoside antibiotics and ribosomal mutations on nonsense and missense suppression and discussed the results in terms of the ability of the variously perturbed ribosomes to discriminate among tRNAs and termination factors. He showed that wild type ribosomes were significantly less accurate than they could become by acquiring certain mutations in ribosomal proteins, and particularly striking was his discovery of mutants with ribosomes more error-prone than the wild type (Rossett and Gorini, 1969). Thus, it was established that the translation error level was readily genetically manipulable and also that the bacterial cell could tolerate an increased level of errors. These systems continue to reveal valuable insights into the mechanisms for control of accuracy (see Chapters 6 and 11).

The recent history of research into questions of molecular accuracy has sounded two principal themes: (1) an ever more detailed analysis of the molecular structures and the kinetics involved in determining accuracy (see Chapters 3, 10 and 11) and (2) a convergence and cross-fertilization of ideas found useful in the various realms of molecular accuracy in biology – transcription (Chapter 7), translation (Chapters 5 and 6), charging of tRNAs (Chapter 4), DNA replication and repair (Chapters 8 and 9), and the substrate specificity of various enzymes (Chapter 3). Attention has also been paid to the integrity of the genetic information transfer system as a whole (Chapter 2 and see Fig. 1.1). As early as 1963, Orgel posed the question: how can the translation process, which itself is mediated by highly specific proteins, be stable against the feedback of mistakes in protein synthesis (Orgel, 1963). The notion of potential instability in the cellular translation process is important to the question of how life, with a stable translation system, emerged in the first place (see below and Chapter 13) and has also been suggested to have relevance to the process of ageing (Orgel, 1963). Since

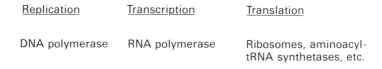

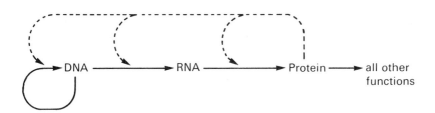

Figure 1.1   Schematic representation of the main pathways of genetic information transfer. The continuous arrows indicate the replication of DNA, the transcription of DNA into RNA, and the translation of RNA into proteins. The dashed arrows indicate the participation of *products* of genetic translation in these processes.

accuracy of information transfer and its associated costs are so fundamental in evolution, the possible role of accuracy breakdown in ageing has stimulated a substantial amount of research (Chapter 12).

1.2 Some preliminary concepts

1.2.1 DEFINITIONS OF ACCURACY

There are several possible ways to define the accuracy of a molecular process. First, the kind of error must be specified. In most of the instances considered in this book, the basic error will be insertion of an incorrect, or noncognate, monomer into a growing polymeric chain in place of the correct, or cognate, monomer. Second, one must be clear whether it is the insertion of a *particular* noncognate monomer which is of interest, or whether all possible misinsertions are to be considered. If the latter, one should be aware that different misinsertions will not usually be made at equal rates. Third, there is the possibility, supported by some experimental evidence (see Chapters 5 and 6), that the error rate for insertion of any given monomer will be influenced by the neighbouring sequence.

Let us assume that the basic error rate for insertion of monomers into a growing polymeric chain is e (misinsertions/insertion), ignoring possible differences in the error frequencies for different monomers. The proportion of correctly synthesized polymers will depend on their lengths. For a polymer made up of N monomeric units, the proportion of correctly synthesized polymers will be $(1 - e)^N$, and the proportion of polymers containing one or more incorrect monomers will be $E_N = 1 - (1 - e)^N$. Table 1.1 shows values of E_N for various values of e and N. Generally, it may be seen that for any given value of e, the proportion of error-containing polymers rises sharply with increasing N. This is likely to be why the larger proteins, for example, are usually made up of smaller subunits. At a basic error rate of $e = 10^{-4}\text{--}10^{-3}$ (Loftfield, 1963), synthesis of a very long protein molecule without error

Table 1.1 Frequencies of error-containing polymers (E_N) as a function of the frequency of inserting incorrect monomers (e) and the length of the polymer (N)

N	e			
	10^{-2}	10^{-3}	10^{-4}	10^{-5}
10	0.096	0.010	0.001	0.000
30	0.260	0.030	0.003	0.000
100	0.634	0.095	0.010	0.001
300	0.951	0.260	0.030	0.003
1000	1.000	0.632	0.095	0.010

would be a rare event, whereas a large molecule made up of subunits could be assembled correctly from a pool of subunits by excluding those subunits which contain errors.

1.2.2 THE COST OF ACCURACY

The cost of synthesizing a polypeptide chain, for example, in terms of the amount of energy directly expended for each amino acid residue incorporated, is known approximately, but the amount of energy that is expended to insure the accuracy which is observed is much more difficult to estimate. To do this requires taking into account the overall balance and rates of change of relevant substrates in the cell (see Chapter 11). A consideration of the total costs for each information transfer system will not be attempted here (see Chapters 11 and 12), but there are simple points of general application that we may observe.

The first point is that complete accuracy, in the sense of reducing the basic error frequency e to zero, is unattainable. If one considers only the fundamental thermodynamic cost in free energy of sorting an arbitrarily large population of monomers into two groups, correct and incorrect, the energy cost of *perfect* accuracy (associated with reducing the so-called entropy of mixing to zero) is infinite. This energy is independent of the energy involved in making the bonds between the monomers.

This consideration determines the absolute theoretical limit, but what of the cost of accuracy in a real system? Consider as an example the template-dependent synthesis of DNA by DNA polymerase. The *in vitro* action of phage T4 DNA polymerase has been studied by several workers (see Bessman *et al.*, 1974; Nossal and Hershfield, 1973; Alberts and Sternglanz, 1977; and references therein). A simplified model for the process of successive insertion and excision of bases has been examined in detail for the case of competition of adenine with the base analogue 2-aminopurine (Galas and Branscomb, 1978). This model, while heavily simplified (see Clayton *et al.*, 1979 and also Chapter 8) provides a clear illustration of accuracy costs.

Figure 1.2 shows the theoretical relationship between cost and error rate, based on the model, and some measured points of the relevant quantities for several different DNA polymerases (Bessman *et al.*, 1974). Cost here is defined as the number of bases inserted in the chain and excised per net base incorporated in the final chain. This cost is realized in two forms: (1) the energy inherent in the synthesis of triphosphate molecules from monophosphates and (2) the reduction in the overall speed of DNA synthesis, since the act of editing out a base requires a subsequent attempt to fill that position. Note, in particular, the concave nature of the curve in Fig. 1.2. As the error rate becomes less than the rate for a wild type polymerase (the middle point), the cost increases at an accelerating pace. The connection with speed is illustrated in Fig. 1.3 in a plot of a simulated base-by-base synthesis of a DNA

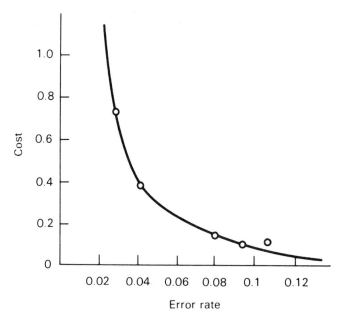

Figure 1.2 The curve is a theoretical fit to the data of a simple model for DNA polymerase action (Galas and Branscomb, 1978) and the data points are from several T4 polymerase alleles studied by Bessman *et al.* (1974). The two leftmost points are anti-mutators, the central point is wild-type, and the two rightmost are mutators.

segment, based on the parameters derived for the *in vitro* reactions. This shows the relative progress of a low-error (anti-mutator) and a high-error (mutator) DNA polymerase. As may be seen, the high-error polymerase is significantly faster.

The relationship between speed and accuracy, or between energy expenditure and accuracy, is not a strict functional relationship (like the curve in Fig. 1.2), but a boundary condition or constraint. A particular enzyme could function in a manner represented by any point above the curve, and the precise nature of the relationship depends on the detailed kinetics of the system. This will be explored in later chapters for several cases of biological interest.

1.3 The accuracy of enzymes

1.3.1 ENZYME TYPES

The primary molecular processes in the transfer of genetic information – the charging of tRNA, the selection of charged tRNA on the ribosome, and the transcription and replication of DNA – illustrate quite well the range of

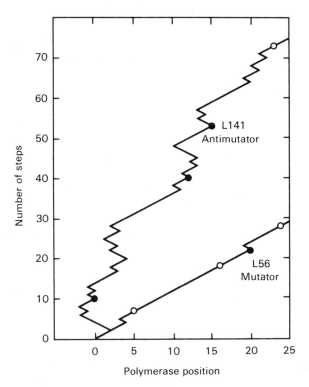

Figure 1.3 Using probabilities for insertion and excision derived from model fits to *in vitro* data (Galas and Branscomb, 1978), the random progress of a mutator (lower trajectory) and anti-mutator (upper trajectory) enzyme was simulated during DNA synthesis. The simulation shows the overall speed difference between the two enzymes. The filled circles indicate misinsertions subsequently removed; the open circles are errors.

accuracy problems for the cell. The molecular selections involved are very different in each case. The charging of tRNA is a fixed, two-fold discrimination problem. It is always the same amino acid and the same tRNA that is to be selected by the enzyme from a large set of competitors. At the ribosome, on the other hand, it is necessary to select a particular charged tRNA from a large set of similar objects only *in response* to the presence of a given codon at the ribosome. At some time or other, all charged tRNAs are selected. The range of molecular structures encountered by the ribosome in its adaptive response to the mRNA is, therefore, almost two orders of magnitude greater than for an aminoacyl-tRNA synthetase.

In transcription and DNA replication, another kind of situation is encountered. The selection process is responsive to a molecular signal (the template DNA strand), as for the ribosome, but the discrimination required is less demanding, involving only a selection among four comparatively simple molecules, the dNTPs (plus a few minor species).

It would be a mistake to underestimate the complexity of the full molecular processes of information transfer as they occur within cells, but it may be equally counter-productive to ignore the possibility of some fundamental, underlying similarities. A general classification of enzyme types may lead us to make significant distinctions between the enzymic mechanisms available to each class.

Let us divide the enzymes for information transfer (for this purpose ribosomes are considered enzymes) into two major categories which we will call 'direct' and 'facultative'. (The choice of these terms is somewhat arbitrary, but they are reasonably descriptive.) By 'direct' we will mean the property of being self-contained in information about the discrimination which is to be performed by the enzyme. In other words, the discrimination is performed the same way each time for these enzymes and no additional information, external to the enzyme structure, is required. For example, the aminoacyl-tRNA synthetases are of this kind. By 'facultative' we will mean the property of needing information external to the enzyme to accomplish the required discrimination. Ribosomes and DNA (or RNA) polymerases are of this kind: they change the selected substrates in response to the messenger RNA and the template DNA strand, respectively. The information contained in the structure of a facultative enzyme itself determines only the class of substrates and external signals that are acceptable for use by the enzyme, and not the specific substrate in this class that is to be accepted. A facultative enzyme is, therefore, a more sophisticated object. (Note that the classification of an enzyme as facultative is not the same as its being allosteric, as these attributes are designated on the basis of completely different measures; a facultative enzyme may, however, be allosteric.)

The distinction between direct and facultative enzymes is useful for several reasons. Outstanding among these is that the requirement for a facultative enzyme to be responsive to an external signal may put severe constraints on the physical means which can be used to make the discrimination between molecules. For example, in one situation, the active site of the facultative enzyme must discriminate sharply *against* a certain substrate, and in another situation must discriminate *in favour of* this same substrate with the same degree of accuracy. A direct enzyme, on the other hand, is always required to make the same discriminations.

1.3.2 CONSTRAINTS

For the moment, let us consider only the simplest of enzyme discrimination systems. A substrate binds to the enzyme complex and an incorporation reaction is catalysed. Even in more complex schemes to be considered later, this constitutes the elemental process. In a Michaelis–Menten type of reaction scheme, for example, the parameters that can be used to discriminate between substrates are the association and dissociation rates (K_m) and the inherent rate of the reaction (v_{max}). How can the requirement of being facultative constrain the use of these parameters?

To provide a background for the ensuing arguments, consider the following essential properties of a facultative enzyme:

(1) It must catalyse the incorporation of the entire set of acceptable substrates at approximately the same rate, in a reaction which is virtually identical for each. (How wide a range of reaction rates exists is not yet clear.)
(2) Cognate and noncognate enzyme–substrate complexes are necessarily similar in their structure in the region of the reactive groups, because all substrates can form complexes with the enzyme and the same groups are involved.
(3) The facultative enzyme must itself bind the substrates with bonds that are common to all substrates. It is, therefore, difficult (though not impossible) for the binding environment of the variable groups on the substrates to transmit changes to the common groups in the active site which could modulate by a large factor the rate of catalysis.

It is now clear that the discussion of enzymic discrimination has brought us face to face with the more general problem of specificity and rate-enhancement in enzymic catalysis, a subject taken up in detail in Chapter 3. The important point is that of all the mechanisms available to enzymes, only a subset of them is available for facultative enzymes because of the above three properties (see also Lipscomb, 1978). A general hypothesis that at once simplifies the problem and unifies our consideration of facultative enzymes can now be made.

This hypothesis is simply that facultative enzymes must make use, as the key to their discriminatory function, of the binding free energy differences between correct and incorrect substrates, rather than differences in inherent reaction rates, while direct enzymes are not so constrained. It is undoubtedly true that differences in the variable moieties of substrates can affect the reaction rates, but the above hypothesis holds that these differences do not provide the essential discrimination. Restating the hypothesis in a different way, it is in the K_m for the correct and incorrect substrates, and not in the v_{max}, that these substrates differ substantially and systematically for facultative enzymes.

Is there any evidence that this hypothesis is correct for known information transfer systems? At present, the evidence is largely indirect. For the tRNA synthetases, which we class as direct enzymes, there are several cases in which no difference at all is seen in the binding of cognate and noncognate tRNAs, but large differences are seen in the reaction rates (Ofengand, 1977). Thus, it appears that v_{max} can be used for the discriminatory function of these enzymes. No such thing has been seen for any facultative enzyme: some binding discrimination has been found in all cases so far examined. In the case of the ribosome, it has been shown by measurement of anti-codon binding between pairs of selected pure tRNA species that the complexes are specific by a factor of a few-fold to a hundred-fold (Grosjean, de Henau and Crothers, 1978, see also Chapter 5). When possible differences in conformation for the ribosome-bound message are taken into account, it seems reasonable to assume that the codon–anticodon binding at the ribosome may yield about a hundred-fold discrimination. While this alone is not enough to explain the accuracy of protein synthesis, it is possible that kinetic enhancement schemes, which re-utilize the basic level of binding energy discrimination (see below and Chapters 10 and 11), may make up the difference. Evidence that DNA polymerase also discriminates on the basis of K_m difference, rather than v_{max}, is reviewed extensively in Chapter 8.

1.3.3 ENZYME SPECIFICITY

To formulate the foregoing simple hypothesis a little more precisely, we need to consider further the intimate relationship between enzyme specificity and rate enhancement. These are, in fact, different facets of the same phenomenon. The specific binding of a substrate brings the catalytic groups of the enzyme and substrate into a precise position, in which the complex is already advanced a substantial distance down the reaction pathway (Jencks, 1975). The observed binding energy determines (or is defined by) the dissociation constant of the complex, and this is therefore always less than the inherent energy (see Chapter 3).

The significance of these general observations about enzyme specificity and rate enhancement for facultative enzymes lies in the fact that the rate-enhancing free-energy change must be derived from the binding of common groups on the substrate to the enzyme, and therefore the reaction velocity will be determined in large part by these common groups. The binding properties of specific groups on the substrate can be used to modulate the dissociation rate and thereby permit the essential discrimination of the reaction. The essential property of the interaction of the selected substrate with the enzyme, which is required to fit the hypothesis, is that the effects of the binding of the variable and common groups are *separable*. This, in turn, means that the binding free-energies derived from the variable and common groups be additive:

$$\Delta G = \Delta G_{var} + \Delta G_{comm}$$

This requires that the total partition function for the complex must be factorizable into two parts, one determined solely by each component of the interaction. In reality, this requirement will never be precisely satisfied as there is always some interaction of the two to be expected, but to the extent that it is a good approximation the hypothesis can be considered a reasonable one.

1.4 The role of kinetics in accuracy

As has been mentioned already, and is extensively discussed in later chapters of this volume, the simple discrimination provided by the interaction free energies of substrate molecules falls short of the discrimination seen in real systems by a substantial factor. This fact underlines the very important point that molecular recognition and the accuracy of information transfer are not at all the same thing. Molecular recognition provides the basis, of course, for discrimination, but responsibility for the real levels of accuracy seen in living cells is now thought to reside in possibly very complex kinetic systems which marshal and co-ordinate the processes of information transfer to extract more specificity in polymer synthesis than could occur at a chemical equilibrium.

General schemes for the kinetic enhancement of accuracy were proposed first by Hopfield (1974) and Ninio (1975) and these are reviewed, with further examples, in Chapters 10 and 11. In essence, the ideas of these schemes are that either the reaction sequence is organized in a series of steps at each of which the inherent molecular discrimination between cognate and non-cognate enzyme–substrate complexes is used again, or the completion of the reaction is delayed to allow a longer time for the differential dissociation rates of cognate and noncognate enzyme–substrate complexes to exert its effect. In order to accomplish this, it is essential that the kinetic process be strictly ordered and isolated from reverse reactions.

A useful analogy is the following. It is as though the candidate enzyme–substrate complex (cognate or noncognate) were being run through a corridor to which there is only one entrance but many side exits arrayed along its length. Arrival at the far end of the corridor corresponds to being accepted by the system, and incorporated in the final product – a polymer. The complex can be rejected through any of the side exits, noncognate complexes being more likely at each exit to be rejected than cognate ones. If the rejected complexes are allowed to leak back through the side exits, particularly the later ones, the whole purpose of the multiple screening system is defeated. If none can leak back through, the system can effectively amplify the discrimination many-fold. A state of kinetic equilibrium corresponds to the free back-flow of rejected complexes, and therefore it is essential to its accuracy-enhancing function that the system be held out of equilibrium. This in turn implies a highly irreversible process and a correspondingly high level of dissipation of energy in accomplishing kinetic enhancement. The fact that dissipation of energy is essential in the enhancement of accuracy brings us back to the important question of the cost of accuracy, discussed above. The question is, how much accuracy is obtained for how much expended energy? The reader is referred to later chapters in this book and to Guéron (1978) for further discussion of kinetic schemes, their structures, costs and efficiencies.

1.5 Molecular accuracy in evolution

Like many aspects of evolution, the evolution of accuracy in the processes of molecular information transfer and the origins of genetic translation are still partly shrouded in speculation and mystery. Nevertheless, considerable insight into the present day systems of accuracy control and the levels at which they operate might be expected if we could identify the selection pressures on these mechanisms and the constraints under which such selection must work.

Ninio (1982) has presented a stimulating discussion on prebiotic metabolism and the origins of life, and further references to this subject are given in Chapter 2. The problems to be solved present themselves at two levels: firstly, how did the earliest self-replicating polymers come into being; secondly, how did these polymers come to code for complex enzymic machinery for facilitating their replication and fostering their survival?

Generally it is believed that the first self-replicating polymers are likely to have been polynucleotides, with replication dependent only on pairing between complementary bases. Although such reactions in the absence of polymerases would have been likely to be quite inaccurate, it is almost certain that for the same reason the rate of phosphodiester bond formation will have been extremely slow. As pointed out by Ninio (see Chapter 10), a very slow rate of polymerization will amplify even small differences in the binding energies of the cognate and noncognate bases. This will, of course, be at the expense of a great loss in efficiency. It is possible, therefore, that replication of polynucleotides in the early days of evolution, before effective catalysis speeded up phosphodiester bond formation, had a respectable degree of accuracy and a very low efficiency.

The origin of genetic translation presents an even greater challenge to our understanding. If one accepts that the first genetic entities were poly-nucleotides, one has to postulate that somehow a primitive type of tRNA arose and with it a mechanism for covalently joining an amino acid to the polynucleotide. We have no information on the number and nature of the amino acids available at that stage. Presumably there will have had to be some selectivity in the joining of amino acids to particular tRNA types. This may have been facilitated again by a slow rate of catalysis. The resulting proteins will, however, almost certainly have been extremely variable in composition and size. Ninio (1982) described these as 'statistical' proteins.

The emergence of primitive tRNAs and mechanisms for joining them to amino acids were essential steps in the evolution of the decoding systems we know today. An additional major step was necessary, however, and that was the development of the grammar needed to make a workable code. Present-day decoding involves a defined start signal and the accurate movement of the machinery in three-base steps along the template. The evolution of the latter mechanism, keeping an accurate reading frame, is particularly difficult for us to grasp, mainly because we still do not know how modern ribosomes perform this function. A number of suggestions have been made, and some are discussed by Kurland and Gallant in Chapter 6. Weiss and Gallant (1983) have found evidence supporting a model originally proposed by Woese (1970), and this has a number of attractions in explaining the possible evolution of message translation.

Woese suggested that it was the anticodon arms of tRNA that moved the message along three bases at a time, by changing their stacking configuration. On this basis, one would not have needed a complex ribosome containing large amounts of specialized RNA and tens of different proteins for sequential reading of a polynucleotide message. In Woese's view, what was required were primitive tRNAs which could form a stem and a loop with seven bases. Three of the seven bases would have been needed to interact with complementary sequences on the putative message. Suppose now that two such tRNAs interacted with successive triplets on the message. A change of stacking interactions would then have resulted in an accurate three-base movement of the polynucleotide relative to the tRNAs. To continue the speculation, if amino acids were joined to a base close to the putative anticodon, they may have been brought sufficiently near each other to allow peptide bond formation. This would have constituted a primitive kind of polypeptide synthesizing system.

Once the synthesis of polypeptides had evolved and been coupled to the sequential reading of a polynucleotide, one must ask what changes such proteins may have initiated. Several of the possibilities are discussed by Ninio (1982). Presumably, at some stage primitive polypeptides would have acquired the ability to catalyse reactions leading to their own synthesis. By this, they would have opened the way for natural selection to improve their own performance and that of the whole polymer synthesizing complex (see Chapter 12). They would also have opened the possibility that random errors in the synthesis of a protein could have consequences of an unusually insidious kind if, by chance, these errors altered the specificity of a part of the protein synthesizing machinery. This possibility, suggested by Orgel (1963), is discussed in the following chapter (see also Kirkwood, Holliday and Rosenberger, 1984).

Over the long evolutionary span since the origins of genetic replication and translation, many factors have changed. One is the enormously increased efficiency of modern organisms. This has been achieved by using many different protein catalysts. For each enzyme, specificity depends on accurate construction and thus the number of sites where errors could be dangerous has multiplied extensively. On the other hand, selective pressures have led to the development of error-correcting mechanisms of various types (see subsequent chapters). Error correction, however, is an energy-consuming process, and each species will have had to strike its own balance between accuracy, and its attendant benefits, and the employment of energy resources for promoting reproductive fitness in other ways (Kirkwood, 1981). Some, such as bacteria, are clearly well buffered against the effects of random errors (Chapter 6). Other, more complex, organisms may have found it more advantageous to concentrate on the survival of small numbers of reproductive cells, rather than to maintain the whole organism (Chapter 12; see also Kirkwood, 1977). In the latter case, errors in decoding proteins of somatic cells may start a feedback cycle which cannot be reversed.

1.6 Accuracy in other information systems

In concluding this introductory chapter, we turn briefly to consider biological information systems other than those directly involved with the processes of genetic information transfer.

Higher eukaryotes contain cells which appear to have identical genomes, but which differ strikingly in their phenotypes. The reasons for this are that cells express only a fraction of their total genetic information and that fraction varies greatly in differently specialized cells. The mechanisms controlling gene expression during differentiation are still not well understood, and a discussion of the data available is outside our present scope. However, one particular aspect indicates that an information system additional to the genetic code is involved.

The differentiation patterns of cells appear to be controlled by their position in the early embryo. This seems to be the determinant for pattern formation in virtually all multicellular organisms. For example, transplanting the zone of polarizing activity from an avian limb bud can give rise to an additional, normally non-existent, wing tip (Wolpert, 1981). Monsters of this kind have been manufactured in many different species. One common explanation is that an informational gradient is set up in the early embryo, presumably by diffusable substances, which can control gene expression. This informational field, whatever its molecular basis, must be highly accurate and reproducible. Since house flies always produce other flies and cows other cows, the origin of the postulated informational gradients must almost certainly be in the genome. But, at least to our present ignorance, it appears to be an additional way of transmitting and decoding information which has developed during the later stages of evolution.

An even more striking biological information storage and handling system

The problem

comprises the instinctive and acquired memories held in brain c
virtually no hard information on how any of these systems wo
personal experiences of their accuracy or otherwise are univers
highly unlikely that the information is stored in the chemical s
polymers, as it is in the genome. A popular, but somewhat anthrop
speculation is that neurones join to form networks similar to the
we ourselves build. Brains are the most complex biological structures that we
know and it seems almost certain that the way they work will be among the last
problems to be clarified in biology.

References

Alberts, B. and Sternglanz, R. (1977) Recent excitement in the DNA replication problem. *Nature*, **269**, 655–661.

Baldwin, A. N. and Berg, P. (1966) Transfer ribonucleic acid-induced hydrolysis of valyl adenylate bound to isoleucyl ribonucleic acid synthetase. *J. Biol. Chem.*, **241**, 839–845.

Bessman, M., Muzyczka, N., Goodman, M. and Schnaar, R. (1974) Studies on the biochemical basis of spontaneous mutation. II The incorporation of a base and its analogue into DNA by wild-type, mutator, and anti-mutator DNA polymerases. *J. Mol. Biol.*, **88**, 409–421.

Brutlag, D. and Kornberg, A. (1972) Enzymatic synthesis of DNA: a proofreading function for the 3' to 5' exonuclease activity in DNA polymerases. *J. Biol. Chem.*, **247**, 241–248.

Clayton, L., Goodman, M., Branscomb, E. and Galas, D. (1979) Error induction and correction by mutant and wild-type T4 DNA polymerases: kinetic error discrimination mechanisms. *J. Biol. Chem.*, **254**, 1902–1912.

Eisley, L. (1958) *Darwin's Century*. Doubleday, New York.

Eldred, E. W. and Schimmel, P. R. (1972) Rapid deacylation by isoleucyl tRNA synthetase of isoleucine specific tRNA aminoacylated with valine. *J. Biol. Chem.*, **247**, 2961–2968.

Galas, D. and Branscomb, E. (1978) The enzymatic determinants of DNA polymerization accuracy: theory of T4 polymerase mechanisms. *J. Mol. Biol.*, **124**, 653–687.

Grosjean, H., de Henau, S. and Crothers, D. (1978) On the physical basis for ambiguity in genetic coding interaction. *Proc. Natl Acad. Sci. USA*, **75**, 610–614.

Guéron, M. (1978) Enhanced selectivity of enzymes by kinetic proofreading. *American Scientist*, **66**, 202–208.

Hopfield, J. J. (1974) Kinetic proofreading: a new mechanism for reducing errors in biosynthetic processes requiring high specificity. *Proc. Natl Acad. Sci. USA*, **71**, 4135–4139.

Jencks, W. P. (1975) Binding energy, specificity and enzyme catalysis: the circe effect. *Advances Enzymol.*, **43**, 219–410.

Kirkwood, T. B. L. (1977) Evolution of ageing. *Nature*, **270**, 301–304.

Kirkwood, T. B. L. (1981) Repair and its evolution: survival versus reproduction. In *Physiological Ecology: An Evolutionary Approach to Resource Use* (eds C. R. Townsend and P. Calow), Blackwell, Oxford, pp. 165–189.

Kirkwood, T. B. L., Holliday, R. and Rosenberger, R. F. (1984) Stability of the cellular translation process. *Int. Rev. Cytol.*, **92**, 93–132.

Lipscomb, W. N. (1978) Intramolecular interactions, enzyme activity and models. In *Molecular Interaction and Activity in Proteins*, Ciba Foundation Symposium 60, Amsterdam.

Loftfield, R. B. (1963) The frequency of errors in protein biosynthesis. *Biochem. J.*, **89**, 82–87.

Ninio, J. (1975) Kinetic amplification of enzyme discrimination. *Biochimie*, **57**, 587–595.

Ninio, J. (1982) *Molecular Approaches to Evolution*, Pitman, London.

Nossal, N. and Hershfield, M. (1973) Exonuclease activity of wild-type and mutant T4 DNA polymerases: hydrolysis during DNA synthesis *in vitro*. In *DNA Synthesis In Vitro* (eds R. Wells and R. Inman), University Park Press, Baltimore.

Ofengand, J. (1977) tRNA and aminoacyl-tRNA synthetases. In *Molecular Mechanisms of Protein Biosynthesis* (eds H. Weissbach and S. Pestka) Academic Press, New York, pp. 7–79.

Orgel, L. E. (1963) The maintenance of the accuracy of protein synthesis and its relevance to ageing. *Proc. Natl Acad. Sci. USA*, **49**, 517–521.

Pauling, L. (1957) The probability of errors in the process of synthesis of protein molecules. In *Festschrift Arthur Stoll*, Birkhauser Verlag, Basel, pp. 597–602.

Rossett, R. and Gorini, L. (1969) A ribosomal ambiguity mutation. *J. Mol. Biol.*, **39**, 95–112.

Schrödinger, E. (1944) *What is Life?*, Cambridge University Press, Cambridge.

Steege, D. and Söll, D. (1979) Suppression. In *Biological Regulation and Development. 1. Gene Expression* (ed. R. F. Goldberger), Plenum, New York, pp. 433–486.

Weiss, R. and Gallant, J. A. (1983) Mechanism of ribosome frameshifting during translation of the genetic code. *Nature*, **302**, 389–393.

Woese, C. (1970) Molecular mechanism of translation: a reciprocating ratchet mechanism. *Nature*, **226**, 817–820.

Wolpert, L. (1981) Positional information and pattern formation. *Phil. Trans. R. Soc. Lond. B*, **295**, 441–450.

Yarus, M. (1972a) Solvent and specificity. Binding and isoleucylation of phenylalanine tRNA (*E. coli*) by isoleucyl-tRNA synthetase from *E. coli. Biochemistry*, **11**, 2352–2361.

Yarus, M. (1972b) Phenylalanyl-tRNA synthetase and ile-tRNA[phe]: a possible verification mechanism for aminoacyl-tRNA. *Proc. Natl Acad. Sci. USA*, **69**, 1915–1919.

2 Errors and the integrity of genetic information transfer

R. F. ROSENBERGER and T. B. L. KIRKWOOD

2.1 Introduction

Accuracy in the synthesis of macromolecules involved with the processes of genetic information transfer, that is with the replication and repair of DNA, the transcription of genes into RNA messages, and the translation of RNA into proteins, is of particular interest in relation to the survival and reproduction of living cells. As was pointed out by Orgel (1963), the ability of a cell to carry out its various vital functions depends not only on its inheritance of an intact complement of genes, but also on its receiving from its parent a viable molecular apparatus for translating them into protein. In particular, Orgel drew attention to a possibility overlooked by the conventional dogma of a strictly unidirectional flow of genetic information from DNA to RNA to protein, namely that errors can be propagated cyclically within the translation apparatus.

The basis of Orgel's hypothesis was the observation that the processes of transcription and translation are mediated by a set of highly specific enzymes which have themselves been produced by a previous round of protein synthesis. Occasionally an error in synthesizing one of these information handling enzymes will create a molecule which retains some or all of its activity, but which has altered specificity. By participating in the next round of translation, this erroneous enzyme can, in turn, generate more faults. In this way, errors might be propagated progressively within the translation apparatus until eventually the cell could no longer carry out essential functions and would succumb to what Orgel graphically termed an 'error catastrophe'.

The risk of error catastrophe is a potential threat to any living cell. In primitive cells, particularly, it would have posed a very serious problem as it seems highly probable that the earliest versions of the translation apparatus were comparatively imprecise. One of the most challenging problems concerning the origins of life is to explain how an accurate translation system evolved out of initially inaccurate parts (Miller and Orgel, 1973; Crick *et al.*, 1976; Eigen and Schuster, 1979). It has been suggested also that error propagation may explain both the clonal senescence of certain types of dividing cells, such as mammalian fibroblasts, and the ultimate loss of function of post-mitotic cells such as neurones (Medvedev, 1962; Orgel 1963, 1973; see also Chapter 12). In the case of dividing cells, the progression to error catastrophe would occur through the inheritance by daughter cells of an increasingly defective translation apparatus.

Over the two decades since the error propagation hypothesis was first formulated, the concept has generated a great deal of interest (for reviews, see Rothstein, 1977; Gershon, 1979; Medvedev, 1980; Gallant, 1981; Laughrea, 1982; Kirkwood, Holliday and Rosenberger, 1984). This interest, mixed with a considerable degree of controversy, reflects not only the breadth of the areas of biology to which the theory applies, but also the difficulty in testing it.

The main difficulty lies in detecting an increase in the rates of making random errors of synthesis of macromolecules, particularly when the error rates are in any case quite low and it is not known what size of increase would be lethal to a cell. It is possible that even quite a small increase in error rates could have a profound effect on cell viability. A second difficulty is that an adequate description of the theoretical dynamics of error propagation has proved more complex than was originally recognized. Without a clear framework of theoretical prediction, the interpretation of experimental results is confused.

For these two reasons, the error propagation hypothesis has stimulated a wide range of both theoretical and experimental studies on the effects of errors on the integrity of genetic information transfer. In this chapter, we review briefly the progress which has been made and assess current evidence concerning the existence, or otherwise, of a significant amount of error propagation. These issues will also be touched upon in several later chapters.

2.2 Theory

The feedback of mistakes into the overall machinery of genetic information transfer can occur at several different levels, with varying severities of effects on the cell. The most serious mistakes are those which lead to a change or mutation in DNA since a change, once made, is permanent (unless the same DNA sequence is mutated again). Mistakes in transcription and translation are, by contrast, transient and last individually only as long as the abnormal mRNA or protein molecule persists in the cell. Transcriptional errors are more serious than translational errors, as a single message may be translated many times and therefore an erroneous message may give rise to a burst of errors in newly synthesized protein. On the other hand, because molecules of protein are generated in greater numbers than molecules of mRNA, errors in protein are likely to be statistically more numerous, even if the basal error rates per monomer (nucleic or amino acid) are the same.

2.2.1 MODELS OF ERROR PROPAGATION IN PROTEIN SYNTHESIS

In modelling the propagation of errors within the genetic translation machinery, a useful start can be made by restricting attention in the first instance to protein errors only. For this purpose, a simplified representation of the translation process will suffice. Figure 2.1 shows the translation apparatus as a series of discrete generations, each of which carries out the synthesis of its successor as well as performing the general translation of DNA into protein. The definition of a generation is here somewhat arbitrary, since the components of the translation apparatus are renewed more or less continuously, but it may be regarded roughly as the mean time to replacement of each constituent molecule. This definition has the advantage that it works equally well in dividing and post-mitotic cells, since protein and RNA turnover occur in both. With this definition, the central issue as regards propagation of translational errors is how the average accuracy q_t (the probability of correctly inserting an amino acid) of any one generation is related to the accuracy q_{t-1} of its predecessor. Obviously, if $q_t < q_{t-1}$ the accuracy of the system is decreasing, while if $q_t > q_{t-1}$ it is increasing. In a stable state, accuracy would be constant with $q_t = q_{t-1} = q_{stable}$.

The earliest, and most simple, model of error propagation which has been suggested (Orgel, 1970) supposes the error level e_t ($= 1 - q_t$) to be composed of two parts, one a residual error frequency E, such as would characterize a translation apparatus containing no erroneous molecules, and

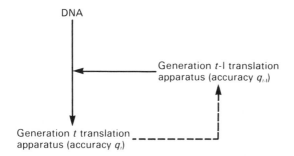

DNA

Generation t-l translation apparatus (accuracy q_{t-1})

Generation t translation apparatus (accuracy q_t)

Figure 2.1 Simplified feedback loop illustrating the potential for propagation of errors during the translation of the genetic message into protein. The stability of translation is determined by the relationship between q_t and q_{t-1}. (From Kirkwood, Holliday and Rosenberger, 1984.)

the other arising through feedback of errors from the previous generation. Orgel assumed the latter to be proportional to the total error frequency of the preceding generation, so

$$e_t = E + \alpha e_{t-1} \qquad (2.1)$$

where α is a constant. If $\alpha \geq 1$, errors in generation $t-1$ result in an even larger number of errors in generation t, and accuracy is rapidly and irreversibly lost. Conversely, if $\alpha < 1$, the number of errors feeding back through the cycle is always less than were there before, and the error level stabilizes to a fixed value $E/(1-\alpha)$. Through its feedback parameter α the model thus predicts an important distinction between limited, convergent error feedback and divergent error amplification.

A later model of the dynamics of error propagation carried the analysis of translational stability further (Hoffmann, 1974; Kirkwood and Holliday, 1975). In this model, each generation of the translation apparatus was assumed to consist of a set of polypeptide 'adaptors', each of which carried out the translation of a given codon into the corresponding amino acid in the nascent polypeptide. The model thus did not attempt to represent the full complexities of the present-day translation process, and we shall return to discuss its limitations later.

For each adaptor, it was assumed there is a subset of m residues in its own sequence whose correct insertion is vital for it to function at all, and a further nonoverlapping subset of n residues whose correct insertion is required for the adaptor to have its normal substrate specificity. Incorrect insertion of one or more of the m residues was assumed to result in complete loss of translational activity, while incorrect insertion of one or more of the n residues was assumed to result in complete non-specificity. In fact, it is more likely that an error in one or more of the n residues would alter specificity, rather than abolish it completely, but if the alteration is random the average effect in a population of adaptors, which is what the model considers, will be the same. Errors in any residues other than the m or n subsets were assumed to be unimportant.

For the purpose of modelling propagation of errors in the translation apparatus, attention can be restricted to those adaptors which do not contain errors in any of the m sites, as any which do contain errors in these sites do not contribute to translation. This is not to say errors in the m sites are unimportant, since synthesis of a high proportion of non-functional adaptors would be a serious burden to the cell. However, this would be a consequence of error propagation, rather than its cause.

The viable adaptors in generation t of the translation apparatus comprise two types: a fraction q_{t-1}^n which are normal (i.e. contain no errors in the n sites) and a fraction $1 - q_{t-1}^n$ which are erroneous (i.e. have lost their specificity). The final ingredient of the model was to specify the different rates of insertion of correct and incorrect amino acids by these normal and erroneous adaptors.

In the more recent version of the model (Kirkwood and Holliday, 1975), the activities of normal and erroneous adaptors were defined as follows:

(1) A normal adaptor makes correct amino acid insertions S times as fast as it makes each of the $\lambda - 1$ possible incorrect insertions, λ being the number of incorrect amino acids available for substitution; S is the *specificity* of a normal adaptor.

(2) An erroneous adaptor makes each of the λ possible insertions at the same rate, with the total rate equal to R times the total rate for a normal adaptor; R is the *residual activity* of an erroneous adaptor.

In the original version of the model (Hoffmann, 1974), R did not feature as a parameter and it was supposed that for an erroneous adaptor the rate of each possible insertion was the same as the rate of each incorrect insertion for a normal adaptor. This was equivalent to setting $R = \lambda/(\lambda + S - 1)$, which for any reasonable values of λ and S amounts to a drastic reduction in total activity. While acknowledging that some reduction in activity is likely, Kirkwood and Holliday (1975) pointed out that single amino acid substitutions are known which alter specificity but do not severely reduce activity, and therefore R was added to the model as a variable parameter to increase its flexibility and realism. The idea that R need not necessarily be very small is supported also by the restriction of attention only to those adaptors which contain no errors in the m subset of sites.

From these assumptions, the average rates for generation t adaptors of making correct and incorrect insertions may be calculated, and the average accuracy q_t determined as the rate of making correct insertions, divided by the total rate of making insertions, both correct and incorrect. This gives the equation

$$q_t = \frac{q_{t-1}^n[\lambda S - (S + \lambda - 1)R] + (S + \lambda - 1)R}{q_{t-1}^n\lambda(S + \lambda - 1)(1 - R) + \lambda(S + \lambda - 1)R} \qquad (2.2)$$

Despite its superficial complexity, Equation (2.2) provides a simple picture of the dependence of q_t on q_{t-1}, from which a number of important conclusions can be drawn. In a plot of q_t against q_{t-1} the equation is represented by an S-shaped curve (Fig. 2.2). If the curve lies entirely below the line $q_t = q_{t-1}$, on which accuracy is constant, the accuracy of translation is always decreasing and error catastrophe is inevitable. Alternatively, if the curve crosses into the upper left region of the graph where $q_t > q_{t-1}$, it intersects the line $q_t = q_{t-1}$ in two points, and the upper of these defines a point of stable accuracy, q_{stable}. In this case, provided accuracy does not fall below the lower point of intersection, $q_{threshold}$, it will always revert to q_{stable} after it is

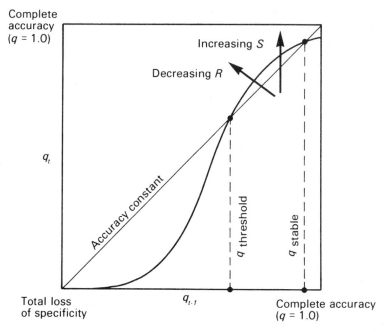

Figure 2.2   The adaptor model of error propagation (see text) predicts an S-shaped relationship between q_t (the average accuracy of generation t adaptors) and q_{t-1}. If the curve crosses the line $q_t = q_{t-1}$, it defines a point of stable accuracy q_{stable}. Provided the accuracy remains above the lower point of intersection $q_{threshold}$, it will always revert to its stable value. Below $q_{threshold}$ accuracy falls progressively, resulting in total loss of specificity. Two principal parameters S (specificity of a normal adaptor) and R (residual activity of an erroneous adaptor) control stability by displacing the curve in the directions indicated. (From Kirkwood, 1980.)

perturbed. If accuracy does drop below $q_{threshold}$, however, it thereafter declines irreversibly towards error catastrophe. The model thus has the important implication that if the margin of safety between q_{stable} and $q_{threshold}$ is small, the cell may exist in a metastable state where there is an appreciable chance that a random increase in the error level will cause it to become irreversibly committed to an error catastrophe.

In summary, taking the parameters λ and n as fixed, the adaptor model defines two variables R and S which together play a critical role in determining the degree of stability in genetic translation. Increasing S (the specificity of normal adaptors) results in a predominantly upward movement of the curve, including the right hand endpoint. Decreasing R (the residual activity of an erroneous adaptor) causes a movement upwards and to the left, leaving both endpoints fixed. Thus, either or both of these changes increases the separation between q_{stable} and $q_{threshold}$. The importance of these parameters is that each of them may be under the direct control of natural selection (Kirkwood, 1977; see also Chapter 12), S through the evolution of enhanced enzyme specificity or of systems for more stringent proofreading of protein synthesis, and R through the evolution of scavenging mechanisms for rapidly degrading abnormal proteins (a shorter lifetime for an erroneous adaptor in the cell being equivalent to a lower net residual activity).

One further model of error propagation in protein synthesis has been described (Goel and Ycas, 1975; Goel and Islam, 1976). However, this model provides only a questionable representation of error propagation (see Kirkwood, 1980) and will not be reviewed further here.

2.2.2 USES AND LIMITATIONS OF THE ADAPTOR MODEL

The adaptor model described in the previous section makes two advances over the simple recognition by Orgel (1970) that the translation process can exist in either a stable or an unstable state. Firstly, it defines stability in a flexible way which allows the possibility that without changing any parameter of the model a cell may jump from the stable to the unstable state. In doing so, it introduces the concept of a threshold error level above which the translation process breaks down but below which it can recover. This is of particular significance to studies in which the accuracy of translation in bacteria has been artificially perturbed with aminoglycoside antibiotics and the subsequent behaviour of the cells followed (for discussion see Kirkwood, 1980; Kirkwood, Holliday and Rosenberger, 1984; and below). Secondly, the model introduces two general parameters which control translational stability and which can be related directly to molecular mechanisms for accuracy control. The model serves, therefore, to refine the predictions of the error propagation hypothesis with regard to the stability of genetic information transfer in different types of cell. For example, bacterial cells are likely to be stable, perhaps highly so, while cells which routinely undergo clonal senescence may be unstable or have such weak stability that it is easily lost.

While recognizing the uses of the adaptor model, however, it is important also to be explicit about its limitations. Obviously, it is a greatly simplified representation of the translation apparatus in that it ignores the separate steps of transcription and translation and it aggregates the functions of ribosomes, synthetases, tRNAs and other regulatory components of the protein synthesizing machinery into a single type of molecule, the polypeptide adaptors. As the error level increases, if error propagation does occur, this lack of realism is likely to become progressively more serious as the probability of successful assembly of an error-free protein synthetic complex decreases sharply. This type of interaction effect is not allowed for in the simple adaptor model, and Equation (2.2) should not therefore be regarded as an exact quantitative prediction. Rather, it may be seen as a qualitative, or semi-quantitative, tool for defining a predictive framework within which to investigate the error propagation hypothesis more closely.

2.2.3 OTHER FACTORS AFFECTING ERROR FEEDBACK

An aspect of accuracy control not so far included in our discussion of error propagation is the role of 'proofreading' activities to enhance the discrimination between correct and incorrect substrates beyond the levels allowed for by simple differences in their binding energies (see Chapters 4 and 10). In essence, these activities involve either the sequential processing of the enzyme–substrate complex through a series of reaction steps at each of which the binding energy discrimination is used again, or the introduction of a time delay so the discrimination produced by differential rates of dissociation of correct and incorrect enzyme–substrate bonds may be amplified (Hopfield, 1974; Ninio, 1975). When these processes are taken into account, the overall machinery of genetic information transfer becomes more elaborate and the effects of simple amino acid substitutions more subtle.

In this connection, it has been pointed out by Ehrenberg and Kurland (1984; and see Chapter 11) that some errors in protein synthesis will not necessarily lead to a reduction in accuracy, and may do the reverse. For example, an error in an information-handling enzyme concerned with regulating the speed of macromolecular synthesis may slow its rate constant, which might have the effect that the reaction it catalyses becomes less efficient but more accurate. Similarly, an error in a ribosomal protein may lower its binding affinity, thereby making it more likely to discard incorrect ternary complexes.

Kurland and Gallant (see Chapter 6) also propose that missense errors at the ribosome may, to some extent, be self-editing, if they also increase the probability of a ribosomal frameshift error. The frameshifted ribosome will then have a high probability of encountering an out-of-frame termination codon, which will result in the early release of the defective polypeptide and thus prevent the formation of a possibly viable but error-prone protein. This concept, termed 'error coupling', which Kurland and Gallant suggest may be how cells come to terms with the presence of defective ribosomes, has some similarity to the earlier idea of 'ribosome editing' (Menninger, 1977, 1983), which proposes that the transfer of peptide to an inappropriate peptidyl-tRNA leads to preferential dissociation of the peptidyl-tRNA and its subsequent destruction by a specific hydrolase.

Two further possibilities can be added to these error-protective systems. Firstly, polymeric enzymes and ribosomal components may have an in-built measure of protection against errors if the structural constraints on their assembly prevent the incorporation of monomers which are abnormal. This would not, however, protect against sequence alterations in monomers which do not affect their structure or binding affinity with other monomers.

Secondly, as ribosomes and RNA polymerases are processive, i.e. they synthesize a whole protein or RNA molecule from start to finish, the more inaccurate they are the greater their chances of making multiple errors in a single molecule. Since about 15% of amino acid substitutions tend to destroy a protein's catalytic activity (Miller and Schmeissner, 1979), multiple errors are quite likely to lead to an inactive product which cannot participate in error feedback.

How important these various factors are to the potential for error propagation remains unclear. The fact that they are not included in simple models of error propagation, such as the adaptor model, does not mean that the general conclusions from these models are wrong. Error coupling, for instance, cannot be a fully effective screen against protein errors, as errors arise during protein synthesis at detectable rates (see below). Although ribosomes may have evolved to damp the contribution to error feedback from error-prone ribosomes, the possibility of an amplification of errors through these or other components of the translation process cannot be ruled out. The effect of error coupling would be to reduce greatly the output of abnormal ribosomes, which is similar (though not identical) to reducing the parameter R in the adaptor model.

Error coupling could also reduce the error level at which error catastrophe occurs, since a small increase in the frequency of errors would result in a substantial increase in the fraction of error-containing ribosomes, which in turn would greatly reduce the overall rate of effective protein synthesis, resulting in slower growth and perhaps even death of the cell. Similarly, the suggestion that some errors may enhance accuracy, instead of lowering it, is obviously not cause for abandoning the idea of error propagation. Firstly, the enhancement in accuracy is suggested to come about through a slowing of the synthesis rate, so the erroneous high accuracy enzymes will be selected against within the cell. Secondly, the proportion of errors which will have an accuracy-enhancing effect, compared to the proportion of errors which will have an accuracy-diminishing effect, for example by a simple change in substrate specificity, is unknown. By the converse of the former point, substitutions which increase the rate of synthesis at the expense of a greater frequency of errors may be selectively favoured, so the overall prediction is not at all clear.

In conclusion, there are several interesting suggestions of factors other than those presently modelled, which may affect the stability of genetic translation. The combined effects of these are likely to modulate, perhaps to a significant degree, the detail of the predictions outlined above. The central concept of a possible feedback of errors through the protein synthesizing machinery remains uncontradicted, however, and while more realistic models are clearly needed, it is only by experiment that the importance of these factors will definitively be established.

2.2.4 GENERAL ERROR THEORY

Models based on protein synthesis alone demonstrate certain principal features of error propagation in a self-replicating translation apparatus. However, as we have remarked already, a realistic theory will have to take account of transcription and DNA replication errors as well (see also Woolhouse, 1969; Lewis and Holliday, 1970; Holliday and Tarrant, 1972; Orgel, 1973; Burnet, 1974; Kirkwood, 1980; Hasegawa *et al.*, 1984). To these can further be added errors in steps such as messenger RNA splicing, DNA methylation, replication of subcellular organelles, such as mitochondria, and in any other process which is involved with, or may affect, some aspect of genetic information transfer.

Although protein errors are likely to be numerically the most common within cells, and are therefore likely to be the primary medium through which error propagation will occur, transcription errors may be of particular significance in the way a stable translation process could become destabilized. Periodically, a transcription error might cause a relatively large increase in the frequency of abnormal protein, and if this occurs in a message for an information-handling enzyme, it may on occasion be sufficient to push the system past the threshold for recovery. Once destabilization has occurred, the error frequency would continue to rise under the impetus primarily of protein errors, until eventually the accuracy of synthesis of RNA and DNA polymerases became so low that transcription errors and mutations would occur quite frequently, and these would then contribute to the final demise of the cell.

The hypothesis that cells may experience an initial build-up of protein errors, followed by progressive deterioration within RNA, DNA, membranes, organelles, and so on, has been termed the *general error theory* (Holliday and Kirkwood, 1983).

2.3 Experimental observations on protein errors and error feedback

Since the original formulation of the error propagation hypothesis (Orgel, 1963), numerous attempts have been made to determine the stability of genetic translation systems experimentally. The questions which experiments needed to answer were not difficult to pose. In the first place, one has to determine the frequency with which different cell types make errors during protein synthesis. To establish the parameters of an error feedback, one needs to measure how much the rates change after sudden perturbations of fidelity or during the lifespan of cells, such as mammalian fibroblasts, which have limited clonal growth. Finally, to evaluate the biological significance of protein errors it is necessary to investigate how different error levels affect cell growth and physiology.

The results obtained from experiments designed to answer these questions have been described in recent reviews and papers (Gallant, 1981; Laughrea, 1982; Kirkwood, Holliday and Rosenberger, 1984) and are considered in several later chapters of this book. Thus, no attempt at another review will be made here, but highly selected aspects of the above problems will be discussed.

2.3.1 ERROR LEVELS DURING PROTEIN SYNTHESIS

Two lines of investigation indicate that in spite of proofreading and other accuracy-enhancing mechanisms a considerable number of random errors escape the protective screens and appear in the finished polypeptide.

One set of data comes from direct estimation of mistakes in protein sequences. A number of factors combine, however, to make the determination of protein errors a difficult and challenging problem (Kirkwood, Rosenberger and Holliday, 1984; see also Chapters 5 and 6). The central problem is that mistakes in translation will introduce errors at random sites in the polypeptide chain and any amino acid may be replaced by one of several alternatives. Detection of errors is therefore much more complex than with mutant proteins, which differ by unique substitutions and can often be resolved into distinct bands or peaks by chromatography or electrophoresis. No such bands or peaks will appear when substitutions are random, and more subtle approaches have been required, each of which has its drawbacks and limitations (Kirkwood, Holliday and Rosenberger, 1984). As also discussed in Chapter 6, measurements of amino acid misinsertions need to be interpreted with care; different codons, and even the same codon in different contexts, cannot be assumed to have identical error frequencies. One cannot, therefore, define a single error rate for a particular organism, nor readily compare error frequencies in cells of widely varying origins. In spite of these reservations, however, it does appear that in *Escherichia coli*, at least, misinsertions can occur as often as 1 in 10^3 amino acids. If similar rates apply to protein synthesis in general, a significant fraction of the larger proteins is likely to be imperfect (see Chapter 1).

A second indication of biologically important levels of error in protein synthesis comes from the apparently universal presence of selective proteases in living organisms, ranging from *E. coli* (Goldberg and St John, 1976; Mount, 1980) to mammals (Hershko and Ciechanover, 1982). One major function of these proteases appears to be to degrade abnormal polypeptides. This activity can be demonstrated by exposing cells to amino acid analogues, which are incorporated as errors in peptide chains, or to antibiotics which reduce the fidelity of protein synthesis. It is found that the resulting abnormal proteins are preferentially degraded (Kemshead and Hipkiss, 1974; Prouty, Karnovsky and Goldberg, 1975; Kelley and Schlesinger, 1978; Hipkiss, 1979; Hightower, 1980; Fong and Poole, 1982; Thomas and Mathews, 1984).

As well as showing very wide phylogenic distribution, the evolutionary conservation of components of the selective proteolysis system is quite remarkable. Eukaryotes cleave many, if not all, of their cytoplasmic proteins by first conjugating them to a small protein, ubiquitin (Hershko and Ciechanover, 1982; Gavilanes *et al.*, 1982; Hershko, 1983). Ubiquitin from yeast and mammalian cells has now been sequenced (Ozkaynak, Finlay and Varshavsky, 1984), and over this range of evolutionary divergence, 73 out of 76 amino acid residues have remained identical, making it the most conserved protein yet described. We do not wish to imply that hydrolysis of error-containing proteins is the only function of ubiquitin-based enzyme reactions. Ubiquitin is also used to mark apparently normal histones for special, but as yet unknown, tasks (Hershko, 1983; Varshavsky *et al.*, 1983) and, presumably, for destroying protein signals needed only at specific points in the cell cycle (Finlay, Ciechanover and Varshavsky, 1984). However, a mutation in the ubiquitin system stops all protein turnover in a eukaryotic cell and is lethal (Finlay, Ciechanover and Varshavsky, 1984; Ciechanover, Finlay and Varshavsky, 1984). We would argue that the universal distribution and evolutionary conservation of selective protease systems is likely to have occurred because errors in the final products of protein synthesis are a real and ongoing threat to living cells.

This argument is strengthened by recent observations that amino acid analogues induce the synthesis of a limited set of proteins in many organisms. In each species, the spectrum of proteins made in response to analogue-containing polypeptides is extremely similar to that induced by other mal-treatments, such as heatshock (Schlesinger, Ashburner and Tissieres, 1982; Neidhardt, van Bogelen and Vaughn, 1984), oxidative damage (Lee, Bochner and Ames, 1983) and severe DNA damage (Neidhardt, van Bogelen and Vaughn, 1984). Again, these reactions appear to occur across the whole evolutionary spectrum, from prokaryotes (Goff, Casson and Goldberg, 1984; Lee, Bochner and Ames, 1983) to mammals (Kelley and Schlesinger, 1978; Hightower, 1980; Schlesinger, Ashburner and Tissieres, 1982). Thus, cells not only have special measures to prevent errors from being made, but they also have complex systems to deal with final, incorrectly assembled products and they appear to have maintained these throughout evolution.

2.3.2 THE OCCURRENCE OF ERROR FEEDBACK

If it is accepted that, in spite of proofreading and other accuracy-promoting devices, proteins do contain an appreciable level of errors, one needs to ask if this leads to a feedback capable of generating an error catastrophe. The arguments for a feedback draw on data for a variety of mutant enzymes which have undergone single amino acid substitutions. Such enzymes lose at least some of their specificity but may retain activity (see Kirkwood and Holliday, 1975). Should this happen through random errors to proteins essential for transcriptional or translational accuracy, then some error feedback is inevitable (see Section 2.2). What cannot be established on theoretical grounds, however, is the extent of any feedback that may occur.

Experiments to measure feedback parameters have, of course, been attempted (for reviews, see Gallant, 1981; Kirkwood, Holliday and Rosenberger, 1984; Holliday, 1986). So far no rigorous proof for error propagation has been obtained. The ideal experiment would be to change the error frequency in cells and to measure error levels after the perturbation. If a significant feedback does take place, the change in misinsertion rates due to the initial perturbation should be followed by further alterations. Such an approach has been attempted by adding aminoglycoside antibiotics to *E. coli* cultures (Branscomb and Galas, 1975; Edelman and Gallant, 1977; Gallant and Foley, 1980; Rosenberger, Foskett and Holliday, 1980; Rosenberger, 1982) or by temperature shifts of a *Neurospora crassa* mutant having a heat-sensitive tRNA charging enzyme (Lewis and Holliday, 1970).

While the kinetics obtained from these perturbation experiments were compatible with error propagation, they could not exclude other interpretations (Gallant, 1981). Proof would depend on demonstrating that some part of the decoding system, which was not affected by the original perturbation, has changed its fidelity. The technical difficulties in doing this are, however, great. The most incisive attempts so far have involved measuring the accuracy of DNA synthesis during the clonal senescence of cultured human fibroblasts (Murray and Holliday, 1981) or in streptomycin-treated *E. coli* (Martin, 1983). In both cases, statistically significant decreases in DNA polymerase fidelity were found, though the differences were not dramatic and proof that they arise through error propagation remains a task for the future.

2.3.3 THE EFFECTS OF PROTEIN ERRORS ON CELL PHYSIOLOGY

The relationship between the magnitudes of protein error rates and their effects on cell growth and metabolism are still quite unclear. The picturesque phrase 'error catastrophe' has done a marked disservice to the subject; it implies an obvious state of disintegration, but even quite modest increases in error levels could have a severe effect on the proportion of error-free proteins. For example, if the error rate increased from 10^{-4} to 10^{-3}, the fraction of erroneous molecules for a small protein with 100 residues would increase from 1 to 10% and for a large protein of 1000 residues from 10 to 63%. This, together with an increase in the rate of protein turnover as scavenging mechanisms attempt to repair the damage, would reduce a cell's efficiency, perhaps to the point where division is no longer possible. Further, there will be protein species with functions which might be disrupted by even a small proportion of faulty molecules. Some examples would be changes in membrane permeases which allow the dissipation of essential gradients, or faulty repressors which bind so tightly that genes are permanently switched off. The possibility that even quite low proportions of erroneous proteins could reduce viability is one that urgently needs clarification.

2.4 Errors in the control of transcription and in the timing of cell cycle events

Almost all studies on error accumulation have concerned themselves with inaccuracies in polymer synthesis and their possible effects on cell physiology. However, errors in two quite different systems could also lead, in a time-dependent manner, to instabilities of cell function.

One of these concerns the control of gene expression by eukaryotic cells. This has been fully reviewed and discussed by Holliday (1984). Briefly, many types of differentiated eukaryotic cells express only a fraction of their genomic capabilities. The controls responsible for such selective gene transcription are inherited in a stable manner during clonal growth. Errors in these transcriptional control mechanisms could lead to a process of de-differentiation (Cutler, 1982); if this occurs in a random manner, growth may be seriously or completely disrupted.

Advances in our knowledge of transcriptional controls have shown the importance of DNA methylation in gene expression (Holliday, 1984). Mistakes in such post-replication modifications of DNA are obviously possible and could accumulate. Methylation of DNA decreases during the *in vitro* lifespan of cultured mammalian fibroblasts (Wilson and Jones, 1983), while experimental inhibition of DNA methylation with azacytidine shortens the lifespan of the cultures. These results indicate that changes in methylation patterns, termed 'epigenetic errors', could be major factors in the instability of cellular growth.

The second system involves the timing of replicon firing in eukaryotic cells. The genome of eukaryotes is organized in a series of many contiguous replicons, each with its own origin of replication. Every replicon is required to duplicate itself once, and only once, in the S phase of a cell generation. Cells must have strict controls to ensure this sequence, although the mechanisms responsible have not yet been characterized. The error under consideration here is when a replicon fires more than once per generation. This would lead to an amplification of the genes in the mistiming region.

Gene amplification is, in fact, a well-characterized phenomenon in living cells. Cells become resistant to antimetabolites (Wahl, Padgett and Stark, 1979) or antitumour agents (Alt *et al.*, 1978) mainly because they overproduce the target enzymes. The most common reason for overproduction is amplification of the genes specifying the enzyme. Approximately one cell in 10^6 appears to have an amplification of a particular operon (Varshavsky, 1981). The causes of amplification have not been established with certainty. One of the most favoured current theories, however, is that mistiming of replicon replication is responsible (Varshavsky, 1981; Schimke *et al.*, 1980; Morrison *et al.*, 1983). The specific replicon is considered to duplicate itself more than once in an *S* phase; this could give rise either to duplications in the chromosome or to small circular DNAs outside of the main genome (Schimke *et al.*, 1980). Both these conditions have been described. Duplications joined covalently to the chromosome are likely to be stable, and if these accumulate the normal balance of enzymes could be greatly disturbed.

2.5 Conclusions

The question of whether a feedback of errors in genetic translation can take place is not in serious dispute. Errors in each step of information transfer are known to occur, and a proportion of these errors will generate active molecules which have altered specificities. What remains unclear, however, is whether under normal biological conditions error feedback can bring about destabilization of cells with sufficient frequency to constitute a serious hazard.

Experimental evidence to date is sufficiently ambiguous as not to provide a clear answer either way. In this situation, there is scope for theoretical approaches to make a significant contribution, both in refining our understanding of what we should expect if error propagation does happen, and in helping to interpret sometimes conflicting and often confusing experimental results. Several of the following chapters in this book deal with aspects of these problems directly, while others touch on them in one respect or another.

Models of error propagation provide a particularly clear way to look at what the theory should predict. As with all mathematical models in biology, however, the difficulty is to incorporate sufficient realism without becoming too complex to be practically useful. The adaptor model, which we have described in detail, illuminates the general behaviour we would predict for a self-replicating genetic translation apparatus. It does not attempt, though, to cover all points. Nonetheless, it does bring out clearly the crucial factor in a protein error feedback, namely the output from error-prone adaptors, represented in aggregate by the parameter R. Since the model was developed, some ten years ago, new concepts, such as ribosome editing and error-coupling, have been proposed. These may help to explain how ribosomal ambiguity mutants of *E. coli* may be viable, despite elevated levels of ribosomal inaccuracy (Kurland *et al.*, 1984; and Chaper 6). Error coupling and ribosome editing do not, however, have a direct bearing on the feedback of mistakes through enzymes, such as those which modify or charge tRNAs, that affect the insertion of each and every residue in polypeptides, whether at normal or abnormal ribosomes. It may be that errors in these systems are particularly significant in generating an error feedback.

Finally, if error propagation can take place, there remains the question of what size of increase in the error level will be required to bring about an error catastrophe. The term 'catastrophe' may, in fact, be misleading in that it suggests a state of gross, and presumably obvious, disorder in the fidelity of synthesis of macromolecules. Early studies sought evidence of major changes in proteins, when it now seems more likely that quite modest increases in error levels may bring about a significant drop in a cell's metabolic efficiency. As error levels begin to rise, there will also be an increasing likelihood of the occasional serious errors which may kill a cell or prevent it from dividing. Studies of ribosomal ambiguity mutants in *E. coli* and of cells perturbed with streptomycin suggest that bacteria can remain viable, even at elevated ribosomal error rates, but these same studies do suggest that a significant feedback of errors may nevertheless be taking place (Gallant and Prothero, 1980). There is good ground to expect bacteria to have evolved to be more stable than, for example, the somatic cells of mammals (see Chapter 12). Primitive cells would almost certainly have been presented with real difficulties from error feedback, and to judge by the investments made in accuracy-promoting devices today, it may well turn out that the risk of error propagation in genetic information transfer is still a potent biological force.

References

Alt, F. W., Kellems, R. E., Bertino, J. R. and Schimke, R. T. (1978) Selective amplification of dihydrofolate reductase genes in methotrexate-resistant variants of cultured murine cell. *J. Biol. Chem.*, **253**, 1357–1370.

Branscomb, E. W. and Galas, D. J. (1975) Progressive decrease in protein synthesis accuracy induced by streptomycin in *E. coli*. *Nature*, **254**, 161–163.

Burnet, F. M. (1974) *Intrinsic Mutagenesis: A Genetic Approach to Ageing*, Wiley, New York.

Ciechanover, A., Finlay, D. and Varshavsky, A. (1984) Ubiquitin dependance of selective protein degradation demonstrated in the mammalian cell cycle mutant ts 85. *Cell*, **37**, 57–66.

Crick, F. H. C., Brenner, S., Klug, A. and Pieczenik, G. (1976) A speculation on the origin of protein synthesis. *Orig. Life*, **7**, 389–397.

Cutler, R. G. (1982) The dysdifferentiative hypothesis of mammalian ageing and longevity. In *The Ageing Brain: Cellular and Molecular Mechanisms of Ageing in the Neurons System* (eds E. Giacobini, G. Filogamo, G. Giacobini and A. Vernadakis), Raven Press, New York.

Edelman, P. and Gallant, J. (1977) Mistranslation in *E. coli*. *Cell*, **10**, 131–137.

Ehrenberg, M. and Kurland, C. G. (1984) Cost of accuracy determined by a maximal growth restraint. *Quart. Rev. Biophys.*, **17**, 45–82.

Eigen, M. and Schuster, P. (1979) *The Hypercycle*. Springer Verlag, Berlin and New York.

Finlay, D., Ciechanover, A. and Varshavsky, A. (1984) Thermolability of ubiquitin-activating enzyme from the mammalian cell cycle mutant ts 85. *Cell*, **37**, 43–55.

Fong, D. and Poole, B. (1982) The effect of canavanine on protein synthesis and protein degradation in IMR-90 fibroblasts. *Biochim. Biophys. Acta*, **696**, 193–200.

Gallant, J. (1981) The error catastrophe theory of cellular senescence: A review. In *Biological Mechanisms of Ageing* (ed. R. T. Schimke), pp. 373–381. Publication no. 81-2194, National Institute of Health, Bethesda, Maryland.

Gallant, J. and Foley, D. (1980) On the causes and prevention of mistranslation. In *Ribosomes: Structure, Function and Genetics* (eds C. Chambliss, C. R. Craven, J. Davis, L. Kahan and M. Nomura), University Park Press, Baltimore, pp. 615–640.

Gallant, J. A. and Prothero, J. (1980) Testing models of error propagation. *J. Theor. Biol.*, **83**, 561–578.

Gavilanes, J. G., de Buitrago, G. G., Perez-Castells, R. and Rodriguez, R. (1982) Isolation, characterisation and amino acid sequence of a ubiquitin-like protein from insect egg. *J. Biol. Chem.*, **257**, 10 267–10 270.

Gershon, D. (1979) Current status of age altered enzymes: Alternative mechanisms, *Mech. Ageing Dev.*, **9**, 189–196.

Goel, N. S. and Islam, S. (1976) Error catastrophe in and the evolution of the protein synthesizing machinery. *J. Theor. Biol.*, **68**, 167–182.

Goel, N. S. and Ycas, M. (1975) Error catastrophe hypothesis with reference to ageing and the evolution of the protein synthesizing machinery. *J. Theor. Biol.*, **55**, 245–282.

Goff, S. A., Casson, L. P. and Goldberg, A. L. (1984) The heatshock regulatory gene, *htpR*, influences rates of protein degradation and expression of the *lon* gene in *E. coli. Proc. Natl Acad. Sci. USA*, **81**, 6647–6651.

Goldberg, A. L. and St John, A. C. (1976) Intracellular protein degradation in mammalian and bacterial cells. *Ann. Rev. Biochem.*, **45**, 747–803.

Hasegawa, M., Yano, T. and Miyata, T. (1984) Evolutionary implications of error amplification and the self-replicating and protein-synthesising machinery. *J. Mol. Evol.*, **20**, 77–85.

Hershko, A. (1983) Ubiquitin: Roles in protein modification and breakdown. *Cell*, **34**, 11–12.

Hershko, A. and Ciechanover, A. (1982) Mechanisms of intracellular protein breakdown. *Ann. Rev. Biochem.*, **51**, 335–364.

Hightower, L. E. (1980) Cultured animal cells exposed to amino acid analogues or puromycin rapidly synthesise several polypeptides, *J. Cell Physiol.*, **102**, 407–427.

Hipkiss, A. R. (1979) Inhibition of breakdown of canavanyl-proteins in *E. coli* by chloramphenicol. *Fems. Microbiol. Lett.*, **6**, 349–353.

Hoffmann, G. W. (1974) On the origin of the genetic code and the stability of the translation apparatus. *J. Mol. Biol.*, **86**, 349–362.

Holliday, R. (1984) The significance of DNA methylation in cellular ageing. In *Molecular Basis of Ageing* (eds A. D. Woodhead and A. D. Blackett), Brookhaven Symposium Volume 33, pp. 1–15.

Holliday, R. (1986) Genes, proteins and cellular ageing. *Benchmark Papers in Genetics* (ed. R. Holliday), van Nostrand Reinhold, Philadelphia.

Holliday, R. and Kirkwood, T. B. L. (1983) Theories of cell ageing: A case of mistaken identity. *J. Theor. Biol.*, **103**, 329–330.

Holliday, R. and Tarrant, G. M. (1972) Altered enzymes in ageing human fibroblasts. *Nature*, **238**, 26–30.

Hopfield, J. J. (1974) Kinetic proofreading: A new mechanism for reducing errors in biosynthetic processes requiring high specificity. *Proc. Natl Acad. Sci. USA*, **71**, 4135–4139.

Kelley, P. M. and Schlesinger, M. J. (1978) The effect of amino acid analogues and heat shock on gene expression in chicken embryo fibroblasts. *Cell*, **15**, 1277–1286.

Kemshead, J. T. and Hipkiss, A. R. (1974) Degradation of abnormal proteins in *E. coli*: relative susceptibility of canavanyl-proteins and puromycin peptides to proteolysis *in vitro*. *Eur. J. Biochem.*, **45**, 535–540.

Kirkwood, T. B. L. (1977) Evolution of ageing. *Nature*, **270**, 301–304.

Kirkwood, T. B. L. (1980) Error propagation in intracellular information transfer. *J. Theor. Biol.*, **82**, 363–382.

Kirkwood, T. B. L. and Holliday, R. (1975) The stability of the translation apparatus. *J. Mol. Biol.*, **97**, 257–265.

Kirkwood, T. B. L., Holliday, R. and Rosenberger, R. F. (1984) Stability of the cellular translation process. *Int. Rev. Cytol.*, **92**, 93–132.

Kurland, C. G., Andersson, D. I., Andersson, S. G. E., Bohman, K., Bouadloun, F., Ehrenberg, M., Jelenc, P. C. and Ruusala, T. (1984) Translational accuracy and bacterial growth. In *Gene Expression*, Alfred Benzon Symposium, Volume 19 (eds B. F. C. Clark and H. U. Peterson), Munksgaard, Copenhagen, pp. 193–207.

Laughrea, M. (1982) On the error theory of ageing: A review of the experimental data. *Exp. Gerontol.*, **17**, 305–317.

Lee, D. C., Bochner, B. R. and Ames, B. N. (1983) ApppA, heat-shock stress and cell oxidation. *Proc. Natl Acad. Sci. USA*, **80**, 7496–7500.

Lewis, C. M. and Holliday, R. (1970) Mistranslation and ageing in *Neurospora*. *Nature*, **228**, 877–880.

Martin, R. (1983) Translational accuracy and the fidelity of DNA replication in *E. coli*. PhD Thesis, CNAA, London.

Medvedev, Zh. A. (1962) Ageing at the molecular level. In *Biological Aspects of Ageing* (ed. N. W. Shock), Columbia University Press, New York, pp. 255–266.

Medvedev, Zh. A. (1980) The role of infidelity of transfer of information for the accumulation of age changes in differentiated cells. *Mech. Ageing Dev.*, **14**, 1–14.

Menninger, J. R. (1977) Ribosome editing and the error catastrophe hypothesis of cellular ageing. *Mech. Ageing Dev.*, **6**, 131–142.

Menninger, J. R. (1983) Computer simulation of ribosome editing. *J. Mol. Biol.*, **171**, 383–399.

Miller, J. H. and Schmeissner, U. (1979) Genetic studies of the *lac* repressor X. Analysis of missense mutations in the *lac* I gene. *J. Mol. Biol.*, **131**, 223–248.

Miller, S. L. and Orgel, L. E. (1973) *The Origins of Life on Earth*, Prentice Hall, New York.

Morrison, P. F., Aroesty, J., Creekmore, S. P., Barker, P. F. and Lincoln, T. L. (1983) A preliminary model of double minutes mediated by gene amplification. *J. Theor. Biol.*, **104**, 71–91.

Mount, D. W. (1980) The genetics of protein degradation in bacteria. *Ann. Rev. Genet.*, **14**, 279–319.

Murray, V. and Holliday, R. (1981) Increased error frequency of DNA polymerase from senescent human fibroblasts. *J. Mol. Biol.*, **146**, 55–76.

Neidhardt, F. C., van Bogelen, R. A. and Vaughn, V. (1984) The genetics and regulation of heat-shock proteins. *Ann. Rev. Genet.*, **18**, 295–329.

Ninio, J. (1975) Kinetic amplification of enzyme discrimination. *Biochimie*, **57**, 587–595.

Orgel, L. E. (1963) The maintenance of the accuracy of protein synthesis and its relevance to ageing. *Proc. Natl Acad. Sci. USA*, **49**, 517–521.

Orgel, L. E. (1970) The maintenance of the accuracy of protein synthesis and its relevance to ageing: A correction. *Proc. Natl Acad. Sci. USA*, **67**, 1476.

Orgel, L. E. (1973) Ageing of clones of mammalian cells. *Nature*, **243**, 441–445.

Ozkaynak, E., Finlay, D. and Varshavsky, A. (1984) The yeast ubiquitin gene: head to tail repeats encoding a polyubiquitin precursor. *Nature*, **312**, 663–666.

Prouty, W. K., Karnovsky, M. L. and Goldberg, A. L. (1975) Degradation of abnormal proteins in *E. coli*. Formation of protein inclusions in cells exposed to amino acid analogues. *J. Biol. Chem.*, **250**, 1112–1122.

Rosenberger, R. F. (1982) Streptomycin-induced protein error propagation appears to lead to cell death in *E. coli*. *IRCS Med. Sci.*, **10**, 874–875.

Rosenberger, R. F., Foskett, G. and Holliday, R. (1980) Error propagation in *E. coli* and its relation to cellular ageing. *Mech. Ageing Dev.*, **13**, 247–252.

Rothstein, M. (1977) Recent developments in the age-related alterations of enzymes: A review. *Mech. Ageing Dev.*, **6**, 241–257.

Schimke, R. T., Brown, P. C., Kaufman, R. J., McGrogan, M., and Slate, D. L. (1980) Chromosomal and extrachromosomal localization of amplified dihydrofolate reductase genes in cultured mammalian cells. *Cold Spring Harbor Symp. Quant. Biol.*, **45**, 785–797.

Schlesinger, M. J., Ashburner, M. and Tissieres, A. (eds) (1982) *Heat Shock: From Bacteria to Man*, Cold Spring Harbor Laboratory, New York.

Thomas, P. G. and Mathews, M. B. (1984) Alterations of transcription and translation in Hela cells exposed to amino acid analogs. *Mol. Cell. Biol.*, **4**, 1063–1072.

Varshavsky, A. (1981) On the possibility of metabolic control of replicon misfiring: Relationship to emergence of malignant phenotypes in mammalian cell lineages. *Proc. Natl Acad. Sci. USA*, **78**, 3673–3677.

Varshavsky, A., Levinger, L., Lundin, O., Barsum, H., Ozkaynak, E., Swerdlow, P. and Finlay, D. (1983) Cellular and SV40 chromatin: replication, segregation, ubiquitination, nuclease hypersensitive sites, HMG-containing nucleosomes and heterochromatin-specific protein. *Cold Spring Harbor Symp. Quant. Biol.*, **47**, 511–528.

Wahl, G. M., Padgett, R. A. and Stark, G. R. (1979) Gene amplification causes overproduction of the first three enzymes of UMP synthesis in *N*(phosphonacetyl)-L-aspartate resistant hamster cells. *J. Biol. Chem.*, **254**, 8679–8689.

Wilson, V. L. and Jones, P. A. (1983) DNA methylation decreases in ageing but not in immortal cells. *Science*, **220**, 1055–1057.

Woolhouse, H. W. (1969) DNA polymerase, genetic variation and determination of the life span. In *Proc. 8th Int. Congr. Gerontol.*, Vol. 1, pp. 162–166. Federation of American Societies for Experimental Biology, Washington, DC.

3 The specificity of enzyme–substrate interactions

M. I. PAGE

3.1 Introduction

Molecules interact with one another. There is continual competition between their kinetic energies to be independent and the forces of intermolecular attraction to bring them together. Kinetic energies may dominate the interaction, such that molecules exist as gases with individual mobilities only slightly impaired by their environment. But even weak forces of attraction, less than 5 kJ mol^{-1}, may be sufficient to encourage the formation of a fragile 'complex' such as an inert gas dimer, held together in loose geometrical configuration by dispersion or Van der Waals forces. At the other extreme, the strong interactions of a chemical bond, with a stabilization of 250 kJ mol^{-1} or more, may hold groups together with a specific, relatively rigid, geometry. Between these two limits are various interactions of intermediate strength which account for the cohesion and yet mobility of liquids, and for the existence of specific biologically important complexes such as those of antibody–antigen, hormone–receptor and enzyme–substrate.

If A is a proton donor and B a proton acceptor then the complex AB may be created by the formation of a hydrogen bond between them. If C is also a proton donor then the complex CB may be formed. In a mixture of A, B and C either both complexes may be formed, or only one may be selected preferentially. Enzymes, the catalysts of life, are seen as successful if they can recognize and select their faithful partner when confronted with a host of enticing infidels.

This chapter will review the principles and define some of the problems associated with understanding enzymic catalysis, and outline the interactions that occur between the reactant 'guest' and enzyme 'host' as the chemical transformation takes place, converting substrate into product.

3.2 Kinetics and thermodynamics

Some interesting conclusions may be drawn simply by examining the kinetics
and thermodynamics of enzyme catalysed reactions.

3.2.1 KINETICS

Experimentally, the initial rate, v, of enzyme catalysed reactions is found to
show saturation kinetics with respect to the concentration of the substrate, S.
That is, at low concentrations of substrate v increases with increasing concentration of S, but it becomes independent of $[S]$ at high or saturating
concentrations of S (Fig. 3.1). This observation was interpreted by Michaelis
and Menten in terms of the rapid and reversible formation of a non-covalent
complex, ES, from the substrate, S, and enzyme, E, which then decomposes
into products, P:

$$E + S \underset{k_{-1}}{\overset{k_1}{\rightleftharpoons}} ES \xrightarrow{k_2} EP \rightleftharpoons E + P \tag{3.1}$$

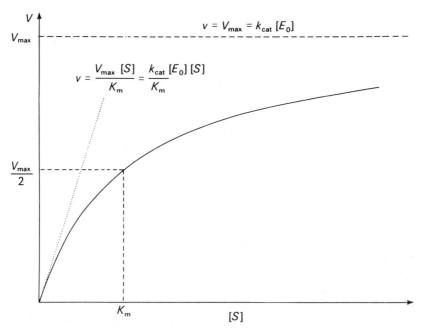

Figure 3.1 Reaction rate–substrate concentration profile for a reaction obeying
Michaelis–Menten kinetics.

This scheme led to the familiar Michaelis–Menten equation:

$$v = \frac{k_2[S]}{(k_{-1}+k_2)/k_1+[S]} = \frac{k_{cat}[E_0][S]}{K_m+[S]} = \frac{V_{max}[S]}{K_m+[S]}, \qquad (3.2)$$

where K_m is the Michaelis constant, which is the concentration of substrate at which the initial rate is *half* the maximal rate at saturation, V_{max}. The Michaelis constant $K_m = (k_{-1}+k_2)/k_1$ if k_2 is comparable with k_{-1}, but $K_m = k_{-1}/k_1$, the true dissociation constant of ES, if $k_{-1} \gg k_2$. The first-order rate constant for the decomposition of ES is commonly called k_{cat} (the turnover number), k_2 in Equation (3.1). At low concentration of substrate, where $[S] \ll K_m$, v is given by:

$$v = \frac{k_{cat}}{K_m}[E_0][S] = \frac{k_1 k_2}{k_{-1}+k_2}[E_0][S] \qquad (3.3)$$

with k_{cat}/K_m being an apparent second-order rate constant. At high concentrations of S, where $S \gg K_m$, v is given by:

$$v = k_{cat}[E_0] = k_2[E_0] = V_{max} \qquad (3.4)$$

and becomes independent of S.

Although the Michaelis–Menten equation (3.2) is valid for most enzyme-catalysed reactions the mechanism (3.1) is not always followed. The measured K_m and k_{cat} values are not always equal, respectively, to the dissociation constant, K_s, for the enzyme–substrate complex and the rate constant for decomposition of ES. The apparent dissociation constant, K_m, can be less than K_s (i.e. apparent tighter binding of substrate) if additional intermediates, covalently or non-covalently bound, are formed during the reaction pathway and the rate limiting step is the reaction of one of these intermediates. Similarly, $K_m > K_s$ if the rate of dissociation of ES to E and S is comparable to or slower than the forward rate of reaction of ES (Briggs–Haldane kinetics). The measured value of k_{cat} may also be a function of several microscopic rate constants (Fersht, 1977), and if the chemical steps occur sufficiently quickly, then either the binding of the substrate or the desorption of the product may become rate limiting. Obedience to Michaelis–Menten kinetics by enzymes in their natural milieu has interesting implications for their mechanism of action.

(a) Working conditions for enzyme and substrate
Individual enzymes *in vivo* have different constraints and requirements. Generally, intracellular enzymes are required to maintain a constant concentration of the various metabolites and this may be achieved by having a wide variation in the reaction flux of material through the various metabolic

pathways. The reaction rate will vary with $[S]$ if the enzyme is working below saturation ($K_m >> [S]$) and the rate is given by Equation (3.3). However, extracellular enzymes are often faced with dramatic changes in the concentration of their substrates and yet are required to maintain a steady flow of material for absorption and use by the cell. The reaction rate will be independent of $[S]$ if the enzyme is working under saturation conditions ($[S] >> K_m$) and the rate is given by Equation (3.4) (Albery and Knowles, 1976).

Given obedience to Michaelis–Menten kinetics, one may make various inferences about the optimal physiological concentration for the substrate. Obviously, the maximal change in rate for a change in substrate concentration is when $[S]$ and v are minimal, ($dv/d[S]$ is maximal at $[S] = 0$, $v = 0$). If there is an optimal physiological concentration for the substrate, enzymes could have evolved to bind the substrate more or less tightly to maximize the changes in rate with respect to changes in substrate concentration. The problem is illustrated in Fig. 3.2, with two substrate concentrations $[S_1]$ and $[S_2]$ having respective rates v_1 and v_2. The minimal fractional change in substrate concentration ($[S_1]/[S_2]$) to obtain a given change in velocity, x,

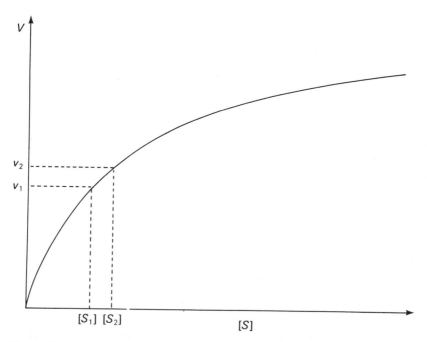

Figure 3.2 A reaction showing saturation kinetics has different rates, v_1 and v_2, at different substrate concentrations $[S_1]$ and $[S_2]$ respectively.

occurs at the K_m for the reaction. This is shown as follows. From Equation (3.2), if $v_2 = v_1 + x$:

$$[S_1] = \frac{v_1 K_m}{V_{max} - v_1} \tag{3.5}$$

and

$$[S_2] = \frac{(v_1 + x) K_m}{V_{max} - (v_1 + x)}. \tag{3.6}$$

Thus,

$$\frac{[S_2]}{[S_1]} = \frac{V_{max} v_1 + V_{max} x - x v_1 - v_1^2}{V_{max} v_1 - v_1^2 - x v_1} \tag{3.7}$$

and

$$\frac{d[S_2]/[S_1]}{dv_1} = \frac{V_{max} x (x + 2v_1 - V_{max})}{(V_{max} v_1 - v_1^2 - x v_1)^2} \tag{3.8}$$

which is zero for $V_{max} = x + 2v_1 = v_1 + v_2$. $\tag{3.9}$

The derivative of $[S_2]/[S_1]$ is therefore zero when $v_1 + v_2 = V_{max}$, i.e. at the K_m of the reaction, and so the optimal physiological concentration of the substrate is at its K_m (Fersht, 1974).

(b) Specificity
The kinetics of enzyme catalysed reactions can also yield interesting conclusions about specificity and accuracy. *In vitro* 'non-specific' substrates are sometimes described as 'poor' because they show a low value of k_{cat} or a high value of K_m. However, *in vivo* discrimination results from a *competition* of substrates for the active site of the enzyme. If two substrates, S and S', compete for the same enzyme wrong conclusions could be reached about the relative rates of product formation if the values of k_{cat} or K_m of the individual substrates are compared instead of their relative values of k_{cat}/K_m (Fersht, 1977).
If the enzyme catalyses the reaction of both S and S':

$$E + S + S' \underset{k_{-1}}{\overset{k_1}{\rightleftharpoons}} ES \xrightarrow{k_2} P \tag{3.10}$$

$$k'_{-1} \updownarrow k'_1$$

$$ES' \xrightarrow{k'_2} P'$$

the relative rates of product formation, $d[P]/d[P']$, may be derived from the usual steady state procedure

$$[ES] = \frac{k_1[E_0][S]}{k_{-1} + k_2 + k_1[S] + k_1'[S](k_{-1} + k_2)/(k_{-1}' + k_2')} \tag{3.11}$$

$$[ES'] = \frac{k_1'[E_0][S']}{k_{-1}' + k_2' + k_1'[S'] + k_1[S](k_{-1}' + k_2')/(k_{-1} + k_2)} \tag{3.12}$$

$$\frac{d[P]}{d[P']} = \frac{k_2[ES]}{k_2'[ES']} \tag{3.13}$$

avoiding any assumptions about the relative values of k_2 and k_{-1} (Briggs–Haldane method). It is given by

$$\frac{v}{v'} = \frac{k_1}{k_1'} \times \frac{k_2}{k_2'} \times \frac{k_{-1}' + k_2'}{k_{-1} + k_2} \times \frac{[S]}{[S']} = \frac{(k_{cat}/K_m)[S]}{(k_{cat}'/K_m')[S']} \tag{3.14}$$

whether the enzyme is working below or above saturation for both substrates. Furthermore, Equation (3.14) is applicable even if the individual K_ms or substrate concentrations are such that the enzyme is working below saturation for one substrate but above saturation for the other. Specificity between competing substrates is therefore given by the relative values of k_{cat}/K_m and not by the individual values of k_{cat} or K_m. The relative rates of product formation depend upon the differences in the free energies of activation of the various substrates. It is incorrect to say that an enzyme catalyses the reaction of A more efficiently than that of B because it binds A more tightly than B. It is also incorrect to say that the difference is attributable to a higher value of k_{cat} (k_2). If for a series of possible substrates some of the microscopic rate constants are similar (e.g. $k_1 \simeq k_1'$ and $k_2 \simeq k_2'$) then discrimination could be discussed in terms of the relative values of k_{-1} and k_{-1}'. However, in general all the microscopic rate constants in Equation (3.14) will contribute to the relative rates of product formation.

3.2.2 THERMODYNAMICS

All chemical reactions are theoretically reversible and many enzymes function in both directions physiologically. In a series of reactions of a metabolic pathway one of the steps may be thermodynamically unfavourable but net synthesis in one direction can still occur if the overall equilibrium constant is favourable. The sequence of events

$$S \rightleftharpoons \overset{w\ \ x}{\underset{}{B}} \rightleftharpoons C \overset{y\ \ z}{\underset{}{\rightleftharpoons}} D \rightleftharpoons P \tag{3.15}$$

may involve the formation of several intermediates, B, C, D, and require the conversion of coenzymes w and y into x and z, respectively. The chemical flux out of S and P may be controlled by (1) changing the concentrations of S or P, (2) changing the concentrations of the coenzymes, or (3) changing the activities of the enzymes involved in the pathway.

Thermodynamically, Equation (3.15) may be described by

$$\Delta G = \Delta G^0 + RT \ln \left(\frac{[P][x][z]}{[S][w][y]} \right) \tag{3.16}$$

where ΔG is the Gibbs free energy change for the reaction under the given conditions and ΔG^0 is the change in Gibbs energy for the reaction with reactants and products at the standard state concentrations. At thermodynamic equilibrium there can be no net flow. Enzymes do not alter the position of equilibrium between *unbound* substrates and products. If any of the intermediates are removed rapidly their concentration will correspond to their steady state concentrations rather than to those at equilibrium. If any of the steps in Reaction (3.15) are near equilibrium the rate of the 'reverse' reaction will be similar to that of the 'forward' direction and none of these steps can therefore limit the rate of production of P.

If in the reaction sequence

$$E + S \underset{K_S}{\rightleftharpoons} ES \overset{K_E}{\rightleftharpoons} EP \overset{K_P}{\rightleftharpoons} E + P \tag{3.17}$$

the dissociation constants for substrate and product, K_S and K_P, respectively, are very different it is conceivable that the equilibrium constant K_E is not a good guide to the value of the overall equilibrium constant

$$K = \frac{K_P}{K_S} \times K_E \tag{3.18}$$

if substrate concentration is not in excess.

3.3 Rates of reaction and accuracy

Because enzymes both increase the rate of reactions and show discrimination between possible substrates there has been a temptation to treat these two phenomena separately. Classically, the rate enhancement is attributed to the chemical mechanism used by the enzyme to bring about transformation of the substrate. The fidelity of enzyme and substrate is accounted for by binding, as in the analogy of a 'lock and key'. It is now apparent that these simple ideas need to be reappraised. Chemical catalysis alone cannot explain the efficiency of enzymes (Page, 1979, 1984). The forces of interaction between the *non-reacting* parts of the substrate and enzyme may also contribute to a lowering of the activation energy of the reaction.

An example illustrates this point (Jencks, 1980). The enzyme succinyl-CoA-acetoacetate transferase catalyses the reaction

$$R_1CO_2^- + R_2COS\text{--}CoA \rightleftharpoons R_1COS\text{--}CoA + R_2CO_2^- \qquad (3.19)$$

($R_1CO_2^-$ = acetoacetate and $R_2CO_2^-$ = succinate) and proceeds by the initial formation of an enzyme–CoA intermediate in which the coenzyme A is bound to the enzyme as a thiol ester of the γ-carboxyl group of glutamate

$$R_2COS\text{--}CoA + Enz\text{--}CO_2^- \rightleftharpoons R_2CO_2^- + Enz\text{--}COS\text{--}CoA \qquad (3.20)$$

In turn, this intermediate is generated by nucleophilic attack of the glutamate carboxylate on succinyl-CoA to give an anhydride intermediate (Fig. 3.3(a)). The second order rate constant for this reaction is 3×10^{13} fold greater than the analogous reaction of acetate with succinyl-CoA (Fig. 3.3(b)). It seems

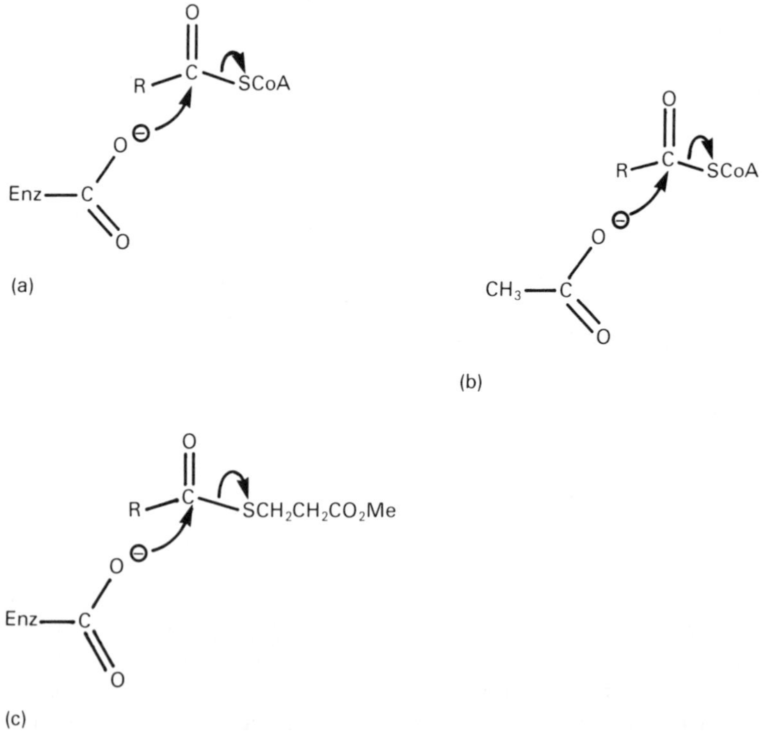

(a)

(b)

(c)

Figure 3.3 A comparison of enzyme and non-enzyme catalysed reactions (see text for details).

unlikely that the chemical reactivity of acetate and the enzyme's glutamate will be vastly different. Similar chemical reactions are therefore being compared and yet the *non-reacting* part of the enzyme lowers the activation energy by 78 kJ mol^{-1} ($RT \ln 3 \times 10^{13}$).

The same example may be used to illustrate specificity. The enzyme also forms an anhydride with the 'non-specific' substrate, succinyl methyl mercaptopropionate (Fig. 3.3(c)). However, the enzyme reacts with this substrate 3×10^{12} fold slower than with succinyl-CoA. The chemical reactivities of the two substrates are similar, for example, towards alkaline hydrolysis. The small substrate, succinyl methyl mercaptopropionate, should be able to fit into the active site. Therefore, the *non-reacting* part of succinyl-CoA of molecular weight *ca* 770 lowers the activation energy by 72 kJ mol^{-1} ($RT \ln 3 \times 10^{12}$).

Enzymes may obviously discriminate against substrates larger than the specific substrate because they are too large to fit into the active site. However, it is not obvious why some small non-specific substrates that presumably can bind to the active site react very slowly. For example, is there enough energy available between the non-reacting parts of the substrate and the enzyme to account for the discrimination between isoleucine (Fig. 3.4(a))

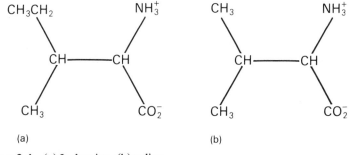

(a)   (b)

Figure 3.4   (a) Isoleucine, (b) valine.

as a specific substrate and valine (Fig. 3.4(b)) as an alternative? Specificity can be reflected in poor binding (high K_s) and/or slow catalytic steps (low k_{cat}), but it was shown earlier (Equation (3.14)) that specificity between competing substrates is controlled by their relative values of k_{cat}/K_m, ($k_1 k_2/(k_{-1} + k_2)$). The second-order rate constant k_{cat}/K_m is directly related to the free-energy difference, $\Delta G^{\ddagger}$, between the enzyme–substrate transition state, $ES^{\ddagger}$, and the *free* unbound substrate and enzyme by

$$E + S \xrightleftharpoons{\Delta G^{\ddagger}} ES^{\ddagger} \qquad\qquad (3.21)$$

$$k_{cat}/K_m = \frac{kT}{h} \exp(-\Delta G^{\ddagger}/RT) \qquad\qquad (3.22)$$

where $\Delta G^{\ddagger}$ is the free energy of activation. The relative rates of an enzyme-catalysed reaction of two substrates S_1 and S_2, of the same chemical reactivity is therefore given by the difference between the free energies of binding the non-reacting parts of the substrates to the enzyme in the transition state and in the ground state. The latter can be important. For example, interchanging hydrophobic and hydrophilic substituents may cause a difference in rates which is attributable to differences in ground state energies. One simple way the latter can be measured is by comparing the solubilities of the two substrates (Bell *et al.*, 1974).

3.3.1 NON-PRODUCTIVE BINDING

It is often stated that discrimination can occur if potentially undesirable substrates are bound preferentially to a site on the enzyme where catalysis cannot occur

$$
\begin{array}{ccc}
& k_1 & k_2 \\
E + S & \underset{k_{-1}}{\overset{}{\rightleftharpoons}} & ES \xrightarrow{} ES^{\ddagger} \\
k'_{-1} \Big\Uparrow\Big\downarrow k'_1 & & \\
ES' & &
\end{array}
\tag{3.23}
$$

The rate is $k_2[ES]$ which in terms of microscopic rate constants is given by

$$
v = \frac{k_2[E][S]}{(k_{-1}+k_2)/k_1 + [S] + k'_1(k_{-1}+k_2)[S]/k'_{-1}k_1}
\tag{3.24}
$$

which may be written as

$$
v = \frac{k_2[E][S]}{K_m^0(1+[S]/K'_s) + [S]}
\tag{3.25}
$$

where K_m^0 is the Michaelis constant for the reaction in the absence of an additional binding site and K'_s, equal to k'_{-1}/k'_1, is the dissociation constant for the non-productive complex ES'. The *observed* binding constant, K_m, is lower than K_m^0 because the additional binding site leads to apparently tighter binding. At saturation, only a fraction of the substrate is productively bound and $k_2[ES]$ is lowered. The *in vitro* conclusion could therefore be that the binding energy between the 'wrong' substrate and the 'wrong' site of the enzyme is used to prevent rapid reaction at the active site; the larger specific substrate can only bind at the active site and is sterically prevented from binding to the 'wrong' site. Although this is an appealing idea, it cannot explain the discrimination between competing substrates *in vivo*. The free-energy difference between the unbound substrate and enzyme, $E + S$, and the transition state, $ES^{\ddagger}$, is unaffected by alternative modes of binding, e.g. to ES' (Equation (3.23)). That is, the free energy of activation, represented by k_{cat}/K_m, is not changed by non-productive binding.

The dominant direct reaction pathway need not necessarily contain the explanation of specificity and there are conceivable, but usually wasteful, types of side reactions which may contribute to the prevention of an undesirable product. There may be an alternative *reactive* site on the enzyme to which the smaller 'unwanted' substrate, intermediate or product *can bind and react* but which is less efficient towards the desired substrate, intermediate or product. This has been suggested to occur with the aminoacyl-tRNA synthetases (Fersht and Kaethner, 1976).

3.3.2 INDUCED FIT

In the induced fit model of discrimination the active site of the free enzyme, E, is in the 'wrong' conformation and is catalytically inactive. The binding of a 'good' substrate induces a conformational change in the enzyme making it catalytically active, E':

$$E \; \overset{K}{\rightleftharpoons} \; E'$$

$$ES \; \rightleftharpoons \; E'S \; \overset{k_{cat}}{\longrightarrow} \; E'S^{\ddagger} \qquad (3.26)$$

A 'poor' substrate does not have enough binding energy to alter the conformational state of the enzyme. On the other hand the 'good' substrate provides enough binding energy to 'pay for' the conversion to the unfavourable, but active, conformation of the enzyme, E'. For this to occur the free energy of binding 'good' substrates must be greater than the free energy of distortion of the enzyme. Induced fit can explain specificity between very good and very poor substrates but it is less successful when applied to a series of substrates that all have sufficient binding energy to compensate for the unfavourable conformational change. The induced-fit enzyme is therefore less efficient than the active enzyme by a factor of K, the ratio of inactive to active molecules of enzyme:

$$(k_{cat}/K_m)_{obs} = k_{cat}K/K_m \qquad (3.27)$$

If the substrates which are required to be discriminated against have sufficient binding energy to compensate for the unfavourable conformational change of the enzyme, then induced fit cannot explain specificity between competing substrates. Compared with the situation where the enzyme is initially in the active conformation, the induced fit mechanism reduces the value of k_{cat}/K_m

for *all* substrates by the *same* fraction and therefore does not affect their relative rates.

Induced fit can therefore only be used to explain the rates of reactions of very poor substrates, e.g. the rate of phosphoryl transfer to water compared with that to glucose catalysed by hexokinase. Although water can almost certainly bind to the active site it must have insufficient binding energy to induce the necessary conformational change in the enzyme (Jencks, 1975).

3.4 Discrimination through binding

The fidelity of enzymes to particular substrates has been discussed in phenomenological terms and this section now reviews the problem in detail. Most reactions, but not all, that are catalysed by enzymes involve a considerable change in geometry and electron density as reactants are converted to products.

The binding energy between substrate and enzyme may be used in a variety of mechanisms to lower the activation energy of the reaction. The main mechanisms are charge (including solvation), geometrical and entropic effects. The electron density distribution in a molecule determines the nuclear configuration and the first two mechanisms are not always easily separable, but as molecules may have similar shapes and yet different charge distributions there are important differences.

It is now generally considered that maximum binding energy, i.e. stabilization, occurs between the substrate and enzyme in the transition state of the reaction. There is an exception, however, for the case in which an enzyme equally stabilizes the ground state and transition state. In this situation, catalysis can only occur if the enzyme is working below saturation conditions (Page, 1984). Below saturation, catalysis would occur if both the ground state and transition state were bound equally tightly (Fig. 3.5). However, there is a limit to this type of catalysis because if the enzyme binds the transition state and ground state very tightly, although the free energy of activation will be reduced, the concentrations of enzyme and substrate required to maintain non-saturation conditions will be decreased. These concentrations may be so low that catalysis would not be observed.

Maximum catalytic efficiency may be achieved by the enzyme stabilizing all transition states and destabilizing, or at least not overstabilizing, all intermediate states in the pathway between reactants and products. The interconversion of S and P may proceed in the thermodynamically favourable direction with a rate that is limited by the diffusion together of the enzyme and, say, S (Fig. 3.6). In the reverse direction the minimum free energy of activation corresponds to the free energy of activation for diffusion plus ΔG_0, the free energy difference between S and P (Fig. 3.6). It is obvious that an efficient enzyme must stabilize the transition state(s) of a reaction. But it is

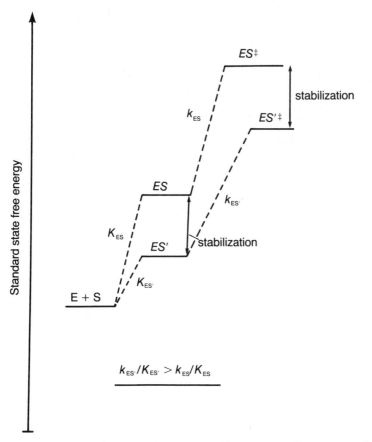

$$k_{ES'}/K_{ES'} > k_{ES}/K_{ES}$$

Figure 3.5 Standard state free energy changes for a series of enzyme-catalysed reactions where the substrate or enzyme is modified so that the ground states and transition states are equally stabilized by the modification.

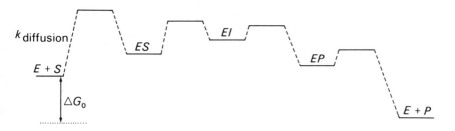

Figure 3.6 Standard state free energy changes for an enzyme-catalysed reaction proceeding via the formation of intermediates. The enzyme should stabilize the transition state but not any of the intermediates.

equally important that the enzyme does not excessively stabilize any intermediate. Stable enzyme intermediate states will bring about saturation conditions at a low concentration of substrate. Valuable enzyme will be then tied up in an energy well.

For a given amount of binding energy between the substrate and enzyme the most effective catalysis will be obtained if this is used to stabilize the transition state, which maximizes the value of k_{cat}/K_m. For a given free-energy of activation $\Delta G^{\ddagger}$ for the process $E + S \rightarrow ES^{\ddagger}$, i.e. for a fixed value of k_{cat}/K_m, and for a given substrate concentration, the maximum rate is obtained if the substrate is bound *weakly*, i.e. a high K_m (Fig. 3.7). A low value of K_m, i.e. the tight binding of the substrate or intermediate state, mediates against catalysis.

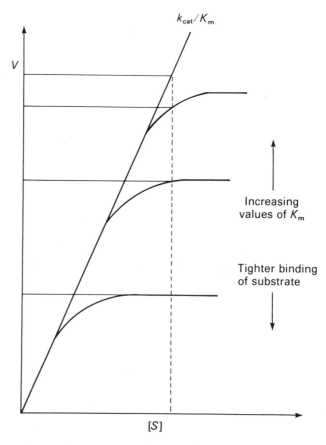

Figure 3.7 Reaction rate–substrate concentration profiles for a given free energy of activation (fixed k_{cat}/K_m) but with varying values of K_m. A higher maximal rate is achieved with weak binding of the substrate.

In agreement with these ideas, the physiological concentrations of most substrates are below their K_m values (Fersht, 1974).

The argument that the enzyme–substrate complex should not be too stable compared with the free enzyme and substrate is sometimes inverted to suggest that the substrate should be *destabilized* in the *ES* complex. This is illustrated in Fig. 3.8 where the energy of the *ES* complex is raised by an unspecified mechanism with the result that k_{cat} is increased. However destabilizing the enzyme–substrate complex has no effect upon k_{cat}/K_m. This mechanism would therefore make the enzyme more effective above saturation but not below saturation. Specificity, in the sense of discrimination between competing substrates, which we have seen depends upon the value of k_{cat}/K_m, would thus not be affected by having the enzyme–substrate complex destabilized. The introduction of strain or any other mechanism of destabilizing the substrate cannot therefore directly affect specificity. However, we have also seen (Fig. 3.7) that for a given k_{cat}/K_m the maximum rate is obtained if the substrate is bound weakly. A high value of K_m, resulting from weak binding or the 'destabilization' of the *ES* complex, favours non-saturation conditions, which is good. If the binding energy used to stabilize the transition state was also used to stabilize the ground state, saturation conditions would set in at very low concentrations of substrate. For given values of k_{cat}/K_m and enzyme and substrate concentrations, a lower rate of reaction results if binding energy

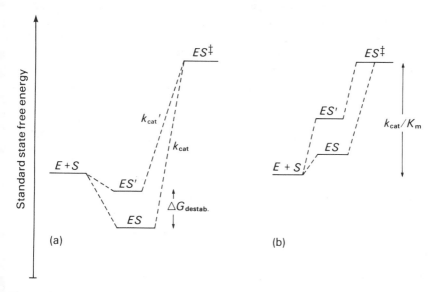

Figure 3.8 Standard state free energy changes for an enzyme-catalysed reaction showing saturation kinetics in which the enzyme–substrate complex is destabilized by an unspecified mechanism. Destabilization of the ground state increases the rate of reaction above saturation (a) but has no effect below saturation (b).

is used to produce tight binding of the ground state, i.e. a stable *ES* complex, and consequently a low value of K_m (Fig. 3.7). Weak binding of the substrate therefore favours efficient catalysis and can contribute to specificity.

A major question in understanding the discrimination between substrates is: how does the system allow the binding energy between the 'non-reacting' parts of the enzyme and substrate to be expressed in the transition state and not in the ground state? If a 'non-specific' and a 'specific' substrate have identical reaction centres (Fig. 3.9) the intrinsic binding energy between the reacting groups undergoing electron density changes and the enzyme should be similar. How can the binding energy of the non-reacting part of the specific substrate stabilize the transition state but not over stabilize the enzyme–

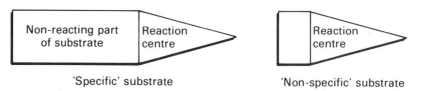

'Specific' substrate   'Non-specific' substrate

Figure 3.9   The difference between specific and non-specific substrates is often in the non-reacting part of the molecule.

substrate or enzyme–intermediate complex? Is the binding energy used to destabilize the substrate or to compensate for thermodynamically unfavourable processes necessary for reaction to occur? That this is a real problem is reflected by the fact that the observed binding constant of a specific substrate is often apparently 'weaker' or no 'tighter' than that for a non-specific substrate (in the experimentally measured sense of the values of K_m). We return to this point in the next section.

In concluding this section, a word of caution is required to prevent a misleading interpretation of the preceding description of enzyme action. To say that 'the enzyme stabilizes the transition state and destabilizes the ground state' should not encourage us to believe that enzymes are present in a fixed state doing 'something' to the substrate. Stabilization is a mutual process – one could equally well regard the substrate in the transition state as stabilizing the enzyme. The important change is the difference in net energy between the ground state – enzyme and substrate in their own environment – and the transition state of the enzyme–substrate complex. A positive charge on an enzyme would probably be 'neutralized' in some way in the ground state of the isolated enzyme, e.g. by anionic charges on a neighbouring part of the protein chain or buffers and by solvation provided by both the rest of the enzyme and water. In the transition state this neutralization could be provided by a negative charge on the substrate and the amount of catalysis would be controlled by the *difference* in free energy provided by these stabilizations in the ground state and the transition state. *If* the enzyme is rigid and *remains undistorted* during the catalytic process the enzyme can be complementary in structure to only one particular molecular geometry and charge distribution during the conversion of the substrate to product.

3.5 Molecular mechanisms

The molecular mechanisms for stabilization of the transition state relative to the ground state at the active site are often equally available for specific and small non-specific substrates. The geometry, 'solvation' and general charge neutralization required by the substrate and enzyme are mutually met in the transition state and are presumably available for the reaction site of both 'specific' and small 'non-specific' substrates. However, for the specific substrate there is in addition the free energy of binding the non-reacting part of the substrate, which may be used to lower the free energy of activation, as indicated by k_{cat}/K_m.

There are two obvious alternative ways in which the binding energy of the non-reacting part of the substrate may be used: (1) it may stabilize the transition state but not the ground state or any intermediate, (2) it may destabilize the substrate by entropy, geometrical, solvation or electrostatic changes in the ground state. This destabilization should be expressed in the transition state, i.e. the binding energy can be used to compensate for unfavourable interactions or processes in the ground state which are relieved or reversed in the transition state.

With alternative (1), the full binding energy is not realized until the transition state is reached. It is simple to visualize this process at the reaction centre: electron density and geometrical changes developed in the substrate on going to the transition state may be stabilized by complementary charges and shape of the active site in the enzyme. However, it is less easy to see how the binding energy of the non-reacting part of the substrate is prevented from being fully expressed in the *ES* complex unless alternative (2) is adopted. If the environment of the enzyme is not compatible with the ground state structure of the substrate, the binding energy from the non-reacting part of the substrate can be used to force the substrate into this 'unwelcoming hole'. This could be manifested in many ways, for example:

(a) a 'rigid' enzyme that has an active site complementary in shape to the transition state but not the ground state;
(b) an active site which is non-polar and conducive to stabilizing a neutral transition state but destabilizing a charged substrate;
(c) desolvation of groups on the substrate or enzyme (not necessarily amino acids concerned with the chemical steps of catalysis);
(d) an active site where electrostatic charges on substrate and enzyme are similar.

If the binding energy is used to compensate for the induction of 'strain' in the substrate, it is essential that the 'strain' is relieved in the transition state in order to increase the rate of reaction. The observed binding energy would be what is 'left over' after the strain has been 'paid for' and may thus *appear* to be weak. The apparent binding constant does not tell us what energy changes have taken place, it only indicates the resultant energy change. It is also essential that the non-reacting group only exhibits its binding energy when the reactive centre of the substrate is bound to the active site. If this does not occur, alternative binding modes may lead to non-productive binding which does not affect k_{cat}/K_m, and hence specificity, but which does decrease K_m leading to saturation conditions at lower concentration of substrate. In fact, the binding energy of the substituent of the specific substrate may be used to prevent non-productive binding.

Probably the most important way that the binding energy is 'used' is to compensate for the unfavourable entropy change that accompanies formation of the *ES* and *ES*$^{+}$ complexes (Page and Jencks, 1971; Page, 1973, 1977). The entropy loss that is required to reach the transition state may already have been partially or completely lost in the *ES* complex. The binding energy of a non-specific substrate may be insufficient to restrict the necessary degrees of freedom of the substrate when bound to the enzyme. Consequently, for this substrate more entropy has to be lost to reach the transition state, which reduces k_{cat}, compared with a specific substrate. For the latter, the binding energy of the additional substituent may compensate for the required loss of entropy which may occur in *ES* or *ES*$^{+}$. Thus, although the free energy change accompanying formation of the *ES* complex may appear to be similar for specific and non-specific substrates, as measured by the observed K_m, the specific substrate may be more 'tightly' bound.

At 25°C for a standard state of 1 M, the complete restriction of medium-sized substrates requires a decrease in entropy and an increase in energy of about 150 kJ mol^{-1}, and unfavourable factor of 10^8. This entropy change is that typically required to form a covalent bond in which the atoms are necessarily confined to a relatively small volume. If the enzyme catalysed reaction requires the formation of a covalent bond then the binding energy of the non-reacting part of the substrate may compensate for this necessary unfavourable entropy change. A non-specific substrate may have insufficient binding energy to compensate for the required entropy loss resulting in a reduced value of k_{cat}/K_m. If the chemical mechanism of catalysis requires the involvement of other functional groups such as general acids or bases, metal-ions or a change in solvation, then a further entropic advantage may be apparent. However, the contribution is smaller than that from covalent catalysis because these secondary effects require less restriction of degrees of freedom, as the 'flexibility' of hydrogen bonds and metal-ion co-ordination is greater than that for covalent bonds.

3.5.1 ESTIMATION OF BINDING ENERGIES

Observed free energies of binding substrates to enzymes, as indicated by K_m, are often less than the true intrinsic binding energy because much of the binding energy is 'used up' in bringing about the required loss of entropy and in inducing any destabilization of the substrate or enzyme. This problem cannot be overcome by measuring enthalpy and entropy changes for binding or activation because of the dominant role that the solvent water plays in determining the observed values. Furthermore the binding energy of a molecule A–B *cannot* be estimated from the values observed for separate A and B analogues.

The intrinsic binding energy of a small group A may be obtained from a comparison of the free energies of binding the larger molecule B and the molecule A–B, where A is a covalently bound substituent (Jencks and Page, 1972). The tight binding of the small substituent or analogue 'molecule' A to an enzyme requires the loss of its translational and rotational entropy which makes an unfavourable contribution to the overall free-energy change. However, when A is a substituent in the molecule A–B, most of this entropy loss is accounted for in the binding of B because the loss of entropy upon binding A–B will be much the same as that upon binding B. This is because the total translational and rotational entropy of the large molecule B will be much the same as that for A–B. The difference in the free energies of binding A–B and B therefore gives an estimation of the true binding energy of A:

$$\Delta G_{AB} - \Delta G_B = \Delta H_{AB} - T\Delta S_{AB} - \Delta H_B + T\Delta S_B \qquad (3.28)$$

$$T\Delta S_{AB} \simeq T\Delta S_B \qquad (3.29)$$

$$\Delta G_{AB} - \Delta G_B = \Delta H_{AB} - \Delta H_B = \Delta H_A \qquad (3.30)$$

The difference, $\Delta G_{AB} - \Delta G_B$, is very much greater than ΔG_A, the observed free energy of binding A, because it is free from the unfavourable entropy term accompanying the binding of A. Consequently, just because $(\Delta G_{AB} - \Delta G_B)$ may give a very favourable negative free energy change for the intrinsic binding energy of A, it would be *incorrect* to conclude that the favourable binding of A–B was mainly due to the favourable binding of A. The binding energy of B in A–B has been 'used' to compensate for the large loss of translational and rotational entropy upon binding B and A–B to the enzyme. The maximum increase in free energy from the entropy loss of binding is *ca* 150 kJ mol^{-1} (a factor of 10^8). If the anchor molecule B is only loosely bound, then the intrinsic binding energy of A in A–B may cause a greater restriction of motion in A–B so that $T\Delta S_{AB} > T\Delta S_B$ and there will be no change, or even a decrease, in the observed binding constant. Intrinsic binding energies estimated in this way are, therefore, likely to be lower limits.

A comparison of ΔG_A with $\Delta G_{AB} - \Delta G_B$, provides a lower limit to the intrinsic loss of entropy upon binding A (Page, 1977):

$$\Delta G_{AB} - \Delta G_B - \Delta G_A = \Delta H_{AB} - \Delta H_B - \Delta H_A - T\Delta S_{AB} + T\Delta S_A + T\Delta S_B$$

$$\simeq T\Delta S_A + T\Delta S_B - T\Delta S_{AB}$$

$$K_A(K_B/K_{AB}) \simeq T\Delta S_A \qquad (3.31)$$

Assuming, as before, that the intrinsic enthalpy of binding A and B approximates to that of AB, the free energy difference $\Delta G_{AB} - (\Delta G_B + \Delta G_A)$, or $K_A(K_B/K_{AB})$, indicates the difference in entropy loss between binding A and B separately compared with A–B. If the entropy loss upon binding the large analogue molecule B is similar to that for binding A–B, then $K_A (K_B/K_{AB})$ approximates the entropy loss of binding the residue A. At a standard state of 1 M, the maximum loss of entropy upon binding is about $- 150$ kJ mol^{-1}, and therefore a comparison of $T\Delta S$, obtained as in (3.31), with this value allows an estimation of the tightness of binding for A, which is true if B binds as tightly as AB.

The equilibrium constant for binding AB to the enzyme may therefore be up to 10^8 greater than that estimated from the product of the binding constants for A and B.

The most interesting problem to arise from the estimation of intrinsic binding energies is the apparently large values associated with small substituents. Examples are 9–16 kJ mol^{-1} for CH_2; 21 kJ mol^{-1} for SCH_3; 34 kJ mol^{-1} for OH and 23–38 kJ mol^{-1} for SH. These intermolecular interactions are much greater than those generally observed between solute and solvent or between molecules within a crystal (Jencks, 1981). Binding energies have recently been determined using site directed mutagenesis (Fersht et al., 1984).

It is often suggested that the interior of a protein is hydrophobic and therefore may be analogous to an organic liquid. The transfer of a group from water to a non-polar liquid is then taken as a model for the transfer of the same group from water to the enzyme:

$$\frac{-CH_2-}{\text{water}} \quad \xrightarrow[\text{kJ mol}^{-1}]{\Delta G = -13} \quad \frac{-CH_2-}{\text{Enzyme}} \qquad (3.32)$$

$$\frac{-CH_2-}{\text{water}} \quad \xrightarrow[\text{kJ mol}^{-1}]{\Delta G = -4.2} \quad \frac{-CH_2-}{\text{Non-polar liquid}} \qquad (3.33)$$

$$\frac{-CH_2-}{\text{water}} \quad \xrightarrow[\text{kJ mol}^{-1}]{\Delta G = -2.1} \quad \frac{-CH_2-}{\text{Dioxane or ethanol}} \qquad (3.34)$$

However, the change in interaction energy estimated in this way for a methylene group, about $- 4.2$ kJ mol^{-1} for transfer to a non-polar liquid and about -2.1 kJ mol^{-1} for transfer to dioxane or ethanol, is at least three times *less* favourable than the change upon transfer to an enzyme. The comparison between the 'enzymic' process (Reaction (3.32)) and the analogous 'non-polar' process (Reaction (3.33)) involves several changes in interaction energies. When a solute is transferred to a non-polar liquid (Reaction (3.33)), there is a gain of solute–solvent interactions and a loss of solvent–solvent interactions, which is equivalent to the free energy of formation of a cavity in the solvent of a suitable size to accommodate the solute molecule.

When a solute is transferred to an enzyme (Reaction (3.32)), there is a gain of solute–enzyme interactions if water is initially absent from the binding site. If water is initially bound to the site of binding and is then displaced by the substituent, there is also a gain of water–water interactions and a loss of water–enzyme interactions. However, the sum of the free energy changes of these latter two interactions must be thermodynamically unfavourable, otherwise water would not initially be bound to the enzyme, and hence, in this situation, the observed free energy of transfer of a substituent from water to an enzyme provides a *lower* limit to the magnitude of enzyme–substrate interactions.

If water is *not* initially bound to the enzyme at the site where the substituent will be bound, then the transfer to the enzyme (Reaction (3.32)) could be thermodynamically more favourable than that to a non-polar liquid (Reaction (3.33)) because the latter process involves the loss of solvent–solvent inter-actions, i.e. the free energy of cavity formation, and thus the free-energy change in Reaction (3.33) provides an underestimation of solute–solvent interactions. This, therefore, may explain, in some cases, part of the dis-crepancy between Reactions (3.32) and (3.33) because the enzyme may have a ready-made 'cavity' and the binding of a substituent to this site would not result in the unfavourable changes which accompany cavity formation in liquids.

An important factor is that the *number* of the favourable dispersion or van der Waals interactions of the substituent with the enzyme is probably greater than with a liquid solvent. The atoms in an enzyme molecule are very closely packed as they are linked together by covalent bonds and, in some cases, the polypeptide chains are cross-linked by covalent disulphide bonds. The fraction of space occupied by the atoms in a molecule of protein is 0.75, which may be compared with 0.74, the value for the most efficient known way of packing spheres, cubic close-packing or hexagonal close-packing. In a liquid the fraction of space occupied is very much less. For example, upper limits for water, cyclohexane and carbon tetrachloride are 0.36, 0.44, and 0.44, respectively. The inside of a protein thus contains little space and is more analogous to a solid than to a liquid. It may be more appropriate, therefore, to compare Reaction (3.32) with the free energy of transfer of a substituent from water to a solid (Page, 1976).

The atoms in an enzyme surrounding the substrate are very compact and most are separated only by their covalent radii. In a liquid, a solute is surrounded by molecules separated by their van der Waals radii. Hence the effective co-ordination number is much greater in the enzyme than it is in a liquid. For example, the van der Waals radii for carbon, nitrogen, and oxygen are about 2.2 times greater than their respective covalent radii and a very simple treatment of the packing of spheres shows that the co-ordination number of a sphere is about 2.5 times greater when surrounded by other spheres separated by such covalent radii than when they are separated by van der Waals radii. In a liquid a large fraction of the surface of a solute molecule is surrounded by empty space between the solvent molecules. Even in a non-polar solid the closest approach of neighbouring molecules is governed by their van der Waals radii.

Physical adsorption of molecules on to solid surfaces exhibits certain similarities to enzyme–substrate binding. The atoms in a solid present a closely packed surface and in some cases, for example graphite, the inter-atomic distance is similar to that found in covalently bonded molecules. Instead of the normal r^{-6} dependence of the London attractive potential between a pair of molecules, there is a r^{-3} dependence for the interaction between a molecule and a solid. This longer range interaction is the basic reason why gases are adsorbed at pressures lower than those at which they condense to liquids or solids. The differential heat of adsorption is often of the order of twice the heat of condensation of the adsorbed vapour (Page, 1977).

3.6 Molecular fit

Both geometrical and electrical compatibility is required between the enzyme and the substrate if the maximum interaction energy is to be realized. To decide whether discrimination between substrates can be achieved by molecular fit, we need to know how the energy of interaction changes with geometry, the magnitude of the binding energy and the accompanying effect upon entropy changes. If a hydrogen bond between the desired substrate and enzyme occurs with optimal geometry, what effect will a less favourable geometry have upon catalysis with an undesired substrate?

Because of the way that we make models and draw reaction mechanisms, we often exaggerate the differences in geometry of ground-state and transition-state structures, whereas the actual distance that atoms move to reach a transition state is not tremendously more than that experienced by normal vibrations. For example, at 25°C mean vibrational amplitudes commonly range from about 0.005 to 0.010 nm and bending amplitudes of $\pm 10°$ are not unusual. The important factor contributing to the increased binding energy of the transition state to the enzyme around the reactant site is that formation of the transition state is often accompanied by large changes in

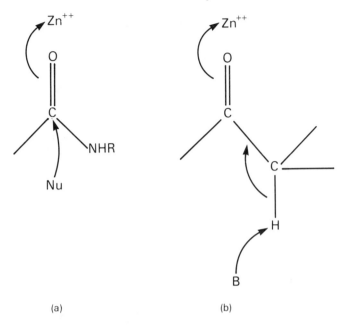

(a)   (b)

Figure 3.10 Enzymes may sometimes catalyse different reactions which involve similar electron density changes.

the electron density surrounding the reacting atoms, which provides the increased favourable interaction between the substrate and enzyme. One consequence of this, of course, is that enzymes can often catalyse different reactions which have similar electron density changes. For example, carboxy-peptidase A catalyses the hydrolysis of amides (Fig. 3.10(a)) and the enolization of ketones (Fig. 3.10(b)). Both reactions involve an increase in electron density on oxygen which favourably interacts with the electrophilic zinc.

To understand the energetics and specificity of enzyme–substrate inter-actions, it is necessary to know the magnitude and geometry of intermolecular and intramolecular forces. It is of interest to know whether the *intermolecular* force field generated by the environment (solvent, crystal, enzyme, etc.) is strong enough to overcome the conformational and geometrical dictates of the *intramolecular* force field of the substrate. Can the enzyme distort the substrate or vice versa? How much does it cost in energy to change the ground state arrangement of atoms in the substrate and enzyme?

The forces controlling the distortion of a molecule, i.e. *intramolecular* effects, may be partitioned into the following contributions: (1) bond stretching; (2) bond angle bending; (3) torsional effects; (4) attractive and repulsive non-bonded interactions; (5) electrostatic interactions such as dipole–dipole and polar effects. Unfortunately, there is by no means universal agreement upon the values of the parameters to be used in the quantitative estimation of these effects, and although they all have a physical reality (probably containing some areas of overlap), there is a tendency to treat them as adjustable parameters (Page, 1973).

Deformation of bond lengths is very difficult and rarely occurs in 'normal' molecules. Bond angle deformation is fairly easy; a $10°$ change costs 11.3 kJ mol^{-1}, and this is the pathway commonly used to relieve non-bonded interaction strain in a molecule. Torsional energy is the 'softest' of all the potential energy terms, and hence distortion of dihedral angles is relatively easy; it generally costs relatively little energy to change conformations of molecules. Unfortunately, probably the most important but the least understood energy function is that describing non-bonded interaction. As a consequence, not only is there a variety of functional forms used to describe this interaction, but there is a range of values reported for the parameters of the same functions. The calculation of meaningful non-bonded interaction energies is beset by a number of complications. Unlike the free atoms, those in molecules do not possess spherical symmetry. To allow for this anisotropic character of non-bonded interactions, it has been suggested that, say, the centre of a hydrogen atom should be shifted along the C–H bond, but the atom still treated as spherical. Another problem is that the effective dielectric constant of the molecule may influence the transmission of the forces involved. Finally, the calculations apply to the gas phase, and in solution the attractive part of the non-bonded interaction would be decreased by the solvent. The relative ease of distortion of a substrate by these parameters is illustrated in Fig. 3.11.

A simplified approach to estimate electrostatic interactions is to use partial charges on the individual atoms, obtained from group dipole moments, and to calculate the electrostatic interaction by Coulomb's law

$$E_{el} = \frac{q_1 q_2}{Dr} \times 138.9 \text{ kJ mol}^{-1} \qquad (3.35)$$

as a function of the distance, r nm, between the partial charges, q, expressed in terms of the electronic charge, in a medium of dielectric constant D. On a qualitative basis, a system of alternate positive and negative partial charges imparts stability to a molecule, while destabilization is associated with adjacent like charges. The use of Coulomb's law should be regarded as a purely empirical procedure since, when two partial charges are not well separated, the solvent molecules and the rest of the solute between and around the two charges do not behave like a continuous medium of constant dielectric constant, and it is also difficult to know where the point dipoles should be located. For two partial charges separated by greater than one width of water layer it has been suggested that the effective dielectric constant

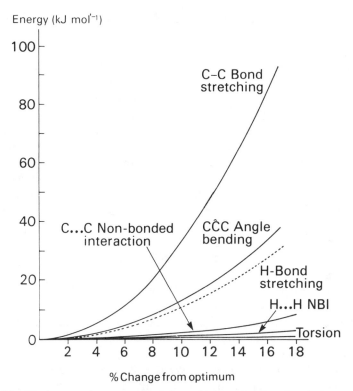

Figure 3.11 The increase in energy that is brought about by changing the parameter indicated from its optimum geometry.

approaches that of bulk water, 80; hence, electrostatic interactions would be negligible at these distances. Electrostatic interactions are discussed further in the next section.

The easiest distortion is that of torsional angle–conformational changes. The *overall* conformational characteristics of a molecule, defined by torsional rotations about bonds that contain large groups on *both* sides of the bond, do *not* appear to be modified by the force field generated by different solvent media or by a crystal. This is not usually the case for the conformational behaviour of small substituents, e.g. methyl groups, which do not generally exhibit conformational specificity.

The origin of attractive *intermolecular* interactions and the factors determining the geometry of molecular complexes is still being sought after decades of experimental and theoretical study (Kollman, 1977; Morokuma, 1977). The interaction energy, ΔE, is the difference between the energy of the species AB and the two isolated molecules A and B. ΔE may be partitioned as

$$\Delta E = E_{es} + E_{pol} + E_{ex} + E_{ct} + E_{disp} \qquad (3.36)$$

where E_{es} is the electrostatic interaction, i.e. the interaction between the undistorted electron distributions of A and B. This contribution includes the interactions of all permanent charges and multipoles such as dipole–dipole, dipole–quadrupole, etc., and may be either attractive or repulsive. E_{pol} is the polarization interaction, i.e. the effect of the distortion of the electron distribution of A by B, and vice versa, and the higher order coupling resulting from such distortions. This component includes the interactions between all permanent charges and induced multipoles and is always an attractive interaction. E_{ex} is the exchange repulsion, i.e. the interaction caused by exchange of electrons between A and B. Physically, this is the short range repulsion due to overlap of the electron distribution of A with that of B. E_{ct} is the charge transfer or electron delocalization interaction, i.e. the interaction caused by charge transfer from occupied molecular orbitals of B to vacant molecular orbitals at A. E_{disp} is the dispersion energy which is a part of the intermolecular correlation energy and is the second-order attraction between fluctuating charges on A and B. It is relatively unimportant for interactions between polar molecules.

The factors determining the equilibrium geometry depend upon the nature of the complex, but the electrostatic interaction has most often been identified as the principal factor. For example, in hydrogen bonding, with a small distance separating the proton donor and acceptor, E_{es}, E_{ct} and E_{pol} can all be important attractive components competing against a large E_{ex} repulsion. At longer distances, for the *same complex*, the short range attractions E_{ct} and E_{pol} are usually unimportant and E_{es} is the only important attraction. However, the importance of an individual component depends on the type of hydrogen bonding, but the electrostatic term is dominant in 'normal' hydrogen bonds between neutral electronegative atoms (Kollman and Allen, 1972).

Two aspects of the geometry of hydrogen bonding are important in determining the specificity of enzyme–substrate interactions. Can the intermolecular force field generated by hydrogen bonding over-ride the intramolecular force field determining the geometry of the substrate? How does the energy of intermolecular hydrogen bonding change with perturbations from optimal geometry? The distance between the electronegative atoms in hydrogen bonding is important, but not critical. The O . . . O and the O . . . N distance in hydrogen bonded systems varies from 0.24 to 0.30 nm and from 0.265 to 0.315 nm, respectively. The force constant for O . . . O stretching in ice is 121 kJ mol^{-1} Å^{-1} and, if distortion from the equilibrium O . . . O distance obeyed a quadratic expression, the energy change as a function of distance would be as illustrated in Fig. 3.11. It is easier to stretch a hydrogen bond than it is to distort a CĈC bond angle. A quadratic probably over-emphasizes the energy dependence upon distance for increasing the distance from the minimum, and theoretical calculations indicate that, for example, the O. . . O distance can vary ±0.3 Å for less than 5 kJ mol^{-1}.

Although at one time it was thought that 'normal' hydrogen bonds were linear, deviations from linearity are now accepted as the normal situation, especially in solids. An examination of the distribution of X–Ĥ . . . Y angles shows that deviations by 15° from linearity occur as frequently as strictly linear hydrogen bonds. The O–Ĥ . . . O and N–Ĥ . . . O angles in intermolecular hydrogen bonded systems vary from 140° to 180° and from 130° to 180°, respectively. Theoretical calculations show that large changes in angle correspond to only 1 or 2 kJ mol^{-1}; for example ±30° for O–Ĥ . . . O=C costs about 3 kJ mol^{-1}. Incidentally, experimental and computational data do not support the hypothesis that the hydrogen bond lies along the maximum of the electron donor lone pair (Olovsson and Jonsson, 1976; Taylor and Kennard, 1984).

Distortion from the optimum hydrogen bond geometry is easy for changes in orientation and slightly less facile for changes in distance. Conversely, the intermolecular force field generated by a hydrogen bond seems unlikely to be able to cause major distortions, except for conformational changes, in the geometry of the substrate.

The electrostatic energy dominates the medium to large range interaction between two molecules and it seems reasonable therefore to emphasize its importance. A promising lead in this direction is the use of electrostatic molecular potentials (Scrocco and Tomasi, 1978; Pullman and Berthod, 1978). This method simulates the coulombic contribution to the intermolecular interaction by substituting appropriately chosen point charges, which may be fractional, to replace actual neighbouring molecules. The potential is a function of the electronic distribution and the position of the nuclei in the molecule and is computed from molecular wave functions. It is the basic premise of the use of electrostatic potentials that E_{pol}, E_{ct} and E_{disp} are much less important than electrostatic effects in determining structural features of complex formation. The deformations of the special charge distribution of a set of atoms bound together to form a molecule are quite complex and the electrostatic potential, like all other one-electron observables, is correct to the second order in the Hartree–Fock approximation. Formally, the electrostatic potential is the value of the interaction between a molecule and a point charge placed at the point k:

$$E_{es}(AB) = \sum_{k} V_A(k)_{qkB} \qquad (3.37)$$

Similar chemical groups have portions of space where the shape of V is reasonably similar. An interpretation of V can be obtained by considering V as the sum of local contributions, transferable from molecule to molecule. This transferability may be exploited to get a rationale of the shape of V in different molecules and to shed light on the effects of chemical substitution on the values assumed by V near a given group in a set of related molecules. The

contributions of the neighbouring groups are sometimes important in determining the shape of V; distant groups can even reverse the sign of V near a
given substituent. A general property of electrostatic potentials is that
different atoms with different local geometries can yield the same electrostatic potential and thus, presumably, the same binding energy. This could be
important in the design of enzyme inhibitors and of drugs in general. The
combined action of several atoms bound together is not at all apparent from a
'ball and stick' model but is an automatic result of electrostatic potentials.

Considerable electrostatic interaction can arise from the dipole moment of
3.5 D for the peptide unit. In an α-helix, the peptide dipole moments are
aligned nearly parallel to the helix axis giving rise to a significant electrical
field. The dipole runs from the C-terminal to the positive end of the N-
terminal. For example, the potential at 0.5 nm from the N-terminus of an
α-helix of 1.0 nm length is about 0.5 V so that, ignoring the effect of solvent, a
negatively charged group at that position involves an attractive energy of
about 50 kJ mol^{-1} (Hol, van Duijnen and Berendson, 1978).

Ideas about the enzyme being complementary to the transition state
structure, particularly with respect to the design of transition state analogues,
have tended to be dominated by geometry. Based on the previous discussions
it is apparent that it is the electrostatic interactions between the transition
state and the enzyme, rather than geometry, which are crucial in determining
the binding energy.

Ideas about the flexibility of enzymes are often dominated by personal
preference for space-filling or skeletal models. In order to bring about the
maximum loss of entropy of substrates, and provide optional alignment of
substrates with catalytic groups at the active site, it is desirable that an enzyme
be as rigid as possible. This does not mean that groups within a protein cannot
undergo internal rotation. In fact, there is much evidence, particularly from
NMR studies, that the interior of some proteins is mobile and that some
proteins 'breathe', i.e. the protein as a whole undergoes some vibronic
motion. Part of the protein may be mobile while another part is not, and these
parts may be of diverse sizes for different proteins. Thermodynamically, of
course, there is no contradiction in a particle the size of an enzyme molecule
having both 'rigid' and 'fluid' properties.

An effect which has not been taken into account in the discussion of
electrostatic interactions is the reaction field of the highly polarizable solvent
water which decreases the field near the molecular surface. The earliest
attempts to take into account solute–solvent interactions focused on varying
the dielectric constant in electrostatic energy terms used to describe pairwise
interactions between species in the solute molecule. In reality, there will be
preferred hydration sites. Because of the intrinsic dielectric properties of the
solute and the voids produced by unfavourable solute group–solvent interactions, the 'local dielectric' behaviour about the solute is heterogeneous and

can be significantly different in magnitude from that of the bulk solvent. No model has been developed in which the local dielectric varies with spatial position. Attempts have been made to treat the dielectric effect as due to the polarization of individual atoms leading to induced dipoles. It is concluded that an enzyme can stabilize charge better than water because, in contrast to water, the enzyme dipoles are kept oriented towards the charge, even when the fluid from the charge is small. In water the dipoles of the first few solvation shells cannot be strongly polarized towards the solute charges because they interact with the surrounding randomly oriented bulk dipoles (Warshel, 1981).

References

Albery, W. J. and Knowles, J. R. (1976) Evolution of enzyme functions and the development of catalytic efficiency. *Biochemistry*, **15**, 5631–5640.

Bell, R. P., Critchlow, J. E. and Page, M. I. (1974) Ground state and transition state effects in the acylation of α-chymotrypsin in organic solvent water mixtures. *J. Chem. Soc. Perkin Trans.*, **2**, 245–249.

Fersht, A. R. (1974) Catalysis, binding and enzyme–substrate complementarity. *Proc. Roy. Soc. London B.*, **187**, 397–407.

Fersht, A. R. (1977) *Enzyme Structure and Mechanism*, Freeman, San Francisco.

Fersht, A. R. and Kaethner, M. M. (1976) Enzyme hyperspecificity. Rejection of threonine by the valyl-tRNA synthetase by misacylation and hydrolytic editing. *Biochemistry*, **15**, 3342–3346.

Ferscht, A. R., Shi, J.-P., Wilkinson, *et al.* (1984) Analysis of enzyme structure and activity by protein engineering. *Angew. Chem. Int. Ed. Engl.*, **23**, 467–538.

Hol, W. G. T., van Duijnen, P. T. and Berendson, H. J. C. (1978) The α-helix dipole and properties of proteins. *Nature*, **273**, 443–446.

Jencks, W. P. (1975) Binding energy, specificity and enzymic catalysis: the Circe effect. *Adv. Enzymology*, **43**, 219–410.

Jencks, W. P. (1980) In *Molecular Biology, Biochemistry and Biophysics*, Vol. 32 (eds F. Chapeville and A. L. Haenni), Springer-Verlag, New York, pp. 3–25.

Jencks, W. P. (1981) On the attribution and additivity of binding energies. *Proc. Natl Acad. Sci. USA*, **78**, 4046–4050.

Jencks, W. P. and Page, M. I. (1972) On the importance of togetherness in enzymic catalysis. *Proc. Eighth FEBS Meeting Amsterdam*, **29**, 45–58.

Kollman, P. A. (1977) Non-covalent interactions. *Accounts Chem. Res.*, **10**, 365–371.

Kollman, P. and Allen, L. C. (1972) Theory of hydrogen bonding. *Chem. Rev.*, **72**, 283–303.

Morokuma, K. (1977) Why do molecules interact? The origin of electron–donor–acceptor complexes, hydrogen bonding and proton affinity. *Accounts Chem. Res.*, **10**, 294–300.

Olovsson, I. and Jonsson, P. (1976) In *The Hydrogen Bond – Recent Developments in Theory and Experiments* (eds P. Schuster *et al.*), North-Holland, Amsterdam, pp. 394–456.

Page, M. I. (1973) The energetics of neighbouring group participation. *Chem. Soc. Rev.*, **2**, 295–323.

Page, M. I. (1976) Binding energy and enzymic catalysis. *Biochem. Biophys. Res. Commun.*, **72**, 456–461.

Page, M. I. (1977) Entropy, binding energy and enzymic catalysis. *Angewandte Chem. Int. Ed.*, **16**, 449–459.

Page, M. I. (1979) The principles of enzymatic catalysis. *Int. J. Biochem.*, **10**, 471–476.

Page, M. I. (1984) *The Chemistry of Enzyme Action*, Elsevier Biomedical Press, Amsterdam.

Page, M. I. and Jencks, W. P. (1971) Entropic contributions to rate accelerations in enzymic and intramolecular reactions and the chelate effect. *Proc. Natl Acad. Sci. USA*, **68**, 1678–1683.

Pullman, A. and Berthod, H. (1978) Electrostatic molecular potentials in hydrogen bonded systems. *Theoret. Chim. Acta*, **48**, 269–277.

Scrocco, E. and Tomasi, J. (1978) Electrostatic molecular potentials. *Adv. in Quantum Chem.*, **11**, 115–193.

Taylor, R. and Kennard, O. (1984) Hydrogen-bond geometry in organic crystals. *Acc. Chem. Res.*, **17**, 320–326.

Warshel, A. (1981) Electrostatic basis of structure–function correlation in proteins. *Accounts Chem. Res.*, **14**, 284–290.

4 The charging of tRNA

A. R. FERSHT

4.1 Introduction

Amino acids are selected in protein synthesis during the aminoacylation of tRNA. This is a two-step reaction composed of activation of the amino acid (Equation (4.1)) followed by its transfer to tRNA (Equation (4.2)).

$$E^{AA} + AA + ATP \longrightarrow E^{AA}.AA-AMP + PPi \qquad (4.1)$$

$$E^{AA}.AA-AMP \xrightarrow{\text{tRNA}^{AA}} AA-tRNA^{AA} + AMP + E^{AA} \qquad (4.2)$$

Both steps are catalysed by the same aminoacyl-tRNA synthetase, each amino acid (AA) having its own specific enzyme (E^{AA}) and family of tRNAs (tRNA^{AA}). The aminoacyl-tRNA synthetases (EC 6.1.1.–) form an extremely diverse family of enzymes differing in molecular weight and subunit structure (for a compilation see Joachimiak and Barciszewski, 1980) as well as kinetic properties.

Measurements so far of the overall accuracy of protein synthesis in *Escherichia coli* suggest an error rate of one mistake per 3000 amino acid residues incorporated (Loftfield, 1963; Loftfield and Vanderjagt, 1972; Edelmann and Gallant, 1977), although a three- to ten-fold lower accuracy has been reported in the synthesis of a bacteriophage coat protein (Parker, Johnston and Borgia, 1980). This figure includes all the errors in transcribing the DNA to mRNA, matching up the codon and anticodon on the ribosome, and the selection of amino acid and tRNA by the amino acid activating enzymes. This last reaction is potentially the weakest link in the chain of accuracy. The fidelity of RNA polymerase *in vitro* is compatible with the overall accuracy (error rates of 1/2400–1/40 000, Springgate and Loeb, 1975). Codon–anticodon interactions are an unknown quantity but the selection of certain amino acids is undoubtedly poor for the following reasons.

4.2 The basic problem in amino acid selection

In principle, two selections are made by an aminoacyl-tRNA synthetase: recognition by the enzyme of first the correct amino acid and then the correct tRNA. In practice, tRNA recognition is not a problem since it is such a large molecule that there is adequate scope for distinctive structural variation (Fersht, 1979). On the other hand, as pointed out by Pauling (1957), the amino acids are sometimes so similar in structure that there are often severe problems in distinguishing between them.

Although Pauling's ideas were formulated prior to our present knowledge of the involvement of tRNA and aminoacyl-tRNA synthetases and were based on the concept of a template constructed to bind the individual amino acids of a polypeptide chain, they highlight the problems involved. Consider the cavity in an enzyme constructed to accept the side chain of a particular amino acid, say valine. The other amino acids which are competing for the valine binding site may be divided up into three classes according to size: those larger than, those smaller than, and those the same size as (*isosteric* with) the correct substrate. The larger substrates may be positively excluded from binding to the cavity by forces of steric repulsion. The smaller substrates, on the other hand, cannot be excluded from binding, but will just bind more weakly. Similarly, an isosteric substrate, such as threonine, must also bind to the active site to some extent as it is not excluded by steric repulsion.

The basic problem of amino acid selection is thus the rejection of smaller or isosteric amino acids. The most severe examples are those of rejection of glycine by E^{Ala}, valine by E^{Ile} and serine by E^{Thr} since the smaller amino acids differ by only one methylene group. Pauling (1957) calculated from a value of about 1 kcal mol^{-1} for the hydrophobic binding energy of a methylene group (the figure usually found from physical chemical measurements) that an error rate of about one in five is expected for glycine replacing alanine, valine replacing isoleucine, etc. Pauling also estimated the errors of alanine replacing glycine ($<1/100$) and phenylalanine replacing tyrosine ($<1/100$). Subsequent experiments have shown these numbers to be severe underestimates, but the ideas are qualitatively correct (Fersht, Shindler and Tsui, 1980).

4.3 The basic kinetic equations of specificity

The extension of Pauling's ideas, based on a template, to enzymes is at first sight complicated because it is well documented that a change in the structure of a substrate can affect both binding and rate constants and need not alter just its affinity for an enzyme. Indeed, the phenomena of 'strain' (where the binding energy of the enzyme and substrate is used to distort the substrate) and 'induced fit' (where the binding energy is used to distort the enzyme into an active conformation) have been proposed to provide specificity. A rigorous analysis of the role of binding energy in enzyme catalysis (Fersht, 1974, 1985), however, shows that strain and induced fit have no effect on specificity in the situation of a mixture of two or more substrates competing for the active site of one enzyme. Differences in binding energy may thus be equated directly with differences in specificity.

The mathematical analysis for enzymes obeying the Michaelis–Menten equation (Equation (4.3) where v is rate, k_{cat} is turnover and K_m is the Michaelis constant) is as follows:

$$v = \frac{k_{cat}[E_0][S]}{K_m + [S]} \tag{4.3}$$

When a mixture of two substrates C ('correct') and I ('incorrect') compete for one active site, their relative rate of reaction is given by:

$$v_C/v_I = (k_{cat}/K_m)_C[C]/(k_{cat}/K_m)_I[I] \tag{4.4}$$

Further, when the enzyme is constructed to bind C, and I is *smaller* because it lacks a group R

$$(k_{cat}/K_m)_C/(k_{cat}/K_m)_I = \exp(-\Delta G_b/RT) \tag{4.5}$$

where ΔG_b is the difference in Gibbs' binding energy of the substrates to the enzyme due to the contribution of R. Thus:

$$v_C/v_I = [\exp(-\Delta G_b/RT)][C]/[I]. \tag{4.6}$$

Equation (4.4) relates that specificity in the sense of competition is controlled by the ratios of the compound kinetic quantity k_{cat}/K_m (the 'discrimination' constant) and not values of k_{cat} and K_m individually (that is, the relative rate of reaction varies as the relative rate constants, relative binding energies and relative concentrations). Application of Equation (4.4) by Fersht (1979) to published data by Yarus (1972b) and Mertes et al. (1972) on the interaction of tRNA[fMet] and tRNA[Phe] with the isoleucyl-tRNA synthetase (all from E. coli) gives discriminations of 6×10^6 against tRNA[fMet] and 2×10^7 against tRNA[Phe] in favour of tRNA[Ile], values far more than adequate to maintain the overall accuracy of protein synthesis.

4.4 The discovery of editing during amino acid selection

Early studies on the isoleucyl-tRNA synthetase showed that valine is activated by the enzyme as predicted by Pauling (1957) but that the value of k_{cat}/K_m is about 200 times lower than that for isoleucine (for example, see Loftfield and Eigner, 1966). However, in the presence of tRNA, no valyl-tRNA^{Ile} is formed and so the overall accuracy is high. The answer to the paradox of how the overall reaction is more accurate than the partial reaction was found by Norris and Berg (1964) and Baldwin and Berg (1966) when they prepared large quantities of the isoleucyl-tRNA synthetase and were able to isolate and examine directly the complexes of the enzyme with its substrates. The enzyme was found to form a relatively stable E^{Ile}.Ile-AMP complex in the presence of ATP and isoleucine. On the addition of tRNA^{Ile} to this complex there is substantial transfer of the isoleucine to the tRNA. A stable E^{Ile}.Val-AMP complex is also formed, but addition of tRNA^{Ile} causes the quantitative hydrolysis of the Val-AMP rather than the formation of Val-tRNA^{Ile}. The sum of these two reactions is that in the presence of tRNA^{Ile} and valine, the isoleucyl-tRNA synthetase acts as an ATP/pyrophosphatase by the sum of the two reactions:

$$E^{Ile} + ATP + Val \longrightarrow E^{Ile}.Val\text{-}AMP + PPi \qquad (4.7)$$

$$E^{Ile}.Val\text{-}AMP + tRNA^{Ile} \longrightarrow E^{Ile} + tRNA^{Ile} + Val + AMP \quad (4.8)$$

This constituted the experimental discovery of the editing mechanism: the enzyme makes a mistake and somehow corrects it. It also provided the basic rules for detection and diagnosis: a partial reaction of a pathway occurs, but the overall reaction occurs only poorly if at all; there is the non-productive hydrolysis of a high energy compound.

The isoleucyl-tRNA synthetase is thus both a synthetase and a hydrolase: the hydrolytic activity being an editing function to remove errors of mis-activation. The term *proofreading* is used synonymously with editing. Editing is now known to be a general phenomenon, occurring also in DNA replication. Many of the key enzymes involved in biosynthetic polymerization are multifunctional, having an additional hydrolytic activity which is used to excise incorrect intermediates as they are formed. This enables the addition of each monomer to be screened twice, once during synthesis and again during proofreading.

4.5 The editing reaction pathway: hydrolysis of mischarged tRNA versus hydrolysis of misactivated amino acid

It was recognized by Baldwin and Berg (1966) that the editing reaction during the rejection of valine by the isoleucyl-tRNA synthetase could occur by the hydrolysis of either Val-AMP or Val-tRNA^{Ile}. It was subsequently found that the aminoacyl-tRNA synthetases are weak esterases towards cognately charged aminoacyl-tRNA but the hydrolytic activity is higher with some mischarged tRNAs. For example, Eldred and Schimmel (1972) found that E^{Ile} deacylates Val-tRNA^{Ile} (*E. coli*) with $k_{cat} = 0.02$ s^{-1} at 3°C, and Yarus (1972a) reported that E^{Phe} deacylates Ile-tRNA^{Phe} with $k_{cat} = 2$ s^{-1} at 37°C.

These authors thus proposed that the deacylation of the mischarged tRNA constituted the editing pathway.

4.5.1 EDITING OF MISCHARGED tRNA (POST-TRANSFER)

The misacylation/deacylation pathway has now been rigorously established for the rejection of threonine (and α-aminobutyrate) by the valyl-tRNA synthetases from *Bacillus stearothermophilus*, *E. coli*, yeast and yellow lupin seeds by using a rapid quenching apparatus to trap and observe the mischarged $tRNA^{Val}$ (Fersht and Kaethner, 1976; Fersht and Dingwall, 1979a; Jakubowski and Fersht, 1981). The relevant observations are that (1) on mixing the E^{Val}.Thr-AMP complex with $tRNA^{Val}$ there is the transient formation of Thr-$tRNA^{Val}$, reaching a maximum after some 20 ms and then disappearing after 100 ms; (2) the deacylation occurs much faster than the dissociation of the mischarged tRNA from the enzyme; (3) on quenching the reaction with phenol, extracting the mischarged tRNA and measuring its rate of enzyme-catalysed deacylation in the rapid quenching machine, the valyl-tRNA synthetase is found to be a very efficient esterase towards the mischarged tRNA ($k_{cat} = 40\,s^{-1}$); (4) this esterase activity is not inhibited by the addition of valyl adenylate which blocks the aminoacylation site, hence suggesting the presence of a distinct hydrolytic site.

4.5.2 EDITING OF MISACTIVATED AMINO ACID (PRE-TRANSFER)

Repetition of the rapid quenching experiments with E^{Ile} (*E. coli*) and valine failed to detect a predicted transient accumulation of Val-$tRNA^{Ile}$ (Fersht, 1977b). Although other explanations could account for this, it was suggested that a pathway involving a tRNA-induced hydrolysis of the E^{Ile}.Val-AMP adenylate must also be considered:

$$
E^{Ile}.Val\text{-}AMP.tRNA \xrightarrow{\ k\ } E^{Ile}.Val\text{-}tRNA^{Ile} \xrightarrow{\ k'\ } E^{Ile} + Val\text{-}tRNA^{Ile}
$$

$$
\Big\downarrow k_h \qquad\qquad\qquad\quad \Big\downarrow k'_h \qquad\qquad\qquad (4.9)
$$

$$
E^{Ile} + Val + AMP \qquad\qquad E^{Ile} + Val + tRNA^{Ile}
$$

That is, a two-step process with major editing by hydrolysis of the Val-AMP followed by the esterase activity mopping the Val-$tRNA^{Ile}$ escaping the first editing step. Jakubowski (1978, 1980) has subsequently shown that in addition to the $tRNA^{Val}$-stimulated hydrolysis of E^{Val}.Thr-AMP, there is a slower tRNA-*independent* reaction. As discussed below, this would not appear to contribute substantially to specificity. More significantly, however, subsequent measurements on the rejection of homocysteine(HCys) by the methionyl-tRNA synthetases from *E. coli* and *B. stearothermophilus* and the isoleucyl-tRNA synthetase from *E. coli* showed that there is very rapid hydrolysis of the E . HCys-AMP complexes in the absence of tRNA and that the rate is stimulated little, if at all, by the presence of tRNA (Jakubowski and Fersht, 1981). Further, the product of the hydrolysis is the homocysteine thiolactone formed by the facile cyclization reaction (4.10).

$$
\begin{array}{ccc}
\overset{+}{N}H_3 & & \overset{+}{N}H_3 \\
| & & | \\
CH \quad O & & CH \\
| & & | \\
CH_2 \quad C & \longrightarrow & CH_2 \qquad C{=}O + AMP \\
| & & | \\
CH_2 \quad AMP & & CH_2 \\
SH & & S
\end{array}
\qquad (4.10)
$$

Coupled with the failure to trap any HCys-tRNA^{Met} during the rejection of HCys by the methionyl-tRNA synthetase (Fersht and Dingwall, 1979c), editing appears to be mainly via the hydrolysis of the aminoacyl adenylate.

Another possibility for the pre-transfer editing is the 'kinetic proofreading' mechanism of Hopfield (1974), shown in Reaction (4.11).

$$
\begin{array}{l}
\quad\quad ATP \quad PP_i \\
E.AA \xrightarrow{\qquad} E.AA{-}AMP \xrightarrow[k_T]{tRNA} E + AA{-}tRNA \\
\;\updownarrow \qquad\qquad\qquad k_1 \updownarrow \qquad\qquad\qquad\qquad + AMP \\
E + AA \qquad\quad E + AA{-}AMP \\
\qquad\qquad\qquad\qquad\;\; \downarrow k_h \\
\qquad\qquad\qquad\quad AA + AMP
\end{array}
\qquad (4.11)
$$

In this, it is postulated that for the misacylation of an incorrect amino acid, the aminoacyl adenylate complex dissociates (via k_1) faster than it is transferred to tRNA (via k_T). However, as pointed out by Mulvey and Fersht (1977) and discussed below, this is an inefficient mechanism.

The scheme represented in Reaction (4.9), with its mixture of pre-transfer and post-transfer editing, was ignored for several years. However, the observations that led to its formulation have now been recently found for several other enzymes and Reaction (4.9) invoked for their editing (Gabius *et al.*, 1983; Lin, Baltzinger and Remy, 1984). It must be stressed, however, that it is difficult to present rigorous proof of pathways in which intermediates do not accumulate.

The relative importance of the three pathways will be discussed in Section 4.8 after a consideration of the costs of editing.

4.6 The double-sieve editing mechanism

An almost bewildering plethora of activations of noncognate amino acids has been reported, summarized by Igloi, von der Haar and Cramer (1978). The number is reduced somewhat, however, after the samples of the noncognate amino acids have been freed from the trace impurities of the cognate amino acid by chemical and enzymic procedures (Fersht and Dingwall, 1979c,d; Fersht, Shindler and Tsui, 1980). The remaining misactivations can be rationalized by one simple model, the double-sieve (Fersht, 1977a; Fersht and Dingwall, 1979d), that predicts the nature of misactivations when editing is required and shows how a large range of amino acids may be edited by the combination of one synthetic and one hydrolytic site.

Table 4.1 Double-sieving by valyl-tRNA synthetase*

Amino acid	Relative rate of activation by E^{Val}	Rate constant for hydrolysis of $AA\text{–}tRNA^{Val}$ catalysed by E^{Val} (s^{-1})
Ile	$<2 \times 10^{-5}$	Not measured[†]
Val	1	4.5×10^{-3}
Thr	4×10^{-3}	40
αBut[‡]	5×10^{-3}	50
Ala	1×10^{-4}	>1
Gly	9×10^{-8}	Not measured

*Data from Fersht and Dingwall (1979d) for valyl-tRNA synthetase from *E. coli*.
[†]In similar examples AA–tRNA is relatively stable (Bonnet and Ebel, 1974).
[‡]α-Amino-L-butyric acid.

The experimental basis of the mechanism comes from measurements on the best characterized example of editing, the rejection of amino acids by the valyl-tRNA synthetase. It is seen in Table 4.1 that isoleucine is activated by E^{Val} at such an extremely low relative rate that if all the isoleucine that is activated is transferred to $tRNA^{Val}$, then the error rate would be less than a tenth of the measured overall error rates of protein synthesis. Thus, amino acids larger than the cognate one can be excluded at a tolerable level by steric repulsion. The smaller amino acids (and the isosteric threonine) are activated by the enzyme at progressively lower rates as their structures differ more and more from that of the cognate substrate. Finally, only the products of the smaller (and natural isosteric) substrates are edited. In this way the whole range of amino acids may be sorted by the two active sites of an aminoacyl-tRNA synthetase according mainly to size and, if necessary, by chemical characteristics in the same way that two sieves of different mesh may be used to isolate particles of just one desired size. Instead of a wire mesh, the sizes of the active site cavities sort the amino acids using the principle that larger substrates may be excluded by steric hindrance from an active site but that smaller substrates must be bound.

The simplest example concerns the selection of an amino acid, such as isoleucine, that has no natural isosteres to cause complications. The activation site of the isoleucyl-tRNA synthetase is presumed to be constructed to be just large enough to accommodate isoleucine but the hydrolytic site is smaller and just large enough to accommodate valine. All amino acids larger than the correct substrate are rejected at a tolerable level from binding at the activation site by steric hindrance. On the other hand, all smaller amino acids have access to the activation site and are activated, albeit at lower rates. But these same amino acids (or rather their products) are accepted by the hydrolytic site. The correct amino acid, on the contrary, is largely excluded from the hydrolytic site by steric hindrance. The principle of steric exclusion is thus used twice to obtain the desired sorting.

The above case is relatively simple since sorting can be performed by size alone. Where isosteric amino acids occur, such as threonine competing with valine, sorting by chemical characteristics must be superimposed upon the sorting by size. For example, the hydrolytic site of the valyl-tRNA synthetase presumably has a hydrogen bond acceptor/donor site to bind specifically the hydroxyl of threonine and repel the corresponding hydrophobic methyl group of valine. Otherwise, the exclusion of all larger amino acids by the activation site and the acceptance of all smaller amino acids by the hydrolytic or editing site takes place as described for the isoleucyl-tRNA synthetase.

Aspects of the double-sieve model may be applied to other mechanisms. There are two cases where the double-sieve breaks down, namely for the phenylalanyl- and alanyl-tRNA synthetases rejecting tyrosine and serine respectively (Igloi, von der Haar and Cramer, 1978; Tsui and Fersht, 1981; Lin et al., 1984). It appears that the CH_3- group is sufficiently large to be rejected at a tolerable level by steric repulsion but the slightly smaller $HO-$ group is not large enough. The relative rates of activation of alanine, serine and α-aminobutyrate by the alanyl-tRNA synthetase are $1:2 \times 10^{-3}:6 \times 10^{-5}$. The activated serine is also rapidly edited (Tsui and Fersht, 1981).

4.6.1 SPECIFICITY OF SORTING BY THE DOUBLE-SIEVE

Measurements of the binding energies of substrates to the aminoacyl-tRNA synthetases, summarized by Fersht, Shindler and Tsui (1980), have provided the maximum contributions different small groups can make to binding. For example, each $-CH_2-$ group can contribute a factor of about 200 to an association constant and an $-OH$ group up to a factor of about 10^5. Cramming an additional $-CH_2-$ group into too small a cavity (e.g. Ile into a Val binding site) can decrease an association constant by a factor of up to 10^6. Measurements of the rates of hydrolysis of Val-tRNA^{Val} and α-aminobutyryl-tRNA^{Val} catalysed by E^{Val} in *E. coli* (Fersht and Dingwall, 1979a) give a specificity of 1.3×10^5 against valine for hydrolytic editing.

4.7 The economics of editing

4.7.1 THE COST–SELECTIVITY EQUATION

Removal by editing of incorrect products implies that some of the correct products are also removed. Thus editing costs energy in two ways: (1) the high energy metabolites used in the formation and subsequent editing of the incorrect products; (2) the fraction of the high energy compounds wasted when some of the correct products are edited. The first of these is clearly necessary and desirable, the second is also necessary, but undesirable and hence should be minimized. There have been some recent detailed analyses of the cost and limits of accuracy including dynamic aspects (Savageau and Freter, 1979a,b; Okamoto and Savageau, 1984a,b), but, for present purposes I wish to use an earlier analysis (Mulvey and Fersht, 1977; Fersht, 1977a) as amplified by Fersht (1981) because this formulation implied some simple equations and predictions in answer to the basic questions: what are the limits of editing, what does it cost and how do these depend on mechanism?

Consider the basic editing mechanism of Reaction (4.12), a Michaelis–Menten mechanism combined with an editing step.

$$E + S \underset{K_m}{\rightleftharpoons} ES \xrightarrow{k_{cat}} EP \begin{array}{c} \xrightarrow{k_s} \text{Synthesis} \\ \xrightarrow{k_d} \text{Destruction} \end{array} \qquad (4.12)$$

The fraction of EP that proceeds to products is given by $k_s/(k_s + k_d)$. Combining this factor with Equation (4.4) for substrates C (correct) and I (incorrect) competing for an enzyme gives:

$$\frac{v_C}{v_I} = \frac{(k_{cat}/K_m)_C}{(k_{cat}/K_m)_I} \times \frac{[k_s/(k_s + k_d)]_C}{[k_s/(k_s + k_d)]_I} \times \frac{[C]}{[I]} \qquad (4.13)$$

where v_C and v_I are now the *fractions* of C and I proceeding to products. The selectivity, S, is defined by:

$$S = \frac{(k_{cat}/K_m)_C}{(k_{cat}/K_m)_I} \times \frac{[k_s/(k_s + k_d)]_C}{[k_s/(k_s + k_d)]_I} \qquad (4.14)$$

that is

$$\frac{v_C}{v_I} = \frac{S[C]}{[I]}.$$
(4.15)

The selectivity may now be defined in terms of two discrimination factors and the cost of editing. First the cost, C, is defined by the fraction of the products of the *correct* substrate that is wastefully hydrolysed. That is:

$$C = [k_d/(k_s + k_d)]_C$$
(4.16)

The discrimination factor in the initial binding reactions, f, is defined by:

$$f = (k_{cat}/K_m)_C/(k_{cat}/K_m)_I$$
(4.17)

Let the discrimination factor in the editing reaction be defined by f'. Combining these equations as described by Fersht (1981) gives the *cost–selectivity* equation:

$$S = f(1 + [f' - 1]C)$$
(4.18)

4.7.2 DEPENDENCE OF THE COST–SELECTIVITY EQUATION ON THE MODE OF RECOGNITION

(a) Single-feature recognition: $f = f'$
The Hopfield (1974) kinetic proofreading mechanism uses the same recognition for editing as it does for initial selection for activation. That is, if the noncognate amino acid dissociates f times faster than the cognate in the initial binding step, then the aminoacyl adenylate complex also dissociates f times faster. Thus, $f = f'$ and the cost–selectivity equation reduces to:

$$S = f(1 - C) + f^2 C$$
(4.19)

(This equation applies also to several models for editing in DNA replication.)

(b) Double-feature recognition: $f' > f$
The double-sieve mechanism is particularly efficient at distinguishing between substrates because it uses different features for discrimination at each site (Fersht and Kaethner, 1976). For example, the valyl-tRNA synthetase discriminates against threonine by first binding its hydroxyl group weakly in a hydrophobic pocket in the activation site and then discriminating against valine in the hydrolytic site by requiring the binding of its relevant methyl group in a hydrophilic locus. Similarly, with the isoleucyl-tRNA synthetase, valine is bound more weakly than isoleucine at the activation site but isoleucine is sterically hindered from entering the hydrolytic site. Experimental evidence on binding energies (Fersht, Shindler and Tsui, 1980), reinforced by measurements of hydrolysis of cognately and noncognately

charged tRNA (Fersht, 1977b; Fersht and Dingwall, 1979a; Table 4.1) shows that $f' \gg f$ in these examples.

Plotted in Fig. 4.1 are values of the selectivity for $f = 200$ (e.g. for E^{Ile}, and Ile versus Val; E^{Val}, and Val versus Thr) and values for f' for (1) the Hopfield mechanism ($f = 200$) and (2) double-feature recognition with $f' = 4 \times 10^4$ and $f' = 2 \times 10^7$, possible values for E^{Val} rejecting Thr and E^{Ile} rejecting Val. It is seen that the Hopfield selection procedure is very inefficient, the selectivity reaching only a value of 4×10^4 ($= f^2$) at the expense of hydrolysing 100% of the correct substrate. The double-feature recognition, on the other hand, gives high values of the selectivity at low cost. Experimentally, the cost of aminoacylation is found to be less than 0.05 (Mulvey and Fersht, 1977).

As f increases, however, the Hopfield proposal becomes more feasible. For example, with the rejection of Ala by E^{Val}, $f = 10^4$ so that a value of $S = 10^6$ may be obtained at $C = 0.01$.

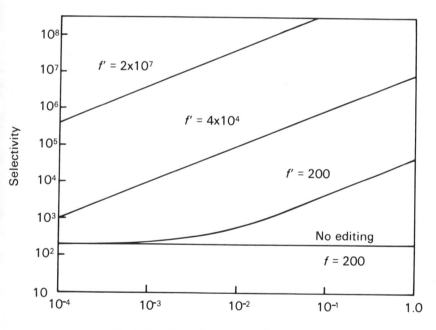

Figure 4.1 Cost and accuracy depend on the method of selection for editing. The cost–selectivity equation (see text) is plotted for a reaction with an initial discrimination (f) of 200. The selectivity is plotted against cost for various values of f', the selectivity in editing. From bottom to top: no editing (or $f' = 1$); $f' = 200$ (single-feature and Hopfield mechanism); $f' = 4 \times 10^4$ and $f' = 2 \times 10^7$ for double-feature selection as for the double-sieve mechanism.

We are now in the position to evaluate semi-quantitatively the relative importance of the different proposed pathways of editing.

4.8 The relative importance of the pre-transfer and post-transfer pathways

The three routes of editing discussed in Section 4.5 may be summarized by Reaction (4.20), where k_p is the rate constant for the dissociation of the AA–tRNA from the enzyme to the elongation factor for protein synthesis.

$$
\begin{array}{ccccc}
& k_T & & k_p & \\
E.AA{-}AMP & \xrightarrow{\quad} & AMP + E.AA{-}tRNA & \xrightarrow{\quad} & E + AA{-}tRNA \\
& \;\;tRNA & & & \\
k_1 \downarrow & \searrow \;\; k_2 & & \downarrow k_3 & \qquad (4.20)\\
& \;\; k_h \searrow & & & \\
E + AA{-}AMP & \rightarrow & E + AA + AMP & &
\end{array}
$$

The fraction edited at any step is given simply by the ratios of the rate constant for editing to that for progress to the next stage of synthesis as follows. The fraction edited (C) for the Hopfield mechanism is given by:

$$C = k_1/(k_T + k_1) \qquad (4.21)$$

Thus for a particular pathway to contribute significantly to editing (using the Hopfield route as an illustration):

$$k_1 \gg k_T \qquad (4.22)$$

and so on for the other pathways.

4.8.1 PRE-TRANSFER PATHWAYS

The cost–selectivity equation shows that the Hopfield (1974) pathway can only be of importance for high values of f, that is, for substrates quite different in structure from the cognate. This mechanism is therefore only useful in amino acid selection when editing is not that crucial because the initial selection is high. The reason for this is that if k_1 for the noncognate substrate is to be much greater than k_T (Equation (4.21)), then k_1 for the correct substrate must also be significant when the two substrates are similar in structure (Mulvey and Fersht, 1977). Equation (4.21) may be used to analyse the significance of the tRNA-independent hydrolysis of the complexes of E^{Val} with noncognate aminoacyl adenylates. The observation that the E^{Val}. Thr-AMP complex hydrolyses at 2% of the rate in the presence of tRNA (Jakubowski, 1980) indicates that $(k_1 + k_2) < 0.02\, k_T$ and so the pre-transfer pathway makes an insignificant contribution to the reduction of errors. Similarly, although the tRNA-independent hydrolysis of E^{Val}. Cys-AMP has

25% of the rate of the tRNA-stimulated reaction (Jakubowski, 1980), the pre-transfer pathway makes little contribution to the overall accuracy although up to 25% of the mistakes are rejected by this route.

In the case of the rejction of Hcys by E^{Met}, and Hcys and Cys by E^{Ile}, however, the hydrolysis rate of the noncognate aminoacyl adenylate complexes is fast in the absence of tRNA and there is little, if any, stimulation by tRNA, so that the pre-transfer pathway is significant.

4.8.2 POST-TRANSFER PATHWAY

The observed values for the rate constant k_3 for the hydrolysis of Thr-tRNA^{Val} and α-aminobutyryl-tRNA^{Val} (>40 s^{-1}) are high compared to rate constants for the dissociation (k_p) of cognately charged tRNAs from aminoacyl-tRNA synthetases (e.g. 12 s^{-1} for E^{Met}.Met-tRNA^{Met} (Mulvey and Fersht, 1977)). Values of k_p for the noncognately charged tRNA could well be far slower if the noncognate aminoacyl moiety binds in the editing site.

4.9 Chemical reaction mechanisms of editing

Virtually nothing is known about the chemistry of editing or the groups involved. The turnover number for the hydrolysis of α-aminobutyryl-tRNA^{Val} is independent of pH between pH 5.8 and 7.8, suggesting that a histidine residue is probably not involved. Inhibition and other kinetic experiments suggest separate aminoacylation and hydrolytic sites (Yarus, 1972a; Schreier and Schimmel, 1972; Fersht and Kaethner, 1976; Güntner and Holler, 1979).

A model invoking the chemical participation of the non-accepting hydroxyl of the tRNA and either a bound water molecule or, in the particular example of the rejection of Thr by E^{Val}, the –OH group of the noncognate substrate, and just one site for both editing and aminoacylation has been proposed (von der Haar and Cramer, 1976; Igloi, von der Haar and Cramer, 1977). There is, however, no direct evidence supporting this scheme. It also conflicts with the evidence supporting the presence of separate sites and also requires different mechanisms for the rejection of threonine and α-aminobutyrate by the E^{Val}. The double-sieve mechanism accounts for the selection for editing of all the competitors of valine by the E^{Val} (Fersht and Dingwall, 1979a).

4.10 Aminoacyl-tRNA synthetases not requiring editing mechanisms

Some amino acids are so different from their possible competitors that adequate selection is possible without recourse to editing – larger competitors are excluded from binding to the active site by the usual process of steric hindrance whilst the smaller competitors are activated too slowly to pose any real problems. A rule of thumb to predict when editing is necessary is to compare the relative rates of activation of the correct amino acid and its isosteric and smaller competitors and calculate whether errors would be introduced at a higher frequency than the observed overall rate of $3/10^4$ for protein synthesis. By this token, the tyrosyl-tRNA synthetase does not need to remove errors of activation of phenylalanine because the activation of tyrosine is favoured by a factor of 10^5 or so (Fersht, Shindler and Tsui, 1980). Similarly, the activation of cysteine is favoured by a factor of 5×10^6 or greater over its isosteric and smaller competitors (Fersht and Dingwall, 1979b). There is no evidence for the participation of a double-sieve type editing mechanism with these enzymes. The glycyl-tRNA synthetase should also face no problems because all other amino acids are larger than glycine. Enzymes which clearly need to possess efficient means of editing are those which have to reject amino acids that are very similar to the correct substrates e.g. the alanyl- (for Gly), threonyl- (Ser and Val), valyl- (Thr and Ala), isoleucyl- (Val), methionyl- (homocysteine), and phenylalanyl- and leucyl- (for the other smaller hydrophobic amino acids) tRNA synthetases.

References

Baldwin, A. N. and Berg, P. (1966) Transfer ribonucleic acid-induced hydrolysis of valyl adenylate bound to isoleucyl ribonucleic acid synthetase. *J. Biol. Chem.*, **241**, 839–845.

Bonnet, J. and Ebel, J. P. (1974) Correction of aminoacylation errors. Evidence for a nonsignificant role of the aminoacyl-tRNA synthetase catalysed deacylation of aminoacyl-tRNAs. *FEBS Letts.*, **39**, 259–262.

Edelmann, P. and Gallant, J. (1977) Mistranslation in *E. coli. Cell*, **10**, 131–137.

Eldred, E. W. and Schimmel, P. (1972) Rapid deacylation by isoleucyl transfer ribonucleic acid synthetase of isoleucine-specific transfer ribonucleic acid synthetase of isoleucine-specific transfer ribonucleic acid aminoacylated with valine. *J. Biol. Chem.*, **247**, 2961–2964.

Fersht, A. R. (1974) Catalysis, binding and enzyme–substrate complementarity. *Proc. R. Soc. London B*, **187**, 397–407.

Fersht, A. R. (1977a) *Enzyme Structure and Mechanism*, W. H. Freeman, San Francisco.

Fersht, A. R. (1977b) Editing mechanisms in protein synthesis. Rejection of valine by the isoleucyl-tRNA synthetase. *Biochemistry*, **16**, 1025–1030.

Fersht, A. R. (1979) In *Transfer RNA: Structure, Properties and Recognition* (eds P. R. Schimmel, D. Soll and J. N. Abelson), Cold Spring Harbor Monograph Series, New York **9A**, pp. 247–254.

Fersht, A. R. (1981) Enzymic editing mechanisms and the genetic code. *Proc. R. Soc. London B*, **212**, 351–379.

Fersht, A. R. (1985) *Enzyme Structure and Mechanism*, 2nd edn (see Chapter 13), W. H. Freeman, New York and Oxford.

Fersht, A. R. and Dingwall, C. (1979a) Establishing the misacylation/deacylation of tRNA pathway for the editing mechanism of prokaryotic and eukaryotic valyl-tRNA synthetases. *Biochemistry*, **18**, 1238–1245.

Fersht, A. R. and Dingwall, C. (1979b) Cysteinyl-tRNA synthetase from *Escherichia coli* does not need an editing mechanism for the rejection of serine and alanine. High binding energy of small groups in specific molecular interactions. *Biochemistry*, **18**, 1245–1250.

Fersht, A. R. and Dingwall, C. (1979c) An editing mechanism for the methionyl-tRNA synthetase in the selection of amino acids in protein synthesis. *Biochemistry*, **18**, 1250–1256.

Fersht, A. R. and Dingwall, C. (1979d) Evidence for the double-sieve editing mechanism for selection of amino acids in protein synthesis: Steric exclusion of isoleucine by valyl-tRNA synthetases. *Biochemistry*, **18**, 2627–2631.

Fersht, A. R. and Kaethner, M. (1976) Enzyme hyperspecificity. Rejection of threonine by the valyl-tRNA synthetase by misacylation and hydrolytic editing. *Biochemistry*, **15**, 3342–3346.

Fersht, A. R., Shindler, J. S. and Tsui, W.-C. (1980) Probing the limits of protein–amino acid side chain recognition with the aminoacyl-tRNA synthetases. Discrimination against phenylalanine by tyrosyl-tRNA synthetases. *Biochemistry*, **19**, 5520–5524.

Gabius, H.-J., Engelhardt, R., Schröder, F. R. and Cramer, F. (1983) Evolutionary aspects of accuracy of phenylalanyl-tRNA synthetase. Accuracy of fungal and animal mitochondrial enzymes and their relationship to their cytoplasmic counterparts and a prokaryotic enzyme. *Biochemistry*, **22**, 5306–5314.

Güntner, C. and Holler, E. (1979) Phenylalanyl-tRNA synthetase of *Escherichia coli* K 10. Multiple enzyme aminoacyl-tRNA complexes as a consequence of substrate specificity. *Biochemistry*, **18**, 2028–2038.

Hopfield, J. J. (1974) Kinetic proofreading: A new mechanism for reducing errors in biosynthetic processes requiring high specificity. *Proc. Natl Acad. Sci.*, **71**, 4135–4139.

Igloi, G. L., von der Haar, F. and Cramer, F. (1977) Hydrolytic action of aminoacyl-tRNA synthetases from baker's yeast. 'Chemical proofreading' of Thr-tRNA^{Val} and amino acid analogues. *Biochemistry*, **16**, 1696–1702.

Igloi, G. L., von der Haar, F. and Cramer, F. (1978) Aminoacyl-tRNA synthetases from yeast: Generality of chemical proofreading in the prevention of misaminoacylation of tRNA. *Biochemistry*, **17**, 3459–3462.

Jakubowski, H. (1978) Valyl-tRNA synthetase from yellow lupin seeds. Instability of enzyme-bound noncognate adenylates versus cognate adenylate. *FEBS Lett.*, **95**, 235–238.

Jakubowski, H. (1980) Valyl-tRNA synthetase from yellow lupin seeds: hydrolysis of the enzyme-bound noncognate aminoacyl adenylate as a possible mechanism of increasing specificity of the aminoacyl-tRNA synthetase. *Biochemistry*, **19**, 5071–5078.

Jakubowski, H. and Fersht, A. R. (1981) Alternative pathways for editing noncognate amino acids by aminoacyl-tRNA synthetases. *Nucl. Acid. Res.*, **9**, 3105–3117.

Joachimiak, A. and Barciszewski, J. (1980) Amino acid: tRNA ligases (EC 6.1.1.–). *FEBS Lett.*, **119**, 201–210.

Lin, S. X., Baltzinger, M. and Remy, P. (1984) Fast kinetic study of yeast phenylalanyl-tRNA synthetase: role of tRNA^{Phe} in the discrimination between tyrosine and phenylalanine. *Biochemistry*, **23**, 4109–4116.

Loftfield, R. B. (1963) The frequency of errors in protein biosynthesis. *Biochem. J.*, **89**, 82–92.

Loftfield, R. B. and Eigner, E. A. (1966) The specificity of enzymic reactions. Aminoacyl-soluble RNA ligases. *Biochim. Biophys. Acta*, **130**, 426–448.

Loftfield, R. B. and Vanderjagt, D. (1972) The frequency of errors in protein biosynthesis. *Biochem. J.*, **128**, 1353–1356.

Mertes, M., Peters, M. A., Mahoney, W. and Yarus, M. (1972) Isoleucylation of transfer RNA^{Met} (*E. coli*) by the isoleucyl-transfer RNA synthetase from *Escherichia coli. J. Mol. Biol.*, **71**, 671–685.

Mulvey, R. S. and Fersht, A. R. (1977) Editing mechanisms in aminoacylation of tRNA: ATP consumption and the binding of aminoacyl-tRNA by elongation factor Tu. *Biochemistry*, **16**, 4731–4737.

Norris, A. T. and Berg, P. (1964) Mechanism of aminoacyl RNA synthesis: Studies with isolated aminoacyl adenylate complexes of isoleucyl tRNA synthetase. *Proc. Natl Acad. Sci.*, **52**, 330–337.

Okamoto, M. and Savageau, M. A. (1984a) Integrated function of a kinetic proofreading mechanism: steady-state analysis testing internal consistency of data obtained *in vivo* and *in vitro* and predicting parameter values. *Biochemistry*, **23**, 1701–1709.

Okamoto, M. and Savageau, M. A. (1984b). Integrated function of a kinetic proofreading mechanism: dynamic analysis separating the effects of speed and substrate competition on accuracy. *Biochemistry*, **23**, 1710–1715.

Parker, J., Johnston, T. C. and Borgia, P. T. (1980) Mistranslation in cells infected with bacteriophage M52: Direct evidence for Lys for Asn substitution. *Mol. gen. Genet.*, **180**, 275–281.

Pauling, L. (1957) The probability of errors in the process of synthesis of protein molecules. In *Festschrift Arthur Stoll*, Birkhauser Verlag, Basel, p. 597.

Savageau, M. A. and Freter, R. (1979a) On the evolution of accuracy and the cost of proofreading in tRNA aminoacylation. *Proc. Natl Acad. Sci. USA*, **76**, 4507–4510.

Savageau, M. A. and Freter, R. (1979b) Energy cost of proofreading to increase fidelity of transfer ribonucleic acid aminoacylation. *Biochemistry*, **18**, 3486–3493.

Schreier, A. A. and Schimmel, P. R. (1972) Transfer ribonucleic acid synthetase catalyzed deacylation of aminoacyl transfer ribonucleic acid in the absence of adenosine monophosphate and pyrophosphate. *Biochemistry*, **11**, 1582–1589.

Springgate, C. F. and Loeb, L. A. (1975) On the fidelity of transcription by *E. coli* ribonucleic acid polymerase. *J. Mol. Biol.*, **97**, 577–591.

Tsui, W.-C. and Fersht, A. R. (1981) Probing the principles of amino acid selection using the alanyl-tRNA synthetase from *Escherichia coli. Nucl. Acid. Res.*, **9**, 4627–4637.

von der Haar, F. and Cramer, F. (1976) Hydrolytic action of aminoacyl-tRNA synthetases from baker's yeast: 'Chemical proofreading' preventing acylation of tRNA with misactivated valine. *Biochemistry*, **15**, 4131–4138.

Yarus, M. (1972a) Phenylalanyl-tRNA synthetase and isoleucyl-tRNA^{Phe}: A possible verification mechanism for aminoacyl-tRNA. *Proc. Natl Acad. Sci.*, **69**, 1915–1919.

Yarus, M. (1972b) Solvent and specificity. Binding and isoleucylation of phenylalanine transfer ribonucleic acid (*Escherichia coli*) by the isoleucyl transfer ribonucleic acid synthetase from *Escherichia coli. Biochemistry*, **11**, 2352–2361.

5 The accuracy of mRNA–tRNA recognition

R. H. BUCKINGHAM and H. GROSJEAN

5.1 Introduction

In order to synthesize functional proteins, mRNA must be translated by the ribosome with a certain level of accuracy. As we will discuss below, our knowledge concerning the levels of precision ultimately attained by the translational machinery, especially at the level of the nascent polypeptide chain, is still incomplete. Furthermore, it will be interesting to compare the sparse data that exist with the results of a naïve calculation of the following type: in order to synthesize correctly at least 50% of the polypeptide chains of an enzyme as large as β-galactosidase which contains 1169 amino acid residues, an average error no larger than 1 in 1700 per amino acid residue can be tolerated (Fig. 5.1). Why should attaining this level of accuracy pose a problem?

The traditional view of aminoacyl-tRNA (AA-tRNA) selection by the acceptor site (A-site) of the ribosome supposes that the only parameter which discriminates between cognate and noncognate AA-tRNA is the stability of the interaction between the three bases of a codon presented in the A-site and the corresponding three bases in the anticodon of AA-tRNA. How this might happen is not yet clear. Our understanding of nucleic acid base pairing equilibria comes very largely from studies in solution. These studies predict that the effects on free energy changes of mispairings are not sufficiently large to explain the discrimination between cognate and noncognate AA-tRNA during protein synthesis on the basis of a one-step selection mechanism (see Chapter 11). Furthermore, studies on the association between tRNAs with complementary anticodons have shown that association between loops is different in several important aspects from association between linear complementary oligonucleotides; we cannot expect, therefore, to be able to apply to the codon–anticodon interaction on the ribosome much of what has been learned about double helix formation between complementary oligoribonucleotides in solution.

$P = (1 - p)^n$

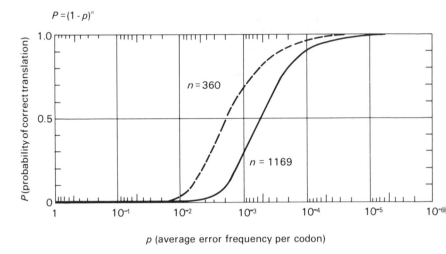

Figure 5.1   Estimates of the probability of synthesizing a correct protein. If p is the average error frequency per codon at each of the n codons of a messenger RNA, the probability (P) of correct translation is $P = (1-p)^n$. The solid curve shows this function for $n = 1169$ (a subunit of β-galactosidase, molecular weight 135 000). The dashed curve shows the function for $n = 360$ which corresponds to the average chain length of prokaryotic and eukaryotic polypeptide chains (molecular weight 40 000, Sommer and Cohen, 1980).

In order to understand AA-tRNA selection we have to take into account several other factors apart from the base–base interactions between codon and anticodon. The whole of the tRNA structure is important, both in the anticodon loop and in other parts of the molecule. In the messenger itself, not only the codon but the sequence surrounding the codon has a significant effect, through a mechanism not yet understood. In addition to the realization that such external factors play an important role in AA-tRNA selection, our knowledge of the base interactions permitted in a cognate tRNA–codon interaction has evolved considerably since the proposal of the wobble hypothesis by Crick (1966). These aspects of AA-tRNA selection will now be discussed in more detail.

5.2 How specific is the process of translation?

5.2.1 MISSENSE ERRORS *in vivo*

Few estimates exist of the number of amino acids misincorporated into cellular proteins; furthermore, these figures are difficult to interpret in terms of the frequency of errors at the level of the nascent polypeptide chain. The direct measurement is difficult because a high degree of purification of the protein, essential for peptide analysis, tends to eliminate all but one or a few molecular species. One strategy the cell appears to use to avoid the build-up of faulty protein is to discard unfinished polypeptides subsequent to mis-incorporation of an amino acid (Manley, 1978a; Menninger, 1983). These products will not normally be detected as finished protein. Finally, degradation of faulty but finished protein may be more rapid than that of correct protein (Goldberg and St John, 1976; Mount, 1980), though this is not observed in the case of many *E. coli* proteins containing errors as a result of histidine or asparagine starvation (Parker, 1981). All systems in which the polypeptide product is subsequently assembled in a larger structure, such as subunit enzymes, the ribosome, cell structures, flagella, etc., may provide further stages of selection for active chains. All these processes tend to mask errors in tRNA selection on the ribosome and lead to underestimates. Other problems may lead to overestimates of such errors; these include failure to eliminate contamination by closely related proteins, unobserved genetic polymorphism or errors in DNA transcription (Rosenberger and Foskett, 1981) or tRNA charging.

It has often been supposed that a single figure can reasonably be used to describe the probability of amino acid misincorporation, but this is misleading. Of the 380 different substitutions possible (20×19), it should be expected, and has indeed been found, that certain types of misincorporation are much more likely than others. Very few have been measured experimentally, and it is not known if these are among the most probable cases of misreading or not. Some are given in Table 5.1, though this is not exhaustive; see also Yarus (1979). The problem is further complicated by an important phenomenon which will be discussed in more detail in Section 5.4.2. Although the mechanism is not yet fully understood, it is clear that the three bases of a codon are not the only bases that affect AA-tRNA selection; those adjacent to the codon also influence this process and, in consequence, the probability of misreading. The selection of sites for the investigation of misincorporation is therefore difficult.

The first documented attempts to measure misreading *in vivo* are those of Loftfield (1963), Loftfield and Vanderjagt (1972) and Edelmann and Gallant (1977). Loftfield measured valine incorporation into non-valine containing peptides of α-globin and chicken oviduct albumin. In one case the precise error involved was identified: valine replaced threonine at a frequency of 2×10^{-4}. The experiment was designed, however, to measure valine incorporation in place of isoleucine; in the threonine-containing peptide analysed in detail this error could not be detected (less than 10^{-5} ?). Furthermore, it is probable that the valine incorporation observed represents mischarging of tRNA^{Thr} with valine (Coons, Smith and Loftfield, 1979).

Edelmann and Gallant (1977) took advantage of the absence of cysteine from the protein flagellin and the ease with which it can be purified; they

Table 5.1 Missense errors *in vivo*

Reference	Protein	Source	Substitution	Misincorporation per site		
				Average over all codons	Average over selected codons	Single site substitution
Loftfield and Vanderjagt (1972)	α-Globin	Rabbit reticulocytes	Val for Ile			$<10^{-5}$
Coons, Smith and Loftfield (1979)	α-Globin	Rabbit reticulocytes	Val for Thr			2×10^{-4}
Harley, Pollard and Stanners (1981)	Actin	W138 Fibroblasts	Gln (?) for His		4×10^{-5} (8 His codons)	
Harley, Pollard and Stanners (1981)	Actin	CHO Cells	Gln (?) for His		1.3×10^{-4} (8 His codons)	
Edelmann and Gallant (1977)	Flagellin	*E. coli*	Cys for Arg	1×10^{-6}	3×10^{-5} (18 Arg codons)	
Bouadloun, Donner and Kurland (1983)	Ribosomal protein L7/L12	*E. coli*	Cys for Arg			1.5×10^{-3}
Bouadloun, Donner and Kurland (1983)	Ribosomal protein S6	*E. coli*	Cys for Trp			4×10^{-3}
Parker *et al.* (1983)	Coat protein	MS2/*E. coli*	Lys for Asn (AAU)		5×10^{-3} (4 AAU codons)	1.5×10^{-3} at AAU12
Parker *et al.* (1983)	Coat protein	MS2/*E. coli*	Lys for Asn (AAC)			2×10^{-4} at AAC3
Khazaie, Buchanan and Rosenberger (1984a)	Coat protein	Qβ/*E. coli*	Trp for ?	1.3×10^{-3}	2×10^{-2} (over 7 codons: single base misreading)	
Khazaie, Buchanan and Rosenberger (1984a)	Coat protein	Qβ/*E. coli*	His for ?	1×10^{-3}	2×10^{-3} (over 41 codons: single base misreading)	
Rice, Libby and Reeve (1984)	0.3 Protein	T7/*E. coli*	Cys for ?	2×10^{-4}		

measured cysteine misincorporation, probably in place of arginine (tryptophan, another site for cysteine misincorporation, is absent from flagellin). They found about 6×10^{-4} mole of cysteine per mole of flagellin. The codons used for the eighteen arginines present are unknown; depending on how many sites are important, the misincorporation per codon could range from 3×10^{-5} to 6×10^{-4}. Analogous experiments on misincorporation of cysteine into the phage T7 0.3 protein indicate one cysteine misincorporated per 5000 codons translated (Rice, Libby and Reeve, 1984).

Some experiments have been reported in which methionine incorporation has been measured in proteins normally lacking this amino acid; however, it must be rigorously demonstrated that all N-terminal methionine has been cleaved from the polypeptide, which has not generally been convincingly achieved (Medvedev and Medvedeva, 1978; Buchanan and Stevens, 1978).

In many experiments on misincorporation, the site of substitution is not precisely determined. Recently, the first measurements have been reported in a prokaryotic system of amino acid misincorporation at specific sites. Bouadloun, Donner and Kurland (1983) determined the level of cysteine substitution for the *unique* arginine in *E. coli* ribosomal protein L7/L12 to be about 10^{-3} in wild type bacteria. In another ribosomal protein, S6, cysteine replaces the *unique* tryptophan at a frequency of 3 to 4×10^{-3}. The strategy in these experiments was ingenious: specific cleavage is possible at the arginine or tryptophan sites in the normal protein but not the erroneous protein, which enables correction to be made for misincorporation at other sites in the protein. No other experiments reported so far employ such a strategy, and their interpretation in terms of errors at specific sites needs caution. Furthermore, ribosomal proteins may be particularly adapted to error measurements. It has been argued that the ribosome is rather tolerant of errors in its structural components (Stöffler, Hasenbank and Dabbs, 1981); these figures may therefore be indicative of error levels in the nascent polypeptide chain.

Misincorporation that changes the charge of a protein leads to satellite spots in isoelectric focusing. This has been used by Parker *et al.* (1983) to estimate misincorporation of lysine for asparagine in MS2 coat protein. If misreading is restricted mostly to the four AAU codons, the misincorporation observed corresponds on average to 5×10^{-3} per codon. Errors at the specific positions 3 and 12 in the polypeptide were 1.5×10^{-3} (an AAU codon) and 2×10^{-4} (an AAC codon), respectively. Measurements of tryptophan and histidine incorporation in Qβ coat protein, which should contain neither of these amino acids, shows misincorporation in both cases of about one amino acid per 10^{3} codons translated (Khazaie, Buchanan and Rosenberger, 1984a, b).

In evaluating the available data on frequencies of misincorporation it is important to know which types of noncognate interaction are the most probable. Errors of misincorporation that have been induced *in vivo* by amino acid starvation or inhibition of tRNA charging with amino acid analogues can be understood as arising by misreading the third codon letter (O'Farrell, 1978; Parker *et al.*, 1978; Parker *et al.*, 1983). Under these conditions, the misincorporation may be as high as 0.1. In missense suppressor tRNAs derived from tRNA^{Gly}, residual recognition of the codon read by the wild type tRNA suggests U.G wobble in the first and second positions of the codon (Murgola and Pagel, 1980). Further information about likely noncognate interactions in prokaryotes comes from the misreading studies conducted in the presence of streptomycin, since it is considered that this drug does not introduce new ambiguities into the translational machinery but merely amplifies those already inherent to the system (Gorini, 1974). Such studies have shown U–U, U–C and G–U anticodon–codon interactions in the 5' codon position and internal positions, and also C–A interactions in the 5' codon position. A C–A interaction, in the wobble position, is also seen in UGA suppression in *E. coli* by tRNA^{Trp} (Hirsh, 1971). The wild type tRNA^{Trp} can misread UGA at a frequency as high as a few percent, in the absence of any antibiotic, in certain codon contexts (Engelberg-Kulka, 1981). UGA suppression in *supK* strains of *Salmonella typhimurium* appears to require a G–G interaction in the internal codon position (Pope, Brown and Reeves, 1978).

5.2.2 MISSENSE ERRORS *in vitro*

Many measurements of missense errors have been made during protein synthesis *in vitro*, most commonly with synthetic polynucleotide messengers, often homopolymers (reviewed in Yarus, 1979). From such results it is generally concluded that pyrimidines, especially uridine, are more easily misread than purines and that alternating copolymers are more accurately translated than the corresponding homopolymers. Unfortunately, much work on missense errors *in vitro* has been performed under conditions, of ionic composition, in particular, that clearly led to artefactually high levels of missense errors and to misleading conclusions. More recently, *in vitro* systems of protein synthesis have been developed that approximate to the performance of the ribosome *in vivo* as regards both the level of missense errors and speed of elongation, at least in the case of synthetic polymer messages (Jelenc and Kurland, 1979). The presence of naturally occurring polyamines (Igarashi *et al.*, 1982; Abraham, 1983) and an efficient ATP regenerating system were found to be important. Under these conditions, even poly U, the most easily misread homopolymer, is translated into polyphenylalanine with a less than 2×10^{-4} error frequency for leucine (Jelenc and Kurland, 1979; Wagner *et al.*, 1982).

One problem in measuring errors in unpurified *in vitro* systems using such messengers is that the preponderance of one or a few codons can seriously deplete the pool of cognate tRNA that is charged and thus raise artefactually the ratio of noncognate to cognate tRNA. Translation then proceeds under conditions of virtual starvation for cognate AA-tRNA. In the case of the strictly repeating messenger poly (U–G), which codes for alternating valine and cysteine residues, errors of tryptophan incorporation in place of cysteine were found to be dominated by this effect; estimates of the real selection error by the ribosome could only be obtained by extrapolation to lower levels of polypeptide synthesis (Carrier and Buckingham, 1984). This problem could be resolved by working in a purified system under different ionic conditions (Andersson, Buckingham and Kurland, 1984). During translation of natural messengers a similar tendency to deplete some AA-tRNA pools might occur if the tRNA population is not adapted to the average codon usage (see Section 5.4.3 for further discussion). Other features of synthetic polynucleotides may militate against accurate translation, including effects of messenger secondary structure. A further problem may arise at the level of initiation; in order to allow initiation on messengers lacking efficient initiation sites, ionic conditions may have to be employed that are deleterious to accurate translation. The development of the *in vitro* system described above (Jelenc and Kurland, 1979) has been highly instrumental in the investigation *in vitro* of ribosomal mechanisms of AA-tRNA selection and proofreading (Ruusala, Ehrenberg and Kurland, 1982a, b).

5.2.3 OTHER TYPES OF ERROR

Errors of frameshifting have been observed both *in vivo* and *in vitro*. The former was first studied in terms of the 'leakiness' of β-galactosidase frameshift mutants (in the range 10^{-5} to at least 10^{-3}) and has been shown to be under the same ribosomal control as missense and nonsense errors (Atkins, Elseviers and Gorini, 1972; Gorini, 1974). This suggests that frameshifting may be associated with errors in AA-tRNA selection (Kurland, 1979; Gallant and Foley, 1979 and Chapter 6 of this volume). As discussed further below, the high frequency of frameshifting derived from leakiness of some frameshift mutants may overestimate the likelihood of these events in translation of wild type messengers, since particularly 'shifty' sequences may be rare in the normal reading frame but common in the other two reading frames.

Addition of purified normal tRNA species to *in vitro* translation systems, which appear to compete with the cognate AA-tRNA at certain sites, induces frameshifts to either the + 1 or − 1 reading frames (Atkins *et al.*, 1979). Similarly, starvation of cells for some amino acids induces leakiness in frameshift mutants (Gallant and Foley, 1979; Gallant *et al.*, 1982). The mechanism by which these frameshifts are induced is not yet clear but some observations suggest that frameshifting arises by the misreading of a codon in the normal reading frame (Weiss and Gallant, 1983), rather than being initiated by the recognition of an out-of-phase codon by its cognate tRNA (Roth, 1981).

In yeast mitochondria, frameshift events seem to occur naturally *in vivo* at a frequency of 10^{-2} to 5×10^{-2} at UUUUC and UUUUUUC in the mitochondrial mRNA, though a polymerase frameshift error rather than a ribosomal error might also explain these observations (Fox and Weiss-Brummer, 1980). So far, little evidence exists that physiologically important products are synthesized that depend upon frameshift events (but see Hensgens *et al.*, 1984). However, it has been suggested that initiation of lysis gene translation in MS2-infected *E. coli* is dependent on a frameshift by ribosomes reading the coat protein cistron, which leads to premature termination just before the start of the lysis cistron (Cody and Conway, 1981; Kastelein and Van Duin, 1982). The remarkable observation has been made that amino acid starvation can lead to suppression not only of frameshift mutations but also of nonsense mutants, an event that would require frameshifting at two sites, one on each side of the nonsense codon, in order to restore normal reading frame (Gallant *et al.*, 1982); double frameshifting had already been observed *in vitro* during MS2 translation when the system was supplemented by specific normal tRNAs (Atkins *et al.*, 1979). This requires a remarkably high probability of frameshifting at the second site, and suggests that such sites may exist and be found frequently in the $+1$ and -1 reading frames. More recent experiments also implicate the sequence UUUC, referred to above, amongst others, as prone to frameshift errors. This occurs in a leaky mutant of the *trpE* gene, *trpE871* (Atkins, Nichols and Thompson, 1983) and in the 'shifty' regions of two phage T7 genes where frameshifting leads to bypassing of the in-frame terminator (Dunn and Studier, 1983). It may be anticipated that evolutionary pressures will minimize 'shifty' sequences in the normal reading frame.

Other types of error, which are difficult to evaluate quantitatively, include incorrect initiation and premature termination (Manley, 1978a, b). Estimates of premature termination indicate a frequency of about 3×10^{-4} per codon which corresponds to a probability of 23% that an initiation will fail to lead to a full length β-galactosidase polypeptide (Manley, 1978a; Caplan and Menninger, 1979). It is difficult to know whether this represents an accident of translation *per se*, or a correction subsequent to an error of misincorporation (Cabanas and Modolell, 1980; Thompson *et al.*, 1981; Hornig, Woolley and Lührmann, 1984), but in the latter case it could be an event of considerable physiological significance. The effect of starving for tryptophan during *in vitro* protein synthesis of MS2 coat protein may be explained by a model that supposes a high probability of premature termination at 'hungry' codons (Goldman, 1982; Menninger, 1983).

It is possible that some apparent errors are physiologically important. In the case of termination, for example, it is known that the growth of some viruses is dependent on the partial readthrough of UGA termination codons, but no case has yet been proven where the host cell employs such a device (see Kohli and Grosjean, 1981; Engelberg-Kulka, 1981; Ryoji, Hsia and Kaji, 1983). Similarly, UAG readthrough leads to extended polypeptides coded by TMV, though again the role is uncertain (see Bienz and Kubli, 1981; Beier, Barciszewska and Sickinger, 1984). The suggestion has been made that cellular differentiation in some organisms may, at some stages, require a relaxation in translational accuracy (Picard-Bennoun, 1982).

At least two strategies may, therefore, be used by the cell in order to counteract the effect of missense errors after their occurrence. An increased probability of a frameshift error subsequent to the misincorporation of an amino acid would normally lead to an early encounter with a termination codon in the new reading frame (see more detailed discussion in Chapter 6 of this volume). Also, this would lead to a lack of adaptation between the codons present in the new reading frame (phases $+1$ and $+2$) and the tRNA population; this might reduce considerably the efficiency of translation in the non-natural reading frame (see Section 5.4.3). A further factor that may reduce the effect of misincorporation is that translocation following such an event leads to a peptidyl-tRNA weakly bound in the P-site, with increased probability that the unfinished polypeptide chain, still attached to tRNA, will dissociate from the ribosome (Menninger, 1983). This mechanism for premature termination is supported by experiments which confirm the decoding potential of the P-site (Lührmann, Eckhardt and Stöffler, 1979; Wurmbach and Nierhaus, 1979; Peters and Yarus, 1979). These strategies reduce the build-up of faulty protein in the cell. They also complicate considerably attempts to measure errors in AA-tRNA selection on the ribosome by looking at finished protein.

Figure 5.2 summarizes the many varied accidents that may occur during the process of translation.

5.3 Decoding of the third codon base

5.3.1 THE WOBBLE HYPOTHESIS

In 1966, Crick rationalized the available evidence on the interactions permitted in decoding in the following way: only the standard base pairs that are found in a Watson–Crick double helical structure are permitted in the first two codon positions, but in the third position, certain additional pairs are permitted in which the geometry is reasonably close to that of the standard pairs. These rules allowed U.G, G.U, I.U, I.C and I.A pairs, but excluded U.C, C.U and U.U pairs and all other pairs that do not form at least two hydrogen bonds. More recently, however, it has become clear that the behaviour of U in the 5' position depends on the presence and the type of post-transcriptional modification, whether or not this changes the geometry of the interaction (for reviews, see Weiss, 1973; Dirheimer, 1983; Björk,

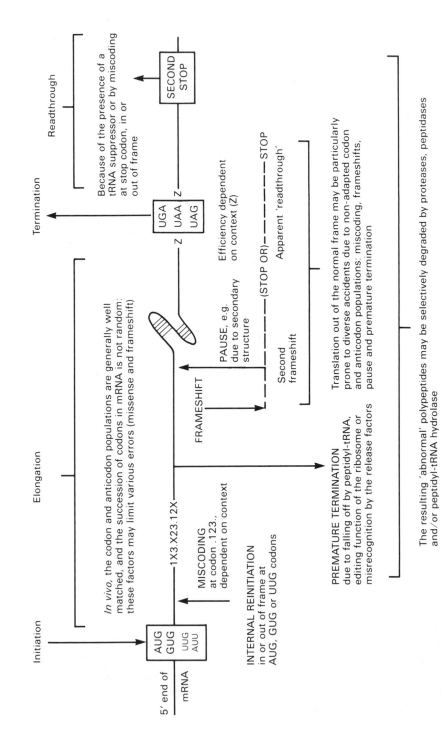

Figure 5.2 Accidents of translation.

1984). In prokaryotes, and in the cytoplasm of eukaryotes, uridine in this position is usually modified: this appears either to increase or decrease the range of possible interactions (Berman et al., 1978; Hillen et al., 1978). Of great interest is the finding that in mitochondria, unmodified uridine is found in the 5' anticodon position of about half of the tRNA species (Heckman et al., 1980; Barrell et al., 1980; Bonitz et al., 1980). These tRNAs are apparently unique for a given amino acid and are used to translate all four codons in the unmixed codon families (i.e. the groups of four codons XYU, XYC, XYA and XYG specifying the same amino acid). Hence, unmodified U in this position on mitochondrial ribosomes can decode all four bases in the third position of the codon. In the mixed codon families (where XYPu and XYPy have different meanings), the two codons XYA and XYG are translated by a single tRNA having a modified uridine (as yet unidentified) in the 5' anticodon position. Thus, modification of U appears to restrict its translational specificity to purines. The two codons in mixed codon families ending with a pyrimidine are read by a single tRNA with G in the wobble position, as predicted by the wobble hypothesis. These relationships are shown in Table 5.2.

These findings suggest that unmodified U in the 5' anticodon position does make an interaction with each of the four bases which contribute to the stability of the codon–anticodon interaction (Barrell et al., 1980). Therefore, this reading does not exactly correspond to a 'two-out-of-three' reading mechanism, to be discussed below, which proposes that the stability of an interaction involving the first two codon bases may be sufficient to decode all four codons $XY^{Pu}/_{Py}$ when they are synonymous. It is not yet known whether mitochondrial tRNAs with an unmodified U in the 5' anticodon position will recognize all four codons of a codon family on prokaryotic or cytoplasmic eukaryotic ribosomes.

5.3.2 THE 'TWO-OUT-OF-THREE' AND OTHER HYPOTHESES

An increasing number of observations has appeared which have been interpreted to show that the wobble hypothesis does not adequately describe the activity of tRNA isoacceptors in decoding synonymous codons. One approach has been the simple triplet-dependent ribosomal binding test which was used initially to elucidate the genetic code (see for example Takemoto et al., 1973; Jank et al., 1977; Kruse et al., 1980; Kawasaki et al., 1980; Diamond, Dudock and Hatfield, 1981). It should be borne in mind in interpreting such experiments that they are done under conditions which may not be able to distinguish between two tRNA species which differ widely in affinity for the ribosome-bound triplet, as was indeed shown by the high levels of noncognate binding observed in early experiments. Furthermore, they are conducted under ionic conditions very different from those needed for accurate protein synthesis, and in the absence of the elongation factors used in vivo. A further

Table 5.2 Codon and anticodon usage in mitochondria. The 64 codons are listed together with their corresponding frequencies of usage in coding sequences in both yeast and man. Frequencies are expressed per thousand codons; the coding sequences available comprised 2168 codons in yeast and 3789 codons in the human mitochondrial genome. The anticodon sequences are placed opposite the codons which are complementary to the Watson–Crick sense; this is not intended to imply any particular decoding properties *in vivo* (see text). Anticodons in italics were deduced from the genomic sequence (*T* is thymine). Information is from Bonitz *et al.* (1980) and Macreadie *et al.* (1983). Many yeast anticodons are known from direct sequencing (see Gauss and Sprinzl, 1984); in these cases data on modified bases in the 5′ anticodon position and 3′ adjacent to the anticodon are included. The latter modification contains an isopentenyl group in tRNA[Tyr] and tRNA[Trp] (yeast). G^1m is 1-methyl guanine, At^6 and Ai^6 are the threonyl carbamoyl and isopentenyl derivatives of A at position 6, U* is an unidentified derivative of U at position 6, iMet is the putative initiator tRNA. Four dashes indicate the absence of a mitochondrial gene coding for a strictly complementary tRNA. Note the different codon usage in yeast and human mitochondria

Codon	Yeast freq.	Yeast anticodon	a.a.	Human freq.	Human anticodon	a.a.
UUU	37	*G AA*.G^1m	Phe	20	*G AA*	Phe
UUC	29	— — — —		37	— — — —	
UUA	127	*U* *AA*.G^1m	Leu	19	*T AA*	Leu
UUG	1	— — — —		4	— — — —	
CUU	2	— — — —		17	— — — —	
CUC	0	— — — —		44	— — — —	
CUA	8	*U AG*.G^1m	Thr	73	*T AG*	Leu
CUG	1	— — — —		12	— — — —	
AUU	88	— — — —		33	— — — —	
AUC	13	*G AU*.At^6	Ile	52	*G AT*	Ile
AUA	17	*C AU*.At^6	Met	44	*C AT*	Met
AUG	33	*C AU*.G^1m	iMet	11	*C AU*	iMet
GUU	22	— — — —		8	— — — —	
GUC	2	— — — —		13	— — — —	
GUA	36	*T AC*.*A*	Val	18	*T AC*	Val
GUG	?			5		

Codon	Yeast freq.	Yeast anticodon	a.a.	Human freq.	Human anticodon	a.a.
UCU	18	— — — —		8	— — — —	
UCC	0.4	— — — —		26	— — — —	
UCA	34	*U GA*.G^1m	Ser	22	*T GA*	Ser
UCG	0	— — — —		2	— — — —	
CCU	20	— — — —		11	— — — —	
CCC	2	— — — —		31	— — — —	
CCA	17	*U GG*.G^1m	Pro	14	*T GG*	Pro
CCG	0.4	— — — —		2	— — — —	
ACU	19	— — — —		13	— — — —	
ACC	0.4	— — — —		41	— — — —	
ACA	24	*T GT*.*A*	Thr	35	*T GT*	Thr
ACG	0	— — — —		3	— — — —	
GCU	30	— — — —		11	— — — —	
GCC	3	— — — —		33	— — — —	
GCA	24	*T GC*.*A*	Ala	21	*T GC*	Ala
GCG	?					

Codon	Yeast (count 1 / count 2)	anticodon	AA	Human (count 1 / count 2)	anticodon	AA
UAU / UAC	48 / 6	GUA.Ai^6	Tyr	12 / 23	GTA	Tyr
UAA / UAG	STOP 0			STOP 0		
CAU / CAC	23 / 2	GUG.G^{1}m	His	5 / 21	GTG	His
CAA / CAG	17 / 2	TTG.A	Gln	21 / 2	TTG	Gln
AAU / AAC	86 / 5	(GTT)	Asn	9 / 35	GTT	Asn
AAA / AAG	31 / 1	U*UU.At^6	Lys	22 / 3	TTT	Lys
GAU / GAC	23 / 1	GTC.A	Asp	4 / 13	GTC	Asp
GAA / GAG	18 / 1	TTC.A	Glu	17 / 6	TTC	Glu

Codon	Yeast (count 1 / count 2)	anticodon	AA	Human (count 1 / count 2)	anticodon	AA
UGU / UGC	6 / 1	GCA.A	Cys	1 / 4	GCA	Cys
UGA / UGG	18 / 0	U*CA.Ai^6	Trp	25 / 3	TCA	Trp
CGU / CGC / CGA / CGG	0.4 / 0 / 0 / 0.4	ACG.A	Arg	2 / 7 / 8 / 0.5	TCG	Arg
AGU / AGC	11 / 0	GCU.A	Ser	4 / 10	GCU	Ser
AGA / AGG	20 / 0	U*CU.At^6	Arg	STOP / STOP		
GGU / GGC / GGA / GGG	45 / 1 / 13 / 3	UCC.Ai^6	Gly	6 / 23 / 18 / 9	TCC	Gly

Table 5.3 Codon and anticodon usage in *E. coli*. The frequency of use of each codon is expressed per thousand codons for two classes of coding sequence: highly expressed genes (column H, calculated over 5253 codons) and little expressed genes (column W, see text and Grosjean and Fiers, 1982; Gouy and Gautier, 1982). Also shown are the anticodon sequences found (or in brackets, assumed), the base 5' adjacent to the anticodon, and the abundance (column A) relative to tRNA^{Leu} (anticodon CAG). Abundance data are from Ikemura (1981 a, b), see also Varenne et al. (1984); major, medium and minor are indications of relative abundance in the absence of precise data (see Post and Nomura, 1980). For anticodon sequences see Gauss and Sprinzl (1984); the anticodons *TGT*, *TAG* and *CCG* are from the corresponding gene sequences only, those marked (?) probably exist but have not yet been demonstrated. Abbreviations are those of Gauss and Sprinzl (1984), except for S(5-methylaminomethyl-2-thiouridine) and C$^+$ (N^4-acetylcytidine). A*, C* and U* are unidentified modified bases. Anticodons are placed opposite the codons to which they show the best Watson–Crick complementarity; this is not intended to imply any particular decoding properties *in vivo* (see text)

Codon	H	W	A	Anticodon	aa	Codon	H	W	A	Anticodon	aa
UUU	7	29	—			UCU	18	7	—		
UUC	22	19	0.35	G AA.Ams$^{2\cdot6}$	Phe	UCC	16	9	Maj.	G GA.A	Ser
						UCA	1	7	0.25	V GA.Ams$^{2\cdot6}$	Ser
UUA	2	14	0.12	A* AA.Ams$^{2\cdot6}$	Leu	UCG	2	12	Min.	C GA.Ams$^{2\cdot6}$	Ser
UUG	3	12	0.12	(CAA)	Leu						
CUU	5	14	—			CCU	4	5	—		
CUC	6	13	0.30	GAG.G'm	Leu	CCC	0.4	9	Min.	G GG.G'm	Pro
CUA	0.6	4	Min.	TAG.G	Leu	CCA	5	9	Med.	V GG.G'm	Pro
CUG	66	56	1.00	CAG.G'm	Leu	CCG	31	19	Maj.	C GG.G'm	Pro
AUU	13	30	—			ACU	20	9	—		
AUC	50	23	1.00	G AU.At^6	Ile	ACC	26	23	0.80	G GU.Amt^6	Thr
AUA	0.4	5	0.05	C* AU.At^6	Ile	ACA	3	6	Med.	*TGT.A*	Thr
						ACG	5	15	?	(?)	Thr
AUG	27	25	0.30	C$^+$ AU.At^6	Met						
GUU	37	21	—			GCU	33	16	—		
GUC	8	13	0.40	G AC.A	Val	GCC	9	34	Med.	G GC.A	Ala
GUA	23	9	1.05	V AC.A^{6}m	Val	GCA	23	20	1.04	V GC.A	Ala
GUG	16	23	—			GCG	25	28	?	(?)	Ala

Codon	H	W	A	Anticodon	aa
UAU	6	18	–	—	
UAC	19	12	0.50	$QUA.Ams^2i^6$	Tyr
UAA	STOP				
UAG	STOP				
CAU	4	18	–	—	
CAC	14	11	0.40	$QUG.A^2m$	His
CAA	7	17	0.30	$SUG.A^2m$	Gln
CAG	32	32	0.40	$CUG.A^2m$	Gln
AAU	2	19	–	—	
AAC	30	19	0.60	$QUU.At^6$	Asn
AAA	49	31	1.00	$SUU.At^6$	Lys
AAG	20	8	?	(?)	
GAU	22	35	–	—	
GAC	39	20	0.80	$QUC.A^2m$	Asp
GAA	63	40	0.90	$SUC.A^2m$	Glu
GAG	20	19	?	(?)	

Codon	H	W	A	Anticodon	aa
UGU	2	6	–	—	
UGC	4	7	Med.	$GCA.Ams^2i^6$	Cys
UGA	STOP				
UGG	5	13	0.30	$CCA.Ams^2i^6$	Trp
CGU	42	19	0.90	$ICG.A^2m$	Arg
CGC	19	25	–	—	
CGA	0.6	5	–	—	
CGG	0.2	8	Min.	$CCG.G$	Arg
AGU	2	11	–	—	
AGC	9	12	0.25	$GCU.At^6$	Ser
AGA	0.6	5	0.02	(?)	Arg
AGG	0.2	3	(?)	(?)	
GGU	43	24	–	—	
GGC	33	27	1.12	$GCC.A$	Gly
GGA	0.8	8	0.15	$U^*CC.A$	Gly
GGG	3	13	0.10	$CCC.A$	Gly

aspect of codon recognition which is lost in triplet-directed binding assays is the effect of codon context: a codon in one reading context might be preferentially translated by one isoacceptor species but in a different context by a different isoacceptor. This must be considered in any discussion of decoding hypotheses. In view of the diversity of factors that can influence AA-tRNA selection, we believe that it is no longer justified to consider as universally applicable any rules that attempt to define a strict correspondence between codons and anticodons.

Much closer to the natural process are the $E.$ $coli$ cell-free systems that have been used to investigate this problem, based on RNA phage messenger translation. Attempts have been made to propose an alternative to the wobble hypothesis for reading within certain groups of codons, where all four codons with the first two bases in common code for the same amino acid (an 'unmixed codon family', as opposed to a 'mixed codon family', where the four codons have more than one meaning, see Table 5.3). These experiments led to the proposal (Lagerkvist, 1978) of the 'two-out-of-three' hypothesis. Essentially, Lagerkvist and his collaborators observed that in the absence of competing species, both $E.$ $coli$ tRNA$_1^{Val}$ (anticodon VAC where V is uridine-5-oxyacetic acid) and $E.$ $coli$ tRNA$_2^{Val}$ (anticodon GAC) were found to read all four valine codons GUX (Mitra et $al.$, 1977); however, when in competition the expected order of specificity is observed: tRNA$_1^{Val}$ (VAC) read GUA and GUG much better than tRNA$_2^{Val}$ (GAC); the latter species was, as expected, much better at reading GUC (Mitra et $al.$, 1979). Similar conclusions were drawn from more recent experiments comparing the relative efficiencies of the three tRNA^{Gly} isoacceptors (anticodons GCC, CCC and U*CC; Samuelsson et $al.$, 1983). Further observations made with yeast tRNA^{Val} (anticodon IAC) in the bacterial in $vitro$ system are difficult to interpret; the general picture is emerging that tRNA structures have evolved over a long period of time in conjunction with the other components of the translational machinery (see below). It is not sufficient to consider that only the anticodon of the molecule is related to its decoding potential, nor is it possible, therefore, to draw conclusions about the normal translational specificity of a tRNA molecule by study of its behaviour in a heterologous system. The same reasoning may be applied to results with yeast tRNA^{Ala} (anticodon IGC), yeast tRNA^{Lys} (anticodon CUU) and $Mycoplasma$ $mycoides$ tRNA^{Gly} (anticodon UCC) translating MS2 RNA in an $E.$ $coli$ cell-free system (Samuelsson et $al.$, 1980, 1983; Lustig et $al.$, 1981). Although these results show a preference for isoacceptor usage in accordance with the same rules as in mixed codon families, it is very difficult to know whether third letter ambiguities occur at a higher level in unmixed codon families than in mixed codon families. Furthermore, it is uncertain whether the mispairings observed in $vitro$ are of significance under normal growth conditions in $vivo$ (see below).

The idea that tRNA–codon specificity is lower in the unmixed codon families, as suggested by Lagerkvist (1978), is clearly an attractive one from the point of view of correlating the occurrence of mixed or unmixed codon families and the base–base interactions involved in decoding. Lagerkvist argues, as others have done previously (see Woese, 1967; Ninio, 1973), that the probability of misreading is related to the stability of the interactions in the first two codon positions; stable interactions with either two G.C pairs in these positions or a G.C followed by a purine in the second position of the codon are supposed to lead to ambiguities in the third position and have therefore become confined to families in which the amino acid is defined by the first two codon bases only. The problem lies in relating stability to base composition in this way.

Although interesting in relation to the evolution of the genetic code (see Eigen and Winkler-Oswatitsch, 1981; Lagerkvist, 1981; Cedergren et al., 1981), such arguments based on the expected stability of codon–anticodon complexes may not be so relevant when considering the present-day architecture of an anticodon loop. Indeed, it is doubtful whether the well-known dependence of stability on base composition shown by complexes between *long* linear complementary oligonucleotides in solution can be applied to the *short* codon–anticodon pair in the complex tRNA–messenger RNA–ribosome (see Section 5.4.1; also Bubienko et al., 1983). In order to understand the role of tRNA isoaccepting species, it is clearly necessary to have genetic studies on the effect *in vivo* of modifying or eliminating the activity of individual isoacceptors. Elegant studies of this type have been described which concern the tRNAs for glycine which, according to the hypothesis of Lagerkvist (1978), should be the most prone to read all four GGX codons. Murgola and his colleagues have demonstrated that the glyT gene product, the only species capable of reading GGA according to the wobble hypothesis, cannot be functionally replaced by any other isoaccepting species of tRNA^{Gly} (Murgola and Pagel, 1980). Furthermore, the growth capability of some strains in which glyT is mutated to yield an AGA/G, UGA/G or GAA/G reading missense suppressor has been shown to be due to residual GGA decoding activity of the suppressor tRNA rather than to 'two-out-of-three' reading by other tRNA^{Gly} (anticodons GCC and CCC) isoacceptors.

Other independent observations *in vivo* on the use of tRNA^{Ser} and tRNA^{Leu} isoacceptors are also difficult to understand on the basis of either the wobble hypothesis or the 'two-out-of-three' hypothesis. Three different anticodon sequences corresponding to the UCN family of serine codons are present in *S. pombe* and *S. cerevisiae* (see Table 5.4): IGA (the major species) U*GA and CGA. Genetic studies indicate that suppressor mutations of the genes coding for the two latter tRNAs are both haplolethal, suggesting that the three gene products are essential for growth (Etcheverry, Colby, and Guthrie, 1979;

Table 5.4 Codon and anticodon usage in yeast. The frequency of use of each codon in highly expressed genes (a total of 2217 codons, column H) and in little expressed genes (a total of 1986 codons, column W) is expressed per thousand codons (see text, and Bennetzen and Hall, 1982); in addition, the genes are included for enolase (H), 3-phosphoglycerate kinase (H), tryptophan synthetase (W), the small subunit of carbamyl phosphate synthetase (W) and of actin (W). The anticodon sequences, including the base 5′ adjacent to it, are from Gauss and Sprinzl (1984). The abundance data of the corresponding tRNA (column A) are from Ikemura (1982); major, medium and minor are indications of relative abundance in the absence of precise data. The anticodons *TAT*, *CGA*, *TTG* and *CCT* are from the corresponding gene sequences, those marked (?) probably exist but have not been demonstrated so far. Abbreviations are those of Gauss and Sprinzl (1984), except for S (5-methoxycarbonylmethyl-2-thiouridine). U* represents modified uridines as stated in Gauss and Sprinzl (1984). Note the differences in codon usage and the corresponding tRNA populations in yeast and in *E. coli*

Codon	H	W	A	Anticodon	aa
UUU	2	18	–	-----	Phe
UUC	34	21	0.80	GmAA.Y	
UUA	5	15	?	(?)	
UUG	76	36	1.00	m⁵CAA.G¹m	Leu
CUU	0	5	?	IAG.G¹m	Leu
CUC	0	3	–	-----	
CUA	2	11	0.50	UAG.G¹m	Leu
CUG	0	4	?	(?)	
AUU	26	29	Maj.	IAU.At⁶	Ile
AUC	31	23	–	-----	
AUA	0	6	?	TAT.A	Ile
AUG	17	20	0.30	CAU.At⁶	Met
GUU	51	36	1.00	IAC.A	Val
GUC	42	37	–	-----	
GUA	0	5	Min.	U*AC.A	Val
GUG	0	7	0.15	CAC.A	Val

Codon	H	W	A	Anticodon	aa
UCU	37	36	1.20	IGA.Ai⁶	Ser
UCC	18	23	–	–	Ser
UCA	1	8	0.40	U*GA.Ai⁶	Ser
UCG	0	2	?	CGA.A	Ser
CCU	2	16	Min.	AGG.G	Pro
CCC	0.4	5	–	–	
CCA	34	24	Med.	U*GG.G¹m	Pro
CCG	0	1	?	(?)	
ACU	24	34	0.90	IGU.At⁶	Thr
ACC	29	17	–	-----	
ACA	0	12	?	(?)	
ACG	0	4	?	(?)	
GCU	79	44	0.95	IGC.I¹m	Ala
GCC	22	24	–	-----	
GCA	0.4	11	?	(?)	
GCG	0	2	?	?	

	H	W	A		
UAU	0.4	14	–		Tyr
UAC	26	22	0.90	GψA.Ai⁶	
UAA	STOP				
UAG	STOP				
CAU	1	14	–		His
CAC	24	13	Maj.	GUG.G¹m	
CAA	21	30	Maj.		Gln
CAG	0	8	?	GGU.At⁶	
AAU	1	13	–		Asn
AAC	40	26	Maj.	GUU.At⁶	
AAA	9	40	0.35	SUU.At⁶	Lys
AAG	77	44	0.80	CUU.At⁶	Lys
GAU	18	28	–		Asp
GAC	49	19	1.30	GUC.G¹m	
GAA	59	40	1.00	SUC.A	Glu
GAG	0.4	10	?	(?)	

	H	W	A		
UGU	7	10	–		Cys
UGC	0	2	0.40	GCA.Ai⁶	
UGA	STOP				
UGG	10	7	0.60	CmCA.A	Trp
CGU	2	6	0.20	ICG.A	Arg
CGC	0	1	–		
CGA	0	0	?	(?)	
CGG	0	0	?	(?)	
AGU	0	7	–		Ser
AGC	1	1	?	(?)	
AGA	30	29	0.90	U*CU.At⁶	Arg
AGG	0	4	?	CCT.A	Arg
GGU	89	61	–		Gly
GGC	2	10	1.40	GCC.A	
GGA	0	4	?	(?)	
GGG	0.4	4	?	(?)	

Munz *et al.*, 1981). With the proviso, difficult to exclude, that the tRNAs might be essential to some function other than protein synthesis (see Björk, 1984), these results imply that the major species of tRNA^{Ser} (anticodon IGA) cannot read UCA *in vivo*, at least in some codon contexts, contrary to the wobble hypothesis. The essential nature of both the UCA and UCG decoding species is also contrary to the 'two-out-of-three' hypothesis. Analogous studies in *S. pombe* indicate that the codon UUA is read efficiently only by the tRNA^{Leu} with anticodon U*AA (Munz *et al.*, 1981), and this contrasts with earlier *in vitro* data for yeast and *E. coli* tRNA which suggested that UUA may be read by more than one leucine isoacceptor (Weissenbach *et al.*, 1977; Holmes *et al.*, 1977; Goldman, Holmes and Hatfield, 1979).

These few examples show that in several cases where 'two-out-of-three' reading, if it occurs, may be expected to allow the survival of strains *in vivo* genetically deleted in a tRNA isoacceptor, it does not do so at the level of efficiency required for survival. Clearly, too much attention has been focused on the coding properties of the first anticodon base to the exclusion of other important parameters, notably other features of the anticodon loop, stem and the rest of the molecule, the messenger, adjacent codons, P-site bound tRNA and the whole population of competing noncognate AA-tRNAs.

5.4 Tuning the codon–anticodon interaction

5.4.1 EFFECTS OF tRNA SEQUENCE AND STRUCTURE

(a) The anticodon loop and stem
Careful analysis of the structural data available on 152 individual tRNA species reveals preferred bases or preferentially avoided bases in certain positions. Furthermore, certain bases may be modified or hypermodified post-transcriptionally. These two features presumably lend to the anticodon loop (see Fig. 5.3) certain structural characteristics or flexibility related to its function (Cedergren *et al.*, 1981; Yarus, 1982; Westhof, Dumas and Moras, 1983). This is particularly evident when all tRNAs having the same decoding properties are compared; this probably arises from their evolution from a common precursor (Larue *et al.*, 1979).

The distribution of purines and pyrimidines in the loop is highly asymmetric, with pyrimidines on the 5′ side of the anticodon and purines on the 3′ side. Related to this asymmetry is the proposal, first advanced by Fuller and Hodgson (1967), that the five bases on the 3′ side of the anticodon form a continuous stack with the anticodon stem. This is supported both by the crystal structures of yeast tRNA^{Phe} (Sussman *et al.*, 1978) and yeast tRNA^{Asp} (Moras *et al.*, 1980), and by physical measurements on tRNA in solution. It should be emphasized, however, that the anticodon loop and the supporting stem are still quite mobile in solution (Johnston and Redfield, 1979; Labuda and Pörschke, 1980; Gorenstein and Goldfield, 1982; Labuda, Striker and

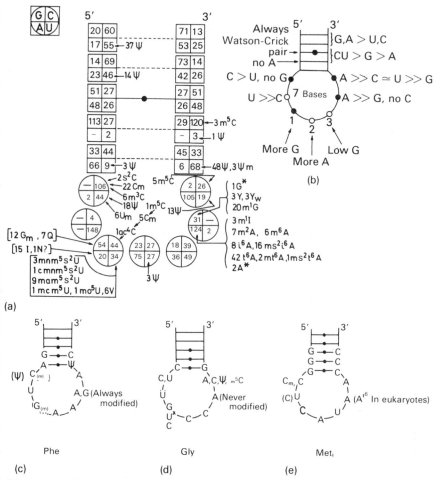

Figure 5.3 (a) A compilation of 152 anticodon loops and stems from tRNA sequences known up to January 1981 (Gauss and Sprinzl, 1981). This includes elongation and initiator tRNAs from prokaryotes and eukaryotes but not from phage T4, mitochondria or chloroplasts. Each position of the anticodon loop is represented by a circle, or by a square in the anticodon stem. The numbers of each of the four bases found are shown in these boxes according to the pattern indicated in the small insert. The fraction of these numbers which corresponds to a modified base is indicated outside the corresponding box. A minus sign means that the corresponding base has never been found at this position. Abbreviations for the modified bases are those used by Gauss and Sprinzl (1981). The dot in the middle of the anticodon stem means that the base pairs are of Watson–Crick type in all the 152 tRNA sequences compiled. (b) Schematic summary of the most interesting characteristics of the compilation shown in (a). (c) Compilation of the anticodon loop and stem of twenty-three species of tRNA^{Phe}. (d) A similar compilation of eleven species of tRNA^{Gly}. (e) A similar compilation of twenty-three species of tRNA$^{Met}_{initiator}$. Only the characteristics that are always present are indicated. A more recent compilation can be found in Grosjean, Cedergren and McKay (1982), and also in Sprinzl, Helk and Baumann (1983) who analyse 256 tRNA sequences; however, the more recent data scarcely modify those presented here.

Pörschke, 1984), and even in the crystal. This mobility may be of crucial importance in translation. The essential nature of a purine in position 37 in a UAG suppressor constructed from yeast tRNA^{Phe} has been shown by anticodon loop replacement studies (Bruce *et al.*, 1982), though G was almost as effective as A in this position. Replacement of U33 by other bases in a different suppressor tRNA showed decreasing efficiency in the order U > C > A > G, but the effect was less striking; even G in position 33 retained one third of the efficiency of U (Bare *et al.*, 1983). U33 replacement by C33 in a UAG suppressor derived from *E. coli* tRNA^{Trp} has also been achieved by manipulation at the tDNA level, and the resulting modified tRNA still suppresses *in vivo* (Thompson, Cline and Yarus, 1982).

Some degree of constraint in the bases found in different positions of the

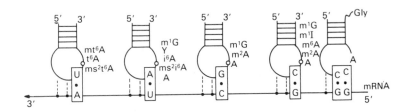

Figure 5.4 Relationship between the third anticodon base and the modification of the following nucleotide. For more details, see Yarus (1982) and Tsang, Buck and Ames (1983).

anticodon stem, in relation to the base in the last anticodon position, has been pointed out by Yarus (1982) who has attempted to use these correlations to explain the relative efficiencies of many tRNA species (the 'extended anticodon theory'). A striking regularity has been observed between the presence of a particular modified purine in position 37 (3' adjacent to the anticodon) and the type of base in the third position of the anticodon (Nishimura, 1979; Tsang, Buck and Ames, 1983; Björk, 1984). However, the precise chemical nature of the modification depends on the origin of the tRNA (prokaryotes, eukaryotes, mitochondria; see Dirheimer, 1983). It is observed (Fig. 5.4) that anticodons ending in A or U are almost always flanked on the 3' side by a highly modified purine (base Y or derivatives in eukaryotic tRNA^{Phe}; m^{1}G, i^{6}A or ms^{2}i^{6}A, t^{6}A, mt^{6}A or ms^{2}t^{6}A in other systems). On the contrary, in the case of anticodons ending in C or G or containing two of these bases, the 3' adjacent base is an unmodified A, or an A or G modified by a simple methylation (m^{2}A, m^{6}A, m^{1}G or m^{1}I). It appears that during evolution a progressive selective advantage has been taken of post-transcriptional base modification, seen clearly in eukaryotic cells

(Kubli, 1980).

There is no clear evidence, however, that the hypermodification of the purine adjacent to the anticodon is an absolute functional requirement for codon–anticodon recognition. Several mutants of *E. coli* and yeast lacking the i^6 or ms^2i^6 modifications have been isolated (for discussions, see Björk, 1984 and Buck and Ames, 1984); these mutants grow almost as well as the wild type strain but lack certain regulatory functions. Even a small difference is, however, highly significant on an evolutionary time scale. Growth of two strains together over many generations may reveal a selective advantage otherwise difficult to demonstrate as in the case of the thymine in position 54 in strains of *E. coli* unable to introduce this base (Björk and Neidhardt, 1975).

Certain tRNA suppressors show a considerable dependence on the presence of a hypermodified base on the 3′ side of the anticodon for their activity. Thus, the decrease in suppressor activity in 'antisuppressor' strains of yeast results from the absence of the i^6A modification (see Kohli, 1983; Laten, 1984). In *E. coli*, an amber suppressor tRNA^{Tyr} (su 3) lacking the ms^2i^6A modification was found to be inefficient in translating an amber mutant of phage f2 *in vitro* (Gefter and Russel, 1969). An analogous observation has been made concerning opal suppression *in vivo* (Petrullo, Gallagher and Elseviers, 1983). Other suppressors are very efficient even in the absence of ms^2i^6 modification to A37 (Murgola *et al.*, 1983). Furthermore, different suppressor species are affected to different degrees by a deficiency in their degree of modification in the wobble position (Colby, Schedel and Guthrie, 1976; Inokuchi *et al.*, 1979; Vacher *et al.*, 1984a; Heyer *et al.*, 1984). Whereas the presence of a modified nucleotide may not always be strictly required for nonsense codon recognition, it may be necessary for maximum translational efficiency in certain codon contexts (see below). This is also true for normal codon recognition (see, for example, Smith *et al.*, 1981, 1984; Meier *et al.*, 1985). The derepression of the *trp* operon enzymes, observed in the *miaA* mutant of *E. coli*, provides a particularly striking example of the importance of hypermodification. This appears to be due to the inability of tRNA^{Trp} lacking ms^2i^6A to translate efficiently tandem tryptophan codons in the *trp* leader mRNA (Eisenberg, Yarus and Soll, 1979). This dependence on hypermodification was, however, not observed in the case of the UGA suppressor tRNA^{Trp} (anticodon CCA, but mutated in the dihydrouridine stem), which has an enhanced ability to translate UGA opal codons in addition to UGG tryptophan codons. The *miaA* mutants also show derepression of the *pheS,T* operon, which is also controlled by attenuation (Springer *et al.*, 1983).

The absence of Ψ at positions 38 and 39 in tRNA^{His} (anticodon QUG.m^2A) of *Salmonella typhimurium* alleviates attenuation of the *his* operon. In an analogous manner to *miaA* mutants, this is thought to arise from a reduced efficiency of the unmodified tRNA to read the seven consecutive His codons in the leader mRNA (see Eisenberg, Yarus and Soll, 1979). Furthermore, the *hisT* mutation reduces the efficiency of suppressor tRNAs which normally have Ψ 38 to an extent which depends on the type of tRNA but also on the sequence around the nonsense codon (see below; see also Bossi and Roth, 1980). Thus, the effect of hypermodification adjacent to the anticodon or Ψ at positions 38–39 on the efficiency of mRNA–tRNA recognition clearly depends on other structural aspects both of the tRNA molecule and the mRNA, which might reinforce or counteract each other on the ribosome.

Studies of complexes between tRNAs with complementary anticodons reveal that the association constant is not related clearly to the base composition of the anticodons (Grosjean, de Henau and Crothers, 1978). Apparently, the architecture of the anticodon loops in natural tRNA species is such that not only is the binding with a complementary nucleotide considerably increased (by several orders of magnitude compared with binding between oligonucleotides), but also the intrinsic difference between G.C and A.U pairs has been considerably reduced. To the extent that the association between complementary tRNAs can be regarded as a model of codon–anticodon binding on the ribosome, we assume that the features of the anticodon which stabilize the anticodon–anticodon complex play the same role on the ribosome. Among such essential features are the loop of seven bases, the presence of a stabilizing dangling purine on the 3′ side of the anticodon and the presence of hypermodification (Grosjean, Söll and Crothers, 1976; Weissenbach and Grosjean, 1981; Vacher *et al.*, 1984b).

However, a standardized energy of binding for the codon–anticodon interaction may not produce a highly accurate system for reading the genetic code. Studies of tRNA binding to programmed ribosomes, as well as of anticodon–anticodon interactions, reveal that there exist stable pairs that are not correct according to the genetic code. Misreading a pyrimidine (especially U) is much more frequent than misreading a purine (especially A); furthermore, the third position of a codon and the corresponding first position in the anticodon are particularly prone to such errors of decoding. The most frequent mistakes are G.U/U.G, U.U, U.C/C.U and C.A; of exceptional occurrence are C.C, A.A, A.G/G.A and G.G. The structural basis of such misbinding is generally explained in terms of hydrogen bonds between nucleotides in the 'wobble' configuration or another unusual configuration (reviewed in Lomant and Fresco, 1975). We believe that misreading is related to a more global property of the decoding system in which other parts of the tRNA molecule, the mRNA, the ribosome and the peptidyl-tRNA that is normally present in the P-site of the ribosome also contribute to the stabilization of AA-tRNA binding in response to a given codon. The relative contribution of all these parameters for decoding the genetic message may not be the same for all codon–anticodon pairs, nor the same in different organisms. During the course of evolution, optimization for accuracy and rate of elongation may have led to appreciable modification of these parameters in the case of certain codons liable to error owing to the existence of noncognate tRNA species effective as competitors for the cognate species. Thus, details in the decoding process are not so universal as at first thought, as shown by the anomalies found in mitochondria (see Fig. 5.3), and very recently, in the cytoplasm of *Paramecium*, where UAA and UAG are sense codons (E. Meyer and F. Caron, personal communication).

(b) Other parts of the tRNA molecule and the role of the ribosome
The effects on tRNA selection of the structure of the tRNA molecule are not limited to the anticodon loop and stem. The most striking demonstration of this so far comes from the UGA suppressor tRNA in *E. coli*, characterized by Hirsh (1971). In this suppressor, derived from tRNA^{Trp}, no change is evident in the anticodon region; suppressor activity is due to a replacement of G by A in the D-stem. The suppressor is also a more effective competitor for tRNA^{Cys} in reading UGU codons than the wild type tRNA^{Trp} (Buckingham and Kurland, 1977). However, these differences in codon specificity have only been observed during elongation; in the absence of ribosomes no difference has been detected in binding to UpGpA (Högenauer, 1974) or the complementary anticodon (Buckingham, 1976; Vacher *et al.*, 1984b).

Hydrogen exchange studies suggest an important difference in the stability of parts of the secondary and tertiary structure involving the D-stem (Ramstein and Buckingham, 1981). These observations have led to the suggestion that the essential difference, seen only during dynamic functioning of the molecule, reflects an altered coupling between interactions at the anticodon and the conformation of other parts of the tRNA (Buckingham and Kurland, 1977, 1980). Some UGA readthrough is apparent even with the wild type tRNA^{Trp} and there appears to be a strong correlation between the efficiency of UGA readthrough and the presence of an adenosine adjacent to 3' side of the nonsense codon (Engelberg-Kulka, 1981). On eukaryotic ribosomes (from rabbit reticulocytes), the suppression of the UGA termination codon of β-globin mRNA is as efficient with the wild type *E. coli* tRNA^{Trp} as with the suppressor species (Grosjean *et al.*, 1980). This suggests that the heterologous system (reticulocyte ribosome/*E. coli* tRNA^{Trp}) is less restrictive than either the homologous *E. coli* system or the homologous reticulocyte system. However, with paromomycin, an antibiotic that binds to eukaryotic ribosomes, appreciable readthrough of the UGA termination codon occurs in the reticulocyte cell-free system. It is stimulated by mammalian and yeast tRNA^{Trp} (anticodon CmCA) for which no suppressor activity has ever been described (Grosjean, H., Parisi, J., Fournier, M. and de Henau, S., unpublished results). Clearly, the accuracy of translation, both in prokaryotic and mammalian systems, depends on subtle structural features of the ribosome. In the examples described above, these effects are apparent when both the tRNA and the ribosomes are of the same origin. This may represent an evolved molecular process for optimizing translational accuracy. So far, there are no other examples comparable to the *E. coli* UGA suppressor of tRNA species showing altered codon response as a result of tRNA modifications outside the anticodon.

A tRNA with remarkable properties has been isolated from the bovine liver cytoplasm by Diamond, Dudock and Hatfield (1981). This species has an anticodon C_mCA, like eukaryotic tRNA^{Trp} and an unusual structure in the D-stem and TΨC stem. The tRNA charges serine, which may be subsequently modified to phosphoserine (Hatfield, Diamond and Dudock, 1982) and can read well to *E. coli* ribosomes in response to the UGA triplet, but not to any serine codon nor, most surprisingly, to the complementary codon UGG. One may speculate that the structural peculiarities of this species, which has two extra nucleotides in the region of the D-stem, lead to the unexpected codon recognition.

It has been understood for many years that the structure of the ribosome has a critical influence on the accuracy of AA-tRNA selection. In particular, the characterization of *strA* and *ram* mutations in ribosomal proteins S12 and S4 or S5 respectively, which are associated with increased and decreased fidelity, has elegantly demonstrated this point (for a review see De Wilde *et al.*, 1977). The same seems to exist with eukaryotic ribosomes (for a review see Picard-Bennoun and Becqueret, 1981). In terms of a proofreading model for AA-tRNA selection, ribosomal mutants appear to affect mainly the probability of rejection of AA-tRNA by the ribosome subsequent to EF-Tu associated GTP hydrolysis, rather than the initial equilibrium between ternary complex AA-tRNA. EF-Tu.GTP and the ribosome (Andersson and Kurland, 1983; Bohman *et al.*, 1984; for further discussion see Chapter 11). Ribosomal control of the rate of elongation is apparent in some mutants of ribosomal proteins which suppress temperature-sensitive aminoacyl-tRNA synthetase mutations. These function by reducing the rate of elongation in relation to the capacity of the defective synthetase (Bückel, Piepersberg and Böck, 1976). Some degree of similarity may exist in the action of ppGpp (O'Farrell, 1978; Wagner and Kurland, 1980).

The environment of the codon on the ribosome must influence the codon–anticodon interaction. Fluorescence experiments suggest that the site is hydrophobic (Robertson and Wintermeyer, 1981); this might have a profound effect on the energetic contribution of hydrogen bonding, which in an aqueous medium contributes little to the stability of complementary base interactions. A further aspect of the site that may be important is an ability to hold the codon in a particular conformation that is sterically incompatible with certain codon–anticodon mispairs (Grosjean, de Henau and Crothers, 1978).

5.4.2 EFFECTS OF CODON CONTEXT IN mRNA

An aspect of mRNA sequence attracting increasing attention is the influence that the sequence of bases surrounding a codon in messenger RNA may have on translational errors at that codon. Compelling evidence exists for the effect of codon context on the efficiency of nonsense codon suppression. Thus, the ratio of the suppression efficiencies at homotopic pairs of amber and ochre mutants by ochre suppressor tRNAs (which can suppress both) has been shown to vary by an order of magnitude from site to site (Feinstein and Altman, 1978). Bossi and Roth (1980) characterized a mutation affecting the base following an amber codon that increased supE-mediated suppression. More recently, using I–Z fusions in the *lac* system, Miller and Albertini (1983) and Bossi (1983) have greatly extended these observations. Nevertheless, the ambiguity remains as to whether the effects of context observed arise because AA-tRNA is sensitive to codon context or because the recognition of nonsense codons by termination factors is also affected. The clearest effects seen in these studies were due to the sequence of the 3' side of nonsense codons.

More direct evidence that codon context affects AA-tRNA selection has come from studies of mistranslation and of missense suppression. Misincorporation of tryptophan in place of cysteine at UGU codons during translation *in vitro* of synthetic polymers indicated a fivefold difference in error depending on the context of the codon (Carrier and Buckingham, 1984). Context effects on missense suppression *in vivo* were observed by comparing suppression efficiencies at two different sites in the *trpA* gene (Murgola, Pagel and Hijazi, 1984). The most immediate difference between the two missense sites was on the 5' side. These experiments point clearly to an effect on tRNA selection and therefore have implications far wider than nonsense codon suppression. The mechanism of the effects is not yet clear. In one sense, the problem is symmetrical with that of the effect of different bases or modified bases in the anticodon loop adjacent to the anticodon on the 3' side (see Section 5.4.1). One possibility is that the conformation of the codon on the ribosome is influenced by interactions between adjacent nucleotides and ribosomal components, either protein or RNA (Sedlacek, Fabry and Rychlik, 1979). Stacking of the nucleotide adjacent to the 3' end of the codon may also play a role (Grosjean, Söll and Crothers, 1976).

The observation by Taniguchi and Weissmann (1978) that a G to A change following the initiation codon of Qβ coat protein cistron increased threefold the binding of fMet-tRNA, gave rise to the suggestion that a fourth base pair might form in the codon–anticodon interaction, using the invariable U on the 5' side of the anticodon (Ganoza, Fraser and Neilson, 1978; Schmitt, Kyriatsoulis and Gassen, 1982). Since all elongator tRNAs have this uridine, however, it is difficult to explain the differential effect on different nonsense suppressors (Bossi and Roth, 1980; Miller and Albertini, 1983; Bossi, 1983) on the basis of such an effect in the A-site.

An alternative explanation for context effects is that interactions occur between tRNA molecules in the A- and P-sites on the ribosome. It is easy to see how tRNA selection at a codon might be affected in this way by bases to the 5' side, though less clear how 3' sequence effects might be mediated in this way. It has been suggested that tRNA–tRNA interactions could affect the efficiency of peptidyl transfer (Feinstein and Altman, 1978; Bossi, 1983), and could thus alter the balance between continued elongation and competing processes such as dissociation of peptidyl-tRNA from the P-site or, perhaps, frameshifting. Nevertheless, it is difficult to see how tenfold variations in nonsense codon suppression could be accounted for in this way.

Apart from the difficulty of accommodating two tRNAs on the ribosome in an appropriate configuration without there being extensive contact, there is little direct reason for invoking tRNA–tRNA interactions at present. However, in the absence of ribosomes, some experiments suggest that tRNA dimerization can occur. When tRNA^{Phe} (yeast) binds the triplet UpUpC at the anticodon, evidence from NMR studies and temperature jump experiments can be intepreted in this way (Geerdes, Van Boom and Hilbers, 1980; Labuda and Pörschke, 1980; Pörschke and Labuda, 1982); however the nature of the complex formed and its general significance remain to be determined. Whatever the mechanism of context effects on tRNA selection, one may speculate that the evolution of messenger sequences may be influenced by the biological need to minimize errors at critical points of the polypeptide chain. This is, of course, only one of many possible evolutionary pressures on the development of the messenger sequence; others include the role of secondary structure in masking spurious initiation sites (see, for example, Gheysen et al., 1982) or modulating the rate of elongation. The possible influence of synonymous codon usage in modulating translation will now be discussed in more detail.

5.4.3 EFFECTS OF tRNA POPULATION AND CODON USAGE IN mRNA

Cells from different organisms generally have different populations of tRNA, both as concerns the total number of tRNA species present, and the relative amounts of acceptors and isoaccepting species (compare Tables 5.3 and 5.4). Furthermore, quantitative variations are found between different tissues from the same organism. Some differences occur because of different degrees of post-transcriptional modification of a tRNA species, whereas others correspond to different gene products. Even more striking are the differences between the cytoplasm and suborganelles such as mitochondria in eukaryotic cells (compare Tables 5.2 and 5.4 for yeast). Some isoaccepting species are essential to the cell, whereas others are not and may be mutated to suppressor species without loss of cell viability (for reviews see Dirheimer, 1983; Björk, 1984).

Similarly, the frequency with which different codons are used varies according to the origin of the cell (Grosjean, 1979; Grantham *et al.*, 1981). This is due partly to varying amino acid composition of the proteins synthesized, and partly to different use of synonymous codons. The former is particularly evident in the case of specialized tissues which make predominantly one product, such as fibroin in the posterior silk gland of the silk worm *Bombix mori* (Chavancy, Daillie and Garel, 1971) or haemoglobin in the reticulocyte (Smith, 1975; Hatfield *et al.*, 1982). Further correlations of this type have been observed in other differentiated tissues such as mammary or salivary gland, lymphocyte, lens, granuloma and pancreas between the amino acid composition of proteins and the amino acid acceptance of the tRNA from the specialized tissue in question (see Garel, 1974; Smith, 1975). Nucleotide sequence data have accumulated greatly in volume and have allowed extensive analysis of synonymous codon usage, which is very non-random and varies according to tissue species and organelle. These data have permitted the correlation to be extended to a relation between the frequency of codon usage and the abundance of the individual tRNA isoacceptors actually thought to be employed (Garel, 1976; Chavancy *et al.*, 1979; Hatfield *et al.*, 1982). Thus, frequently used codons in a given mRNA for abundant proteins correspond to major tRNA isoacceptors in the cell in which the mRNA will be translated, whereas rarely used codons generally correspond to minor isoacceptors in the same cell. This correlation is not peculiar to differentiated animal cells but is observed in *E. coli* and in yeast particularly in the case of abundant proteins such as ribosomal proteins, the major cell wall lipoproteins and the enzymes involved in basic metabolism such as glycolysis and the tricarboxylic acid cycle (Post and Nomura, 1980; Ikemura, 1981a, b, 1982; Bennetzen and Hall, 1982). Such mutual adaptation of tRNA population and codon usage needs to be considered from two points of view, which are closely related: the rate of elongation and the occurrence of errors of translation.

(a) The rate of translation
One factor that has been thought to be important in the relationship between the tRNA species present in a cell and the codons used in mRNA is the possibility that the speed of elongation may be regulated by the use of rare codons and corresponding minor tRNA isoaccepting species (Ames and Hartman, 1963; Stent, 1964; Fiers, 1979; Konigsberg and Godson, 1983). According to this theory, ribosomes would not move regularly along the mRNA, but with 'pauses' at codons corresponding to rare AA-tRNA species. These would slow down the overall rate of elongation and encourage premature termination due to the peptidyl-tRNA falling off the ribosome at these locations (see Menninger, 1983). Recent experimental data from Pedersen (1984) suggest that those mRNAs in *E. coli* with a majority of codons corresponding to abundant tRNAs are, on average, translated slightly

faster than those mRNAs which frequently use rare codons. Furthermore, Varenne *et al.* (1984) were able to demonstrate clearly that soliciting rare tRNAs does indeed induce pauses, while not excluding a contribution from other factors, such as mRNA secondary structure. The presence in high copy number of a gene into which the rare AGG codon had been inserted was shown to reduce translational efficiency (Robinson *et al.*, 1984). For a more detailed discussion of pauses during translation, see Yarus and Thompson (1984).

Although 'pauses' do indeed occur, there is little evidence that the cell uses this systematically as a mechanism for slowing down translation of mRNA (see Chavancy and Garel, 1981 and references therein). These authors, as others (Follon *et al.*, 1979; Kim and Warner, 1983), argue that the efficiency of elongation *in vivo* is dependent primarily on mRNA availability, tRNA species being produced in balanced amounts so as to maximize the mean rate of protein synthesis. In the case of proteins produced in small amounts, where a less close correlation is observed between isoacceptor tRNA abundance and codon usage (Ikemura, 1981a, b, 1982; Bennetzen and Hall, 1982), this argument is less evident and the modulation hypothesis of elongation rate should not be discarded. It may be of some interest that rare codons occur frequently in the $+1$ and -1 reading frames (Grosjean and Fiers, 1982; Konigsberg and Godson, 1983). Thus translation of mRNA following a frameshift is likely to be inefficient.

The rate of elongation may be modulated, not only by the use of specific rare codons, but also through weak or short-lived interactions of an AA-tRNA, even if abundant, with certain codons at specific locations in mRNA. As discussed above (Sections 5.4.1 and 5.4.2), a slight modification of the structure of the tRNA either in the anticodon loop (base change by mutation or a deficiency in hypermodification) or in the dihydrouridine stem, or of the mRNA (by modifying the codon context) is sufficient to modulate the affinity of an AA-tRNA for its cognate codons. The differential utilization of iso-accepting tRNAs and the corresponding codons depends on the messenger considered and may change with the physiological state of the cell. Such phenomena may be important, for example, in attenuator control of the transcription of certain operons (see Yanofsky, 1981) or of the synthesis of specific proteins particularly prone to such phenomena (see Tamura, Nishimura and Ohki, 1984).

(b) The accuracy of translation
The other factor which would be expected to exert an evolutionary pressure on the adaptation between messenger sequence and tRNA population is the effect of errors in protein synthesis. As discussed above (see Section 5.2.3), extreme imbalance in AA-tRNA species produced, for example, by amino acid starvation leads to frequent errors of misincorporation and frameshifting

in both eukaryotic cells and relaxed strains of *E. coli* (see Gallant and Foley, 1979; Atkins *et al.*, 1979; Weiss and Gallant, 1983). However, some evidence also exists for an effect on accuracy of less extreme disruptions of the normal AA-tRNA population. For example, translation of tissue-specific mRNA *in vitro* often appears to suffer from excessive premature termination and other errors unless tRNA from the appropriate tissue is present (Sharma, Beezley and Roberts, 1975; Sharma and Kuchino, 1977). A further aspect of tRNA adaptation to the messenger population is that potential competitor tRNAs will frequently be rare, or even missing (Ninio, 1971); likewise, abundant tRNA species will often be noncognate competitors only for codons which are seldom used.

Accurate and efficient translation is, therefore, dependent upon the whole tRNA population as well as the codon preference and the adaptation between these sets of parameters both in phase and out of the reading frame (see above). It can be seen readily that the effects of imbalance on precision are likely to be exacerbated by the recycling of tRNA: a certain proportion of any tRNA species will be sequestered in a form other than AA-tRNA to an extent dependent on its rate of utilization. Thus, imbalance in the total RNA population will be reflected in a more extreme way in the AA-tRNA population. This effect may be compared to the artefactual increase in errors seen during synthetic polymer translation *in vitro* (see Section 5.2.2).

Analysis of synonymous codon usage tends to show a strong, often extreme, preference for certain codons, particularly in the case of abundant proteins (Grantham *et al.*, 1981; Gouy and Gautier, 1982; Bennetzen and Hall, 1982). Furthermore, unlike amino acid usage in protein, the pattern of codon usage differs markedly from one class of living cell to another (see Tables 5.2 to 5.4). This is particularly striking in the case of thermophilic and halophilic bacteria, where a preference for G.C or C.G pairs in the third codon position has been observed (Dunn *et al.*, 1981; Kagawa *et al.*, 1984). The complete set of sequences of the 41 tRNAs is known, and shows interesting differences compared to those from eubacteria, eukaryotic cytoplasm and organelles (Gu *et al.*, 1983; Gupta, 1984). Thus, at least some elements of gene evolution are independent of the protein product. Indeed, the degeneracy of the genetic code is clearly essential in order to allow this measure of independent evolution of protein and nucleic acid sequences. Many different constraints may act at the level of the nucleic acid sequence. One of these, closely related to accuracy of translation, concerns again the energy of the codon–anticodon interaction and the choice of synonymous codons. For *E. coli* and most prokaryotes, it has been suggested that intermediate strength codon–anticodon interactions are preferred over very strong or very weak interactions (Grosjean *et al.*, 1978; Ikemura, 1981a, b; Gouy and Gautier, 1982; Grosjean and Fiers, 1982). Thus, if the first two interactions involve G.C pairs, the third interaction tends to be the wobble pair G . U rather than G . C. Conversely, the interaction in the wobble position tends to be G . C if it is preceded by two A . U pairs. This is seen clearly for many genes corresponding to abundant proteins, such as the ribosomal proteins, RNA polymerase subunits, elongation factors, the *recA* protein and outer membrane lipoprotein, but not for some other, less expressed, genes such as those for the repressor proteins. The same rule seems to apply in yeast (Bennetzen and Hall, 1982; Ikemura, 1982). This supports the idea of a cellular strategy in codon usage related to optimum translational efficiency.

5.5 Concluding remarks

An impressive amount of knowledge has been accumulated over the last twenty years concerning the components involved in protein synthesis, and the general mechanism of this process is now quite well understood. More recently, growing attention has been paid to the possible aberrations of the system and this has highlighted aspects of the translational apparatus previously neglected. We have attempted to stress the areas in which these new perspectives have been opened. We hope it has become clear that the overall magnitude of the problem is still poorly understood because data concerning the real levels of misincorporation and frameshifting on the ribosome remain fragmentary. Context effects on misincorporation and frameshifting clearly occur, and may be linked, but the mechanism(s) underlying context effects have not yet been elucidated. Proofreading appears to be used by the ribosome to amplify the specificity of tRNA–mRNA recognition or accelerate accurate AA-tRNA selection.

Changes in many components of the translational apparatus have repercussions on the levels of errors. Some of these effects are more selective than others. Thus, the structure of the ribosome and the levels of magic spot have rather general effects. The evolution of the whole structure of each tRNA species governs its decoding properties and also its potential for introducing errors. Post-transcriptional modification of the tRNA molecule is a further element in this process and plays an important role in regulating the specificity of codon–anticodon interactions, particularly in the third codon position. Considerable doubt remains concerning the allowed inter-base interactions in decoding this codon position. Recent genetic experiments in yeast suggest that interactions involving inosine may be more restricted than allowed by the wobble hypothesis. Clearly, more experiments of this kind *in vivo* are needed to define the permitted interactions and investigate the influence of codon context in mRNA.

Degeneracy in the genetic code allows the cell sufficient flexibility to avoid some error-prone situations, without necessarily imposing restrictions on the amino acid sequences that may be synthesized. Thus, evolution has further opportunities to avoid sequences prone to either misincorporation or frame-shifting. This may be of special importance in long mRNAs, although statistical evidence of such a strategy has not yet been produced. Co-evolution of the tRNA population and the choice of different codons appears to be an important means of reducing errors. Even more striking is the mechanism for adapting the population of tRNA species to the special needs of highly differentiated tissues which synthesize predominantly one protein.

Finally, the cell has further options once errors have occurred. Without relieving the ribosome of the burden of precision, faulty unfinished chains may nevertheless be discarded by the ribosome and released from tRNA by peptidyl hydrolase. These peptides, like finished but faulty polypeptides, are then substrates for the complex degradative mechanisms of the cell which are able to act upon them with some measure of selectivity.

References

Abraham, A. K. (1983) The fidelity of translation. *Prog. Nucl. Ac. Res. Mol. Biol.*, **28**, 81–100.

Ames, B. N. and Hartman, P. E. (1963) The histidine operon. *Cold Spring Harbor Symp. Quant. Biol.*, **28**, 349–356.

Anderson, S., Bankier, A. T., Barrell, B. G. *et al.* (1981) Sequence and organization of the human mitochondrial genome. *Nature*, **290**, 457–464.

Andersson, D. I. and Kurland, C. G. (1983) Ram ribosomes are defective proof-readers. *Mol. Gen. Genet.*, **191**, 378–381.

Andersson, S. G. E., Buckingham, R. H. and Kurland, C. G. (1984) Does codon–anticodon composition influence ribosome functions? *EMBO J.*, **3**, 91–94.

Atkins, J. F., Elseviers, D. and Gorini, L. (1972) Low activity of β-galactosidase in frameshift mutants of *Escherichia coli*. *Proc. Natl Acad. Sci. USA*, **69**, 1192–1195.

Atkins, J. F., Gesteland, R. F., Reid, B. R. and Anderson, C. W. (1979) Normal tRNAs promote ribosomal frameshifting. *Cell*, **18**, 1119–1131.

Atkins, J. F., Nichols, B. P. and Thompson, S. (1983) The nucleotide sequence of the first externally suppressible −1 frameshift mutant, and of some nearly leaky frameshift mutants. *EMBO J.*, **2**, 1345–1350.

Bare, L., Bruce, A. G., Gesteland, R. and Uhlenbeck, O. C. (1983) Uridine-33 in yeast tRNA not essential for amber suppression. *Nature*, **305**, 554–556.

Barrell, B. G., Anderson, S., Bankier, A. T. *et al.* (1980) Different pattern of codon recognition by mammalian mitochondrial tRNAs. *Proc. Natl Acad. Sci. USA*, **77**, 3164–3166.

Beier, H., Barciszewska, M., Sickinger, H. D. (1984) The molecular basis for the differential translation of TMV RNA in tobacco protoplasts and wheat germ extract. *EMBO J.*, **3**, 1091–1096.

Bennetzen, J. L. and Hall, B. D. (1982) Codon selection in yeast. *J. Biol. Chem.*, **257**, 3026–3031.

Berman, H. M., Marcu, D., Narayana, P., Fissekis, J. D. and Lipnick, R. L. (1978)

Modified bases in transfer-RNA. Structures of 5-carbamoylmethyl uridine and 5-carboxymethyl uridine. *Nucl. Acid Res.*, **5**, 593–903.

Bienz, M. and Kubli, E. (1981) Wild-type tRNA$_G^{Tyr}$ reads the TMV-RNA stop codon, but Q base-modified tRNA$_Q^{Tyr}$ does not. *Nature*, **294**, 188–190.

Björk, G. R. (1984) Modified nucleosides in RNA: their formation and function. In *Processing of RNA* (ed. D. Apirion) CRC Press Inc., Boca, Raton, Florida, pp. 291–330.

Björk, G. R. and Neidhardt, F. C. (1975) Physiological and biochemical studies on the function of 5-methyluridine in the tRNA of *E. coli. J. Bacteriol.*, **124**, 99–111.

Bohman, K., Ruusala, T., Jelenc, P. C. and Kurland, C. G. (1984) Kinetic impairment of restrictive streptomycin resistant ribosomes. *Mol. Gen. Genet.*, **198**, 90–99.

Bonitz, S. G., Berlani, R., Coruzzi, G. *et al.* (1980) Codon recognition rules in yeast mitochondria. *Proc. Natl Acad. Sci. USA*, **77**, 3167–3170.

Bossi, L. (1983) Context effects: translation of UAG codon by suppressor tRNA is affected by the sequence following UAG in the message. *J. Mol. Biol.*, **164**, 73–87.

Bossi, L. and Roth, J. R. (1980) The influence of codon context on genetic code translation. *Nature*, **286**, 123–127.

Bouadloun, F., Donner, D. and Kurland, C. G. (1983) Codon-specific missense errors *in vivo. EMBO J.*, **2**, 1351–1356.

Bruce, A. G., Atkins, J. F., Wills, N., Uhlenbeck. O., Gesteland, R. F. (1982) Replacement of anticodon loop nucleotides to produce functional tRNA: amber suppressors derived from yeast tRNA-Phe. *Proc. Natl Acad. Sci. USA*, **79**, 7129–7131.

Bubienko, E., Cruz, P., Thomason, J. F. and Borer, P. N. (1983) Nearest-neighbor effects in the structure and function of nucleic acids. *Progr. Nucl. Ac. Res. Mol. Biol.*, **30**, 41–90.

Buchanan, J. H. and Stevens, A. S. (1978) Fidelity of histone synthesis in cultured human fibroblasts. *Mech. Ageing Dev.*, **7**, 321–334.

Buck, M. and Ames, B. N. (1984) A modified nucleotide in tRNA as a possible regulator of aerobiosis: synthesis of cis-2-methyl-thioribosylzeatin in the tRNA of *Salmonella. Cell*, **36**, 523–531.

Bückel, P., Piepersberg, W. and Böck, A. (1976) Suppression of temperature-sensitive aminoacyl-tRNA synthetase mutations by ribosomal mutations: a possible mechanism. *Mol. Gen. Genet.*, **149**, 51–61.

Buckingham, R. H. (1976) Anticodon conformation and accessibility in wild-type and suppressor tryptophan tRNA from *E. coli. Nucl. Acids Res.*, **3**, 965–975.

Buckingham, R. H. and Kurland, C. G. (1977) Codon specificity of UGA suppressor tRNA^{Trp} from *E. coli. Proc. Natl Acad. Sci. USA*, **74**, 5496–5498.

Buckingham, R. H. and Kurland, C. G. (1980) Interactions between UGA suppressor tRNA^{Trp} and the ribosome: mechanisms of tRNA selection. In *Transfer RNA, Biological Aspects* (eds D. Söll, J. N. Abelson and P. R. Schimmel), Cold Spring Harbor Laboratory, New York, monograph 9B, pp. 421–426.

Cabanas, M. J. and Modolell, J. (1980) Non-enzymatic translocation and spontaneous release of non-cognate peptidyl-tRNA from *E. coli* ribosomes. *Biochemistry*, **19**, 5411–5416.

Caplan, A. B. and Menninger, J. R. (1979) Tests of the ribosomal editing hypothesis: amino acid starvation differentially enhances the dissociation of peptidyl-tRNA from the ribosome. *J. Mol. Biol.*, **134**, 621–637.

Carrier, M. J. and Buckingham, R. H. (1984) An effect of codon context in the mistranslation of UGU codons *in vitro. J. Mol. Biol.*, **175**, 29–38.

Cedergren, R. J., La Rue, B., Sankoff, D. and Grosjean, H. (1981) The evolving tRNA molecule. In *CRC Critical Reviews in Biochemistry*, Vol. 11, CRC Press, Florida, pp. 35–104.

Chavancy, G. and Garel, J. P. (1981) Does quantitative tRNA adaptation to codon content in mRNA optimise the ribosomal translation efficiency? Proposal for a translation system model. *Biochimie*, **63**, 187–195.

Chavancy, G., Daillie, J. and Garel, J. P. (1971) Adaptation fonctionnelle des tRNA à la biosynthèse protéique dans un système cellulaire hautement différencié. IV – Evolution des tRNA dans la glande séricigène de *Bombyx mori* L. au cours du dernier âge larvaire. *Biochimie*, **53**, 1187–1197.

Chavancy, G., Chevallier, A., Fournier, A. and Garel, J. P. (1979) Adaptation of iso-tRNA concentration to mRNA codon frequency in the eukaryotic cell. *Biochimie*, **61**, 71–78.

Cody, J. D. M. and Conway, T. W. (1981) Defective lysis of streptomycin-resistant *E. coli* cells infected with bacteriophage f2. *J. Virology*, **37**, 813–820.

Colby, D. S., Schedel, P. and Guthrie, C. (1976) A functional requirement for modification of the wobble nucleotide in the anticodon of a T4 suppressor tRNA. *Cell*, **9**, 449–463.

Coons, S. F., Smith, L. F. and Loftfield, R. B. (1979) The nature of amino acids errors in *in vivo* biosynthesis of rabbit hemoglobin. *Fed. Proc.*, **38**, 328.

Crick, F. H. C. (1966) Codon–anticodon pairing: the wobble hypothesis. *J. Mol. Biol.*, **19**, 548–555.

De Wilde, M., Cabezon, T., Herzog, A., Bollen, A. (1977) Apport de la génétique à la connaissance du ribosome bactérien. *Biochimie*, **59**, 125–140.

Diamond, A., Dudock, B. and Hatfield, D. (1981) Structure and properties of a bovine liver UGA suppressor serine tRNA with a tryptophan anticodon. *Cell*, **25**, 497–506.

Dirheimer, G. (1983) Chemical nature, properties, location and physiological variations of modified nucleosides in tRNAs. In *Recent Results in Cancer Research*, Vol. 84, Springer Verlag, Berlin and Heidelberg, pp. 15–46.

Dunn, J. J. and Studier, F. W. (1983) Complete nucleotide sequence of bacteriophage T7 DNA and the location of T7 genetic elements. *J. Mol. Biol.*, **166**, 477–535.

Dunn, R., McCoy, J., Simsek, M., Majumdar, A., Chang, S. H., RajBhandar, U. L. and Khorana, H. G. (1981) The bacteriorhodopsin gene. *Proc. Natl Acad. Sci. USA*, **78**, 6744–6748.

Edelmann, P. and Gallant, J. (1977) Mistranslation in *E. coli. Cell*, **10**, 131–137.

Eigen, M. and Winkler-Oswatitsch, R. (1981) Transfer-RNA, an early gene? *Naturwissenschaften*, **68**, 282–292.

Eisenberg, S. P., Yarus, M. and Soll, L. (1979) The effect of an *E. coli* regulatory mutation on tRNA structure. *J. Mol. Biol.*, **135**, 111–126.

Engelberg-Kulka, H. (1981) UGA suppression by normal tRNA[Trp] in *E. coli*: codon context effects. *Nucl. Acids Res.*, **9**, 983–991.

Etcheverry, T., Colby, D. and Guthrie, C. (1979) A precursor to a minor species of yeast tRNA[Ser] contains an intervening sequence. *Cell*, **18**, 11–26.

Feinstein, S. I. and Altman, S. (1978) Context effects on nonsense codon suppression in *E. coli. Genetics*, **88**, 201–219.

Fiers, S. (1979) Structure and function of RNA bacteriophages. In *Comprehensive Virology* (eds H. Fraenkel-Conrat and R. R. Wagner), Vol. 13, Plenum Publishing Corp., New York, pp. 69–203.

Follon, A. M., Jinks, C. S., Strycharz, G. D. and Nomura, M. (1979) Regulation of ribosomal protein synthesis in *E. coli* by selective mRNA inactivation. *Proc. Natl Acad. Sci. USA*, **76**, 3411–3415.

Fox, T. D. and Weiss-Brummer, B. (1980) Leaky +1 and −1 frameshift mutations at the same site in yeast mitochondrial gene. *Nature*, **288**, 60–64.

Fuller, W. and Hodgson, A. (1967) Conformation of the anticodon loop in tRNA. *Nature*, **215**, 817–821.

Gallant, J. and Foley, D. (1979) On the causes and prevention of mistranslation. In *Ribosomes, Structure, Function and Genetics* (eds G. Chambliss, G. R. Craven, J. Davies, K. Davis, L. Kahan and M. Nomura), University Park Press, Baltimore, pp. 615–638.

Gallant, J., Ehrlich, H., Weiss, R., Palmer, L. and Nyari, L. (1982) Nonsense suppression in aminoacyl-tRNA limited cells. *Mol. Gen. Genet.*, **186**, 221–227.

Ganoza, M. C., Fraser, A. R. and Neilson, T. (1978) Nucleotides contiguous to AUG affect translational initiation. *Biochemistry*, **17**, 2769–2776.

Garel, J. P. (1974) Functional adaptation of tRNA population. *J. Theor. Biol.*, **43**, 211–225.

Garel, J. P. (1976) Quantitative adaptation of isoacceptor tRNAs to mRNA codons of alanine, glycine and serine. *Nature*, **260**, 805–806.

Gauss, D. H. and Sprinzl, M. (1981) Compilation of tRNA sequences. *Nucl. Acids Res.*, **9**, r1–r42.

Gauss, D. A. and Sprinzl, M. (1984) Compilation of tRNA sequences and of tRNA genes. *Nucl. Acids Res.*, **12**, r1–r131.

Geerdes, H. A., Van Boom, J. H. and Hilbers, C. W. (1980) Codon–anticodon interaction in tRNA[Phe]: NMR study of the binding of the codon UUC. *J. Mol. Biol.*, **142**, 219–230.

Gefter, M. L. and Russel, R. L. (1969) Role of modifications in tyrosine tRNA: a modified base affecting ribosome binding. *J. Mol. Biol.*, **39**, 145–157.

Gheysen, D., Iserentant, D., Derom, C. and Fiers, W. (1982) Systematic alteration of the nucleotide sequence preceding the translation initiation codon and the effects on bacterial expression of the cloned SV 40 small-T antigen gene. *Gene*, **17**, 55–63.

Goldberg, A. L. and St John, A. C. (1976) Intracellular protein degradation in mammalian and bacterial cells: Part 2. *Ann. Rev. Biochem.*, **45**, 747–803.

Goldman, E. (1982) Effect of rate-limiting elongation on bacteriophage MS$_2$ RNA-directed protein synthesis in extracts of *E. coli. J. Mol. Biol.*, **158**, 619–636.

Goldman, E., Holmes, W. M. and Hatfield, G. W. (1979) Specificity of codon recognition by *Escherichia coli* isoaccepting species determined by protein synthesis *in vitro* directed by phage RNA. *J. Mol. Biol.*, **129**, 567–585.

Gorenstein, D. G. and Goldfield, E. M. (1982) High-resolution phosphorus nuclear magnetic resonance spectroscopy of tRNAs: multiple conformations in the anticodon loop. *Biochemistry*, **21**, 5839–5849.

Gorini, L. (1974) Streptomycin and misreading of the genetic code. In *Ribosomes* (eds

M. Nomura, A. Tissières and P. Lengyel), Cold Spring Harbor Laboratory, New York, pp. 791–803.

Gouy, M. and Gautier, R. (1982) Codon usage in bacteria: a correlation with gene expressivity. *Nucl. Acids Res.*, **10**, 7055–7074.

Grantham, R., Gautier, C., Gouy, M., Jacobzone, M. and Mercier, R. (1981) Codon catalogue is a genome strategy for gene expressivity. *Nucl. Acids Res.*, **9**, r43–r74.

Grosjean, H. (1979) Codon usage in several organisms. In *Transfer RNA: Biological Aspects* (eds D. Söll, J. N. Abelson and P. Schimmel), Cold Spring Harbor Laboratory, New York, monograph 9B, pp. 565–569.

Grosjean, H. and Fiers, W. (1982) Preferential codon usage in prokaryotic genes: the optimal codon–anticodon energy hypothesis. *Gene*, **18**, 199–209.

Grosjean, H., Cedergren, R. J. and McKay, W. (1982) Structure in tRNA data. *Biochimie*, **64**, 387–397.

Grosjean, H., de Henau, S. and Crothers, D. M. (1978) On the physical basis for ambiguity in genetic coding interactions. *Proc. Natl Acad. Sci. USA*, **75**, 610–614.

Grosjean, H., Söll, D. and Crothers, D. M. (1976) Studies of the complex between tRNA with complementary anticodons. *J. Mol. Biol.*, **103**, 499–518.

Grosjean, H., Sankoff, D., Jou, M. J., Fiers, W. and Cedergren, R. J. (1978) Bacteriophage MS2 RNA: a correlation between the stability of the codon:anticodon interaction and the choice of code words. *J. Mol. Evol.*, **12**, 113–119.

Grosjean, H., de Henau, S., Houssier, C. and Buckingham, R. H. (1980) Wild-type *E. coli* tRNA[Trp] efficiency suppresses UGA opal codon in an eukaryotic cell-free protein synthesis: evolutionary implications. *Arch. Internat. Physiol. Biochim.*, **88**, 168–169.

Gu, X., Nicoghosian, K., Cedergren, R. J. and Wong, J. T. (1983) Sequence of halobacterial tRNAs and the paucity of U in the first position of their anticodons. *Nucl. Acids Res.*, **11**, 5443–5450.

Gupta, R. (1984) *Halobacterium volcanii* tRNAs: identification of 41 tRNAs covering all amino acids, and the sequences of 33 class I tRNAs. *J. Biol. Chem.*, **259**, 9461–9471.

Harley, C. B., Pollard, J. W. and Stanners, C. P. (1981) Model for messenger RNA translation during amino acid starvation applied to the calculation of protein synthetic error rates. *J. Biol. Chem.*, **256**, 10 786–10 794.

Hatfield, D., Diamond, A. and Dudock, B. (1982) Opal suppressor serine tRNAs from bovine liver form phosphoseryl-tRNA. *Proc. Natl Acad. Sci. USA*, **79**, 6215–6219.

Hatfield, D., Varrichio, F., Rich, M. and Forget, B. G. (1982) The aminoacyl-tRNA population of human reticulocytes. *J. Biol. Chem.*, **257**, 3183–3188.

Heckman, J. E., Sarnoff, J., Alzner-Deweerd, B., Yin, S. and Rajbhandary, U. L. (1980) Novel features in the genetic code and codon reading patterns in *Neurospora crassa* mitochondria based on sequences of six mitochondrial tRNAs. *Proc. Natl Acad. Sci. USA*, **77**, 3159–3163.

Hensgens, L. A., Brakenhoff, J., de Vries, B. F., Sloof, P., Tromp, M. C., Van Boom, J. H. and Benne, R. (1984) The sequence of the gene for cytochrome oxydase sub-unit I, a frameshift containing gene for cytochrome oxydase II and seven unassigned reading frames in *Trypanosoma brucei* mitochondrial maxi circle DNA. *Nucl. Acids Res.*, **12**, 7327–7344.

Heyer, W. D., Thuriaux, P., Kohli, J., Ebert, P., Kersten, H., Gehrke, C., Kuo, K. and Agris, P. F. (1984) An antisuppressor mutation of *S. pombe* affects the post-transcriptional modification of the 'wobble' base in the anticodon of tRNAs. *J. Biol. Chem.*, **259**, 2856–2862.

Hillen, W., Egert, E., Lindner, H. J. and Gassen, H. G. (1978) Restriction or amplification of wobble recognition: the structure of 2-thio-5-methylaminomethyl-uridine and the interaction of odd uridines with the anticodon loop backbone. *FEBS Lett.*, **94**, 361–364.

Hirsh, D. (1971) Tryptophan transfer RNA as the UGA suppressor. *J. Mol. Biol.*, **58**, 439–458.

Högenauer, G. (1974) Binding of UGA to wild type and suppressor tryptophan tRNA from *E. coli. FEBS Lett.*, **39**, 310–316.

Holmes, W. M., Goldman, E., Miner, T. A. and Hatfield, G. W. (1977) Differential utilization of leucyl-tRNAs by *Escherichia coli. Proc. Natl Acad. Sci. USA*, **74**, 1393–1397.

Hornig, H., Woolley, P. and Lührmann, R. (1984) Decoding at the ribosomal A-site: the effect of a defined codon–anticodon mismatch upon the behaviour of bound aminoacyl-tRNA. *J. Biol. Chem.*, **259**, 5632–5636.

Igarashi, K., Hashimoto, S., Miyake, A., Kashiwagi, K. and Hirose, S. (1982) Increase of fidelity of polypeptide synthesis by spermidine in eukaryotic cell-free system. *Eur. J. Biochem.*, **128**, 597–604.

Ikemura, T. (1981a) Correlation between the abundance of *E. coli* tRNAs and the occurrence of the respective codons in its protein genes. *J. Mol. Biol.*, **146**, 1–21.

Ikemura, T. (1981b) Correlation between the abundance of *E. coli* tRNAs and the occurrence of the respective codons in its protein genes: a proposal for a synonymous codon choice that is optimal for the *E. coli* translational system. *J. Mol. Biol.*, **151**, 389–409.

Ikemura, T. (1982) Correlation between the abundance of yeast tRNAs and the occurrence of the respective codons in protein genes. *J. Mol. Biol.*, **158**, 573–597.

Inokuchi, H., Yamao, F., Sakano, H. and Ozeki, H. (1979) Identification of tRNA suppressors in *E. coli* 1. Amber suppressor su$^+$2, an anticodon mutant of tRNA$_2^{Glu}$. *J. Mol. Biol.*, **132**, 649–662.

Jank, P., Shindo-Okado, N., Nishimura, S. and Gross, H. J. (1977) Rabbit liver tRNA$_1^{Val}$. I. Primary structure and unusual codon recognition. *Nucl. Acid Res.*, **4**, 1999–2008.

Jelenc, P. C. and Kurland, C. G. (1979) Nucleotide triphosphate regeneration decreases the frequency of translation errors. *Proc. Natl Acad. Sci. USA*, **76**, 3174–3178.

Johnston, P. D. and Redfield, A. G. (1979) Proton FT-NMR studies of tRNA structure and dynamics. In *Structure, Properties and Recognition* (eds P. Schimmel, D. Söll and J. N. Abelson), Cold Spring Harbor, monograph 9A, pp. 191–206.

Kagawa, Y., Nojima, H., Nukiwa, N., Ishizuka, M., Nakajima, T., Yasuhara, T., Tanaka, T. and Oshima, T. (1984) High guanine plus cytosine content in the third letter of codons of an extreme thermophile. *J. Biol. Chem.*, **259**, 2956–2960.

Kastelein, R. and Van Duin, J. (1982) Ribosomal frameshift errors control the expression of an overlapping gene in RNA phage. In *Interaction of Translational and*

Transcriptional Controls in the Regulation of Gene Expression (eds M. Grunberg-Manago and B. Safer), Elsevier Biomedical, New York, pp. 221–240.

Kastelein, R. A., Berkhout, B. and Van Duin, J. (1983) Opening the closed ribosome-binding site of the lysis cistron of bacteriophage MS2. *Nature*, **305**, 741–744.

Kawasaki, M., Tsonis, P. A., Nishio, K. and Takemura, S. (1980) Abnormal codon recognition of glycyl-tRNA from the posterior silk glands of *Bombix mori*. *J. Biochem. (Tokyo)*, **88**, 1151–1157.

Khazaie, K., Buchanan, J. H. and Rosenberger, R. F. (1984a) The accuracy of Qβ RNA translation. I. Errors during the synthesis of Qβ protein by intact *Escherichia coli* cells. *Eur. J. Biochem.*, **144**, 485–489.

Khazaie, K., Buchanan, J. H. and Rosenberger, R. F. (1984b) The accuracy of Qβ RNA translation. II. Errors during the synthesis of Qβ proteins by cell-free *Escherichia coli* extracts. *Eur. J. Biochem.*, **144**, 491–495.

Kim, C. H. and Warner, J. R. (1983) Messenger RNA for ribosomal protein in yeast. *J. Mol. Biol.*, **165**, 79–89.

Kohli, J. (1983) The genetics concerning modified nucleotides in relation to their influence on tRNA function. In *The Modified Nucleosides of tRNA*, Vol. II (eds P. F. Agris and R. A. Kopper), Alan R. Liss, New York.

Kohli, J. and Grosjean, H. (1981) Usage of the three termination codons: compilation and analysis of the known eukaryotic and prokaryotic translation termination sequences. *Mol. Gen. Genet.*, **182**, 430–439.

Konigsberg, W. and Godson, G. N. (1983) Evidence for use of rare codons in the DNA C-gene and other regulatory genes of *E. coli*. *Proc. Natl Acad. Sci. USA*, **80**, 687–691.

Kopelowitz, J., Schoulaker-Schwarz, R., Lebanon, A. and Engelberg-Kulka, H. (1984) Modulation of *E. coli* tryptophan attenuation by the UGA readthrough process. *Mol. Gen. Genet.*, **196**, 541–545.

Kruse, T. A., Clark, B. F. C., Appel, B. and Erdmann, V. A. (1980) The structure of the CCA end of tRNA, aminoacyl-tRNA and aminoacyl-tRNA in the ternary complex. *FEBS Lett.*, **117**, 315–318.

Kubli, E. (1980) Transfer RNA modification in eukaryotes: an evolutionary interpretation. *Trends in Biochem. Sci.*, **5**, 190–191.

Kurland, C. G. (1979) Reading frame errors on ribosomes. In *Nonsense Mutations and tRNA Suppressors* (eds J. E. Celis and J. D. Smith), Academic Press, New York, pp. 98–108.

Labuda, D. and Pörschke, D. (1980) Multistep mechanism of codon recognition by tRNA. *Biochemistry*, **19**, 3799–3805.

Labuda, D., Striker, G. and Pörschke, D. (1984) Mechanism of codon recognition by tRNA and codon-induced tRNA association. *J. Mol. Biol.*, **174**, 587–604.

Lagerkvist, U. (1978) 'Two out of three': an alternative method of codon reading. *Proc. Natl Acad. Sci. USA*, **75**, 1759–1762.

Lagerkvist, U. (1981) Unorthodox codon reading and the evolution of the genetic code. *Cell*, **23**, 305–306.

Larue, B., Cedergren, R. J., Sankoff, D. and Grosjean, H. (1979) Evolution of methionine initiator and phenylalanine tRNA. *J. Mol. Evol.*, **14**, 287–300.

Laten, H. M. (1984) Antisuppression of class I suppressors in an isopentenylated-tRNA deficient mutant of *S. cerevisiae*. *Current Genetics*, **8**, 29–32.

Loftfield, R. B. (1963) The frequency of errors in protein biosynthesis. *Biochem. J.*, **89**, 89–92.

Loftfield, R. B. and Vanderjagt, D. (1972) The frequency of errors in protein biosynthesis. *Biochem. J.*, **128**, 1353–1356.

Lomant, A. J. and Fresco, J. R. (1975) Structural and energetic consequences of non-complementary base oppositions in nucleic acid helix. *Progr. Nucl. Acids Res. Mol. Biol.*, **15**, 185–218.

Lührmann, R., Eckhardt, H. and Stöffler, G. (1979) Codon–anticodon interaction at the ribosomal peptidyl-site. *Nature*, **280**, 423–425.

Lustig, F., Elias, P., Axberg, T., Samuelson, T., Tittawella, I. and Lagerkvist, U. (1981) Codon reading and translation error: reading of the glutamine and lysine codons during protein synthesis *in vitro*. *J. Biol. Chem.*, **256**, 2635–2643.

Macreadie, I. G., Novitski, C. E., Maxwell, R. J., John, U., Ooi, B. G., McMullen, G. L., Lukins, H. B., Linnane, A. W. and Nagley, P. (1983) Biogenesis of mitochondria; the mitochondrial gene aap 1 coding for mitochondrial ATPase subunit 8 in *S. cerevisiae*. *Nucl. Acids Res.*, **11**, 4435–4451.

Manley, J. L. (1978a) Synthesis and degradation of termination and premature-termination fragments of β-galactosidase *in vitro* and *in vivo*. *J. Mol. Biol.*, **125**, 407–432.

Manley, J. L. (1978b) Synthesis of internal re-initiation fragments of β-galactosidase *in vitro* and *in vivo*. *J. Mol. Biol.*, **125**, 449–466.

Medvedev, Z. A. and Medvedeva, M. N. (1978) Use of H1 histone to test the fidelity of protein biosynthesis in mouse tissues. *Biochem. Soc. Trans*, **6**, 610–612.

Meier, F., Suter, B., Grosjean, H., Keith, G. and Kubli, E. (1985) Modification of the wobble base in tRNA[His] influences *in vivo* decoding properties. *EMBO J.*, **4**, 823–827.

Menninger, J. R. (1983) Computer simulation of ribosome editing. *J. Mol. Biol.*, **171**, 383–399.

Miller, J. H. and Albertini, A. M. (1983) Effects of surrounding sequence on the suppression of nonsense codon. *J. Mol. Biol.*, **164**, 59–71.

Mitra, S. K., Lustig, F., Akesson, B. and Lagerkvist, U. (1977) Codon:anticodon recognition in the valine codon family. *J. Biol. Chem.*, **255**, 471–478.

Mitra, S. K., Lustig, F., Akesson, B., Axberg, T., Elias, P. and Lagerkvist, U. (1979) Relative efficiency of anticodons in reading the valine codons during protein synthesis *in vitro*. *J. Biol. Chem.*, **254**, 6397–6401.

Moras, D., Comarmond, M. B., Fischer, J., Weiss, R., Thierry, J. C., Ebel, J. P. and Giégé, R. (1980) Crystal structure of yeast tRNA[Asp]. *Nature*, **288**, 669–674.

Mount, D. W. (1980) The genetics of protein degradation in bacteria. *Ann. Rev. Genet.*, **14**, 279–319.

Munz, P., Leupold, U., Agris, P. and Kohli, J. (1981) *In vivo* decoding rules in *Schizosaccharomyces pombe* are at variance with *in vitro* data. *Nature*, **294**, 187–188.

Murgola, E. and Pagel, F. T. (1980) Codon recognition by glycine transfer RNAs of *Escherichia coli in vivo*. *J. Mol. Biol.*, **138**, 833–844.

Murgola, E. J., Pagel, F. T. and Hijazi, K. A. (1984) Codon context effects in missense suppression. *J. Mol. Biol.*, **175**, 19–27.

Murgola, J., Prather, N. E., Mims, P. H. and Ishigazi, K. A. (1983) Missense and nonsense suppressors derived from a glycine tRNA by nucleotides insertion and deletion *in vivo. Mol. Gen. Genet.*, **193**, 76–81.

Ninio, J. (1971) Codon–anticodon recognition: the missing triplet hypothesis. *J. Mol. Biol.*, **6**, 63–82.

Ninio, J. (1973) Recognition in nucleic acids and the anticodon families. *Progr. Nucl. Acids Res. Mol. Biol.*, **13**, 301–337.

Nishimura, S. (1979) Modified nucleosides in tRNA. In *Transfer RNA: Structure, Properties and Recognition* (eds P. R. Schimmel, D. Söll and J. B. Abelson). Cold Spring Harbor Laboratory, New York, monograph 9A, pp. 59–79.

O'Farrell, P. H. (1978) The suppression of defective translation by ppGpp and its role in the stringent response. *Cell*, **15**, 545–547.

Parker, J. (1981) Mistranslated protein in *E. coli. J. Biol. Chem.*, **256**, 9770–9773.

Parker, J., Pollard, J., Friesen, J. D. and Stanners, C. P. (1978) Stuttering: high level mistranslation in animal and bacterial cells. *Proc. Natl Acad. Sci. USA*, **75**, 1091–1095.

Parker, J., Johnston, T. C., Borgia, P. T., Holtz, G., Remaut, E. and Fiers, W. (1983) Codon usage and mistranslation. *J. Biol. Chem.*, **258**, 10007–10012.

Pedersen, S. (1984) *Escherichia coli* ribosomes translate *in vivo* with variable rate. *EMBO J.*, **3**, 2895–2898.

Peters, M. and Yarus, M. (1979) Transfer RNA selection at the ribosomal A and P sites. *J. Mol. Biol.*, **134**, 471–491.

Petrullo, L. A., Gallagher, P. J. and Elseviers, D. (1983) The role of 2-methylthio-*N*6-isopentenyladenosine in readthrough and suppression of nonsense codons in *E. coli. Mol. Gen. Genet.*, **190**, 289–294.

Picard-Bennoun, M. (1982) Does translational ambiguity increase during cell differentiation? *FEBS Lett.*, **149**, 167–170.

Picard-Bennoun, M. and Becqueret, J. (1981) Genetic analysis of cytoplasmic ribosomes in fungi. *Trends in Biochem. Sci.*, **6**, 272–274.

Pope, W. T., Brown, A. and Reeves, R. H. (1978) The identification of the tRNA substrates for the supK tRNA methylase. *Nucl. Acids Res.*, **5**, 1041–1057.

Pörschke, D. and Labuda, D. (1982) Codon induced tRNA association: quantitative analysis by sedimentation equilibrium. *Biochemistry*, **21**, 53–56.

Post, L. E. and Nomura, M. (1980) DNA sequences from the str operon of *E. coli. J. Biol. Chem.*, **255**, 4660–4666.

Quigley, G. H. and Rich, A. (1976) Structural domains of transfer RNA molecules. *Science*, **194**, 796–806.

Ramstein, J. and Buckingham, R. (1981) Tritium exchange on tRNA: slowly exchanging protons sensitive to a change in the dihydrouridine stem. *Proc. Natl Acad. Sci. USA*, **78**, 1567–1571.

Rice, J. B., Libby, R. J. and Reeve, J. N. (1984) Mistranslation of the mRNA encoding bacteriophage T7-0.3 protein. *J. Biol. Chem.*, **259**, 6505–6510.

Robertson, J. M. and Wintermeyer, W. (1981) Effect of translocation on topology and conformation of anticodon and D-loops of tRNA^{Phe}. *J. Mol. Biol.*, **151**, 57–69.

Robinson, M., Lilley, R., Little, S., Emtage, J. S., Yarranton, G., Millican, A., Eaton, M. and Humphreys, G. (1984) Codon usage can affect efficiency of translation of genes in *E. coli. Nucl. Acids Res.*, **12**, 6663–6672.

Rosenberger, R. F. and Foskett, G. (1981) The estimate of the frequency of *in vivo* transcriptional errors at a nonsense codon in *E. coli. Mol. Gen. Genet.*, **183**,

561–563.

Roth, J. R. (1981) Frameshift suppression. *Cell*, **24**, 601–602.

Ruusala, T., Ehrenberg, M. and Kurland, C. G. (1982a) Catalytic effects of elongation factors Ts on polypeptide synthesis. *EMBO J.*, **1**, 75–78.

Ruusala, T., Ehrenberg, M. and Kurland, C. G. (1982b) Is there proofreading during polypeptide synthesis? *EMBO J.*, **1**, 741–745.

Ryoji, M., Hsia, K. and Kaji, A. (1983) Read-through translation. *Trends in Biochem. Sci.*, **8**, 88–90.

Samuelsson, T., Elias, P., Lustig, F. *et al.* (1980) Aberrations of the classic reading scheme during protein synthesis *in vitro*. *J. Biol. Chem.*, **255**, 4583–4588.

Samuelsson, T., Axberg, T., Boren, T. and Lagerkvist, U. (1983) Unconventional reading of the glycine codons. *J. Biol. Chem.*, **258**, 13 178–13 184.

Schmitt, M., Kyriatsoulis, A. and Gassen, H. G. (1982) The context theory as applied to the decoding of the initiator tRNA by *E. coli* ribosomes. *Eur. J. Biochem.*, **125**, 389–394.

Sedlacek, J., Fabry, M. and Rychlik, I. (1979) The arrangement of nucleotides in the coding regions of natural templates. *Mol. Gen. Genet.*, **172**, 31–36.

Sharma, O. K. and Kuchino, Y. (1977) Infidelity of translation of encephalomyocarditis viral RNA with tRNA from human malignant trophoblastic cells. *Biochem. Biophys. Res. Commun.*, **78**, 591–595.

Sharma, O. K., Beezley, D. N. and Roberts, W. K. (1975) Limitation of reticulocyte tRNA in the translation of heterologous mRNAs. *Biochemistry*, **15**, 4313–4318.

Smiley, B. L., Lupski, J. R., Svec, P. S., McMacken, R. and Godson, G. N. (1982) Sequences of the *E. coli* dnaG primase gene and regulation of its expression. *Proc. Natl Acad. Sci. USA*, **79**, 4550–4554.

Smith, D. W. E. (1975) Reticulocyte tRNA and hemoglobin synthesis: tRNA availability may regulate hemoglobin synthesis in developing red blood cells. *Science*, **190**, 529–534.

Smith, D. W. E., McNamara, L., Rice, M. and Hatfield, D. L. (1981) The effects of posttranscriptional modification on the function of tRNA^{Lys} isoaccepting species in translation. *J. Biol. Chem.*, **256**, 10 033–10 036.

Smith, D. W. E., McNamara, A. L., Mushinski, J. F. and Hatfield, D. L. (1984) Tumor-specific hypomodified Phe-tRNA is utilised in translation in preference to the fully-modified isoacceptor of normal cells. *J. Biol. Chem.*, **260**, 147–151.

Sommer, S. S. and Cohen, J. E. (1980) The size distribution of protein, mRNA and nuclear RNA. *J. Mol. Evol.*, **15**, 37–57.

Springer, M., Trudel, M., Graffe, M., Plumbridge, J. A., Fayat, G., Mayaux, J. F., Sacerdot, C., Blanquet, S., Grunberg-Manago, M. (1983) *E. coli* Phe-tRNA synthetase operon is controlled by attenuation *in vivo*. *J. Mol. Biol.*, **171**, 263–279.

Sprinzl, M., Helk, B. and Baumann, U. (1983) Structural and functional studies on the anticodon and T-loop of tRNA. In *Gene Expression, the Translation Step and its Control* (eds B. Clark and H. U. Petersen), Munksgaard, Copenhagen, pp. 235–254.

Stent, G. S. (1964) The operon: on its third anniversary. *Science*, **144**, 816.

Stöffler, G., Hasenbank, R. and Dabbs, E. R. (1981) Expression of the L1–L11 operon in mutants of *E. coli* lacking the ribosomal protein L1 or L11. *Mol. Gen. Genet.*, **181**, 164–168.

Sussman, J. F., Holbrook, S. R., Warrant, R. W., Church, G. M. and Kim, S. H.

(1978) Crystal structure of yeast phenylalanine tRNA. *J. Mol. Biol.*, **123**, 607–630.

Takemoto, T., Takeishi, S., Nishimura, S. and Ukita, T. (1973) Transfer of valine into rabbit haemoglobin from various isoaccepting species of valyl-tRNA differing in codon recognition. *Eur. J. Biochem.*, **38**, 489–496.

Tamura, F., Nishimura, S. and Ohki, M. (1984) The *E. coli* div-E mutation which differentially inhibits synthesis of certain proteins, is in tRNA$_1^{Ser}$. *EMBO J.*, **3**, 1103–1107.

Taniguchi, T. and Weissmann, C. (1978) Site-directed mutagens in the initiator region of the bacteriophage Qβ coat cistron and their effect on ribosome binding. *J. Mol. Biol.*, **118**, 533–565.

Thompson, R. C., Cline, S. W. and Yarus, M. (1982) Site-directed mutagenesis of the anticodon region: the universal U is not essential to tRNA synthesis and function. In *Interaction of Translational and Transcriptional Controls in the Regulation of Gene Expression* (eds M. Grunberg-Manago and B. Safer), Elsevier Biomedical, New York, pp. 189–220.

Thompson, R. C., Dix, D. B., Gerson, R. B. and Karim, A. M. (1981) A GTPase reaction accompanying the rejection of Leu-tRNA$_2$ by UUU-programmed ribosomes. *J. Biol. Chem.*, **256**, 81–86.

Tsang, T. H., Buck, M. and Ames, B. N. (1983) Sequence specificity of tRNA-modifying enzymes: an analysis of 258 tRNA sequences. *Biochim. Biophys. Acta*, **741**, 180–196.

Vacher, J., Grosjean, H., de Henau, S., Finelli, J. and Buckingham, R. H. (1984a) *Eur. J. Biochem.*, **138**, 77–81.

Vacher, J., Grosjean, H., Houssier, C. and Buckingham, R. H. (1984b) The effect of point mutations affecting *E. coli* tryptophan-tRNA on anticodon:anticodon interactions and on UGA suppression. *J. Mol. Biol.*, **177**, 329–342.

Varenne, S., Buc, J., Lloubes, R. and Lazdunski, C. (1984) Translation is a non-uniform process: effect of tRNA availability on the rate of elongation of nascent polypeptide chains. *J. Mol. Biol.*, **180**, 549–576.

Wagner, E. G. H. and Kurland, C. G. (1980) Translation accuracy enhanced *in vitro* by (p)ppGpp. *Mol. Gen. Genet.*, **180**, 139–145.

Wagner, E. G. H., Jelenc, P. C., Ehrenberg, M. and Kurland, C. G. (1982) Rate of elongation of polyphenylalanine *in vitro*. *Eur. J. Biochem.*, **122**, 193–197.

Weiss, G. B. (1973) Translational control of protein synthesis by tRNA unrelated to changes in tRNA concentration. *J. Mol. Evol.*, **2**, 199–204.

Weiss, R. and Gallant, J. (1983) Mechanism of ribosome frameshifting during translation of the genetic code. *Nature*, **302**, 389–393.

Weissenbach, J. and Grosjean, H. (1981) Effect of threonylcarbomoyl modification (t^{6}A) in yeast tRNA^{Arg} on codon:anticodon and anticodon:anticodon interactions: a thermodynamic and kinetic evaluation. *Eur. J. Biochem.*, **116**, 207–213.

Weissenbach, J., Dirheimer, G., Falcoff, R., Sanceau, J. and Falcoff, E. (1977) Yeast tRNA^{Leu} (anticodon UAG) translates all six leucine codons in extracts from interferon treated cells. *FEBS Lett.*, **82**, 71–76.

Westhof, E., Dumas, P. and Moras, D. (1983) Loop stereochemistry and dynamics in tRNA. *J. Biomol. Struct. and Dyn.*, **1**, 337–355.

Woese, C. (1967) *The Genetic Code: the Molecular Basis for Genetic Expression*. Harper and Row, New York.

Wurmbach, P. and Nierhaus, K. H. (1979) Codon–anticodon interaction at the ribosomal P (peptidyl-tRNA) site. *Proc. Natl Acad. Sci. USA*, **76**, 2143–2147.

Yanofsky, C. (1981) Attenuation in the control of expression of bacterial operon. *Nature*, **289**, 751–758.

Yarus, M. (1979) The accuracy of translation. *Progr. Nucl. Acids Res. Mol. Biol.*, **23**, 195–225.

Yarus, M. (1982) Translational efficiency of tRNAs: uses of an extended anticodon. *Science*, **218**, 646–652.

Yarus, M. and Thompson, R. C. (1984) Precision of protein biosynthesis. In *Gene Function in Prokaryotes*, Cold Spring Harbor Laboratory, New York, pp. 23–63.

6 The secret life of the ribosome

C. G. KURLAND and J. A. GALLANT

6.1 Introduction

We begin by admitting that it is not yet possible to describe in molecular detail how ribosomes carry out the accurate translation of a messenger RNA. Indeed, there is still some question about how accurate protein synthesis really is. Nevertheless, some recent developments have provided unexpected insights into the origin and containment of translation errors. The principal thrust of recent work has been to explore the interpretation that the errors of translation and the strategies to contain them are of an essentially kinetic character. This notion can be motivated on at least two levels.

The first of these levels, namely that determined by the physical constraints imposed on translation, was delineated nearly a quarter of a century ago, and then forgotten. Thus, Crick (1958) in his discussion of 'the essence of the problem' referred to three headings under which knowledge about protein synthesis could usefully be organized. These three headings corresponded to 'the flow of matter, the flow of energy and the flow of information'. Our discussion will also be concerned with the flows over the ribosomes, however, we will want to amend Crick's list.

The flow of information referred to by Crick was more narrowly defined as a flow of sequence information. Thus, it is usual to think of the synthesis of a particular protein as being accompanied by a flow of information from the nucleotide sequence of the corresponding cistron to the amino acid sequence of the final product. Nevertheless, there are problems in this intuitive picture. Not the least of these problems is that while the flows of matter and energy are relevant physical fluxes, the flow of information is not really the parameter of choice to evaluate the fidelity of translation. For this reason we will replace the intuitive notion of information flow with a more narrowly defined parameter that is both directly measurable and directly related to the problem at hand, namely that of accuracy.

Accordingly, we will focus our attention on the two physical fluxes, those of mass and energy, and we will characterize the mass flow of amino acids into proteins by a discrimination ratio. This index is simply the ratio of the flow into peptide of cognate amino acid compared to that of the noncognate amino acids under the influence of codon-programmed ribosomes. Obviously, the accuracy of the proteins accumulated by a cell can be influenced by post-translational events such as the enzymic degradation of defective polypeptides. However, for simplicity, and unless it is otherwise specified, we will mean the discrimination ratio when we refer to the accuracy of ribosome function.

One of our objectives is to outline a physical theory that describes the necessary connections between the mass flow and energy flow in protein synthesis as well as the accuracy of these flows. We then use this physical theory to motivate the discussion of three more distinctly biological aspects of the problem. These are the evaluation of *in vitro* data concerning the accuracy of the translation mechanism, the description of the elements of a metabolic strategy to control translation errors under conditions of physiological stress, and the introduction of the notion that the translation apparatus may be constructed in such a way that it compensates for the errors of its own construction. This latter suggestion is directed to the more general kinetic problem of the long-term stability of translational accuracy (see Chapter 2), a problem that was also noted previously by a distinguished protagonist.

Thus, Orgel (1963) recognized that the accuracy of gene expression might be sensitive to the destructive effects of errors accumulated in the proteins responsible for DNA replication, RNA transcription and protein synthesis. Here we draw attention to two aspects of the error feedback problem. In one part of this discussion we stress the potential value of tight coupling between the errors of tRNA selection and those of mRNA movement. Such error coupling can function as an editing mechanism through which the potentially deleterious effects of error-prone ribosomes can be contained. We review the recent experimental evidence indicating the existence of such an error coupling mechanism on the ribosome.

In the other part of our discussion of the error feedback problem, we consider the statistical effects of errors of ribosome construction on the accuracy of aminoacyl–tRNA selection. The conditions under which these statistical effects would tend to damp the error feedback are described, as is the evidence that ribosomes are, indeed, constructed so that errors in their biosynthesis do not in general lead to an error catastrophe.

6.2 Missense error frequencies

One unquestionable constraint on the accuracy of protein synthesis must be that the majority of proteins synthesized are functional. Studies of the *lac* repressor suggest that roughly half of the random single amino acid replacenents obtained in this protein completely inactivate it (Miller, 1979). This suggests that at least half of the proteins must be perfect translation products if the biosynthetic system is to function effectively. With this as the starting-point we can estimate what the average error rate per translated codon might be.

We can take 300 as the average chain length of the proteins to be synthesized and we calculate the fraction of perfect copies synthesized when the missense error frequencies are fixed at certain values. Thus, with a 10^{-2} missense error frequency per codon, the fraction of perfect copies of chain length 300 would be $(1 - 0.01)^{300} = 0.05$. Similarly, with an error frequency of 10^{-3}, this fraction is 0.74; with an error frequency of 10^{-4}, it is 0.97. Clearly, an error frequency much above 10^{-3} is incompatible with a largely homogeneous primary structure for proteins of average length. Since there are many proteins (and protein complexes) containing many more than 300 amino acids, we might expect the average missense error frequency to be below 10^{-3}. This supposition is confirmed by data on thermal inactivation of a variety of enzymes. Since more than half of the random amino acid substitutions generate a significant increase in thermolability (Langridge, 1968), a protein which exhibits an exponential, single-hit inactivation curve must be largely homogeneous in primary structure. Many proteins exhibit such single-hit inactivation kinetics, including some composed of great, hulking protometers larger than 1000 amino acids (reviewed by Gallant and Palmer, 1979). This evidence, considered in terms of the calculation illustrated above, sets an upper limit on the average error frequency in the vicinity of 3×10^{-4}.

Direct measurements of the error frequency, of which there are surprisingly few, are consistent with this expectation. Loftfield's classic measurements of isoleucine–valine mistakes in rabbit globin yield estimates of $2-6 \times 10^{-4}$ (Loftfield and Vanderjagt, 1972). Likewise, Edelmann and Gallant (1977) found that E. coli flagellin contained trace quantities of cysteine at a frequency of about six residues per 10 000 flagellin molecules, leading to an upper bound of 6×10^{-4} for missense errors involving cysteine. Consideration of the categories of decoding errors which could bring illegitimate cysteine residues into flagellin leads to an estimate of about 10^{-4} for their average frequency (Edelmann and Gallant, 1977). Parker et al. (1981) and Ellis and Gallant (1982) have used electrophoretic heterogeneity, detected by two-dimensional electrophoresis, to estimate the average frequency of charge substitution errors. Both reports indicate an average of about two to five errors per 10^4 codon translation cycles.

Each of these studies aggregates errors at an undefined collection of codons, and each depends upon rough-and-ready approximations as well as normalization procedures to arrive at an average value. Recently, Bouadloun, Donner and Kurland (1983) have developed a more refined procedure for estimating the frequency of codon-specific missense errors at defined positions. They found frequencies of $1-3 \times 10^{-3}$ for specific substitutions in two ribosomal proteins. These studies are roughly consistent with the maximum value we inferred above, from the homogeneity of large proteins, and hint at considerable variability for the error frequency at individual positions. Indeed, it seems unlikely from the outset that missense error rates could be the same for all organisms, all codons, and all messenger RNA contexts.

The leaky synthesis of active gene products in nonsense mutants provides an opportunity to measure missense events of a special sort. Here we can study the occasional misreading of a termination codon by a tRNA species with an anticodon that is presumably complementary to the nonsense triplet at two out of three positions. The drawback in these measurements is that while the nonsense suppression event depends on the outcome of a competition between a tRNA species and a termination factor, the normal missense error represents the outcome of a competition between different tRNA species. It is, therefore, a little uncertain whether these two sorts of missense errors are directly comparable.

The leakiness of twenty-five UAA and UAG mutants in *phoA* and *lacZ* (data from Garen and Siddiqi (1962) and unpublished results of J. Gallant) is distributed over a broad range from less than 10^{-5} to 8.5×10^{-4} of the activity found in isogenic wild type strains. The mean value of this collection of estimates is 2.65×10^{-4} with a standard deviation of 2.75×10^{-4}. Thus, the mean value of the spontaneous nonsense suppression frequency is reasonably close to the missense frequency noted above, but the range of two orders of magnitude demonstrates considerable variability. To repeat, the 'canonical' estimate of about 3×10^{-4} should be understood as the mean of a very heterogeneous collection of events.

One source of this variability could be the presence of different near-neighbouring sequences to the nonsense codons in the mRNA. Clearly, the structure to which a tRNA associates is not a ribosome site *per se*, but rather a ribosome associated with a considerable length of mRNA as well as another tRNA molecule. The structure of the complex may therefore be affected by the particularities of mRNA sequence throughout the region associated with the ribosome. Indeed, detailed studies of the T4 $r_{II}B$ gene show that such context effects vary the efficiency of suppression over three orders of magnitude (Fluck, Salser and Epstein, 1977). The sequencing revolution holds the promise of discovering the rules of these context effects. The first step in this direction, by Bossi and Roth (1980), has demonstrated that suppressibility is strongly affected by the base immediately 3' of a nonsense allele, suggesting perhaps a form of quadruplet reading. More recent studies (Bossi, 1983; Miller and Albertini, 1983) have confirmed the primary importance of the 3' neighbour, and suggest additional context rules extending two or even three bases to the 3' side. The physical basis of these context rules is a complete mystery. At the least, they demonstrate that the specificities of translation are not quite so simple as the triplet code taught in biology texts and comic books.

Another complication is that the frequency of mistranslation is dependent on ribosome structure. For example, certain mutant alleles of the *E. coli* gene coding for ribosome protein S12 (*rpsL*, formerly called *strA*) lead to a reduced error frequency while certain mutant alleles of the gene for protein S4 (*rpsD*, formerly called *ram*) enhance the error frequency (Gorini, 1974). Accuracy is also affected by mutational alteration of ribosome proteins S5 and S17 (de Wilde *et al.*, 1977) and L6 (Kuhlberger *et al.*, 1979). Similarly, in the eukaryote *Podospora*, accuracy is affected by mutation in several ribosome genes (Picard-Bennoun, 1976; Bennoun, personal communication). Accordingly, there is the possibility that the ribosome of different cell types may have evolved in such a way that their average frequencies of translational errors are significantly different.

In the case of *E. coli*, the major source of errors in gene expression seems to be the selection of aminoacyl-tRNA by the codon-programmed ribosome. This could be inferred from a comparison of the missense errors measured at two particular codon sites in wild type bacteria and in a restrictive strepto-mycin resistant mutant (Bouadloun *et al.*, 1983). The restrictive ribosome mutant supports error rates that were roughly one-quarter those of the wild type. If these results are truly representative, they suggest that the aminoacyl-tRNA synthetases and RNA polymerase function at a higher accuracy than do the ribosomes. A corollary that will have greater significance when we discuss the error loop is that the existence of restrictive mutants shows that wild type ribosomes work far from the maximum attainable accuracy of translation.

It should be clear by now that the missense error frequency is not a universal constant. Considerable descriptive work will be required before the patterns of variability from site to site can be formulated. Nevertheless, all of the available information is consistent with the conclusion that most missense errors of translation occur with frequencies in the range of 10^{-4} to 10^{-3} per codon under normal conditions.

Error frequencies are of course influenced by the concentrations of the substrates involved in protein synthesis, such as the aminoacyl-tRNA and the guanine nucleotide pools. We will discuss these relationships in detail later. For the moment, however, it should be borne in mind that error frequencies can be driven up to very high values by imbalances in the charging levels of competing tRNAs. Were the charging level of one tRNA reduced to near zero, then in theory the error frequency at codons calling for that species would effectively approach unity. Indeed, Parker, Johnston and Borgia (1980) have detected error frequencies as high as 0.3 for replacement of asparagine by lysine in cells specifically starved of asparaginyl-tRNA.

The dependence of error frequency upon aminoacyl-tRNA concentration has obvious implications for the recombinant DNA industry. Strongly expressed *E. coli* genes have a particular pattern of codon preference (see Chapter 5), and these preferences are generally mirrored in the concentrations of the corresponding isoacceptor tRNAs in *E. coli* cells. For example, *E. coli* genes use the CCU proline codon very rarely, and the tRNA which reads this codon is very scarce. Eukaryotic genes have a different pattern of codon preference in this and several other cases. Accordingly, efficient translation of a eukaryotic gene insert cloned into a multi-copy plasmid expression vector places unusual demands upon the protein synthesis apparatus of the host *E. coli* cells. We conjecture that the scarcity of some rarely used *E. coli* tRNA species may lead to unusually high error frequencies at codons in cloned eukaryotic gene inserts which call for them. Indeed, we suspect that practitioners of cloning have often, although inadvertently, carried out in this way interesting experiments on translational accuracy.

Thus far we have catalogued some of the structural elements that influence the accuracy of aminoacyl-tRNA selection during translation. This, however, is only part of the picture, though it must be added that it is the part most representative of the conventions of molecular biology. The missing part of the picture concerns the dynamics of the interactions between these structural elements. Although it is somewhat unconventional in molecular biology to be concerned with kinetics, by so doing we in fact return to Crick's original picture of translation as a composite flow.

6.3 Bioenergetics of translation

According to the accepted view of protein synthesis (Lipmann, 1969; Kaziro, 1978), the partial reactions that make up a peptide elongation cycle on the ribosome are arranged as follows. First, a ternary complex consisting of elongation factor Tu (EF-Tu), GTP and aminoacyl-tRNA (AA-tRNA) is bound at the A-site of the ribosome. Following complex formation, the GTP is hydrolysed and EF-Tu is released from the ribosome leaving the AA-tRNA bound at the A-site. At this point the selection of the AA-tRNA to match the codon in the A-site is thought to be completed, and peptide bond formation is catalysed by a ribosomal centre. This reaction results in the transfer of the growing polypeptide chain to the tRNA at the A-site and it leaves a deacylated tRNA in position at the P-site.

The next phase of the cycle is thought to begin with the binding to the ribosome of a complex containing elongation factor G (EF-G) and GTP. The release of the deacylated tRNA from the P-site, the translocation of peptidyl-tRNA from the A- to the P-site and the advance of the mRNA relative to the ribosome by one codon are all mediated in some way by EF-G. The cycle is completed when the EF-G, following GTP hydrolysis, is released from the ribosome.

Although this flow scheme has the virtue of imposing some order on the complexities of the system, it is silent about the relationships between the functions of these different bits of machinery and the accuracy of the translated product. Most surprising in retrospect is the long time it has taken to focus attention on the relationship between the factor-dependent GTPase reactions and the accuracy of translation. The delay in appreciating the significance of the partial reactions involving GTP hydrolysis can be attributed in part to a failure to appreciate the physical constraints under which the translation of the genetic code must take place. Two opposing viewpoints that surfaced during the mid-seventies illustrate this problem.

Until recently the prevailing view has been that the structural rules governing the codon–anticodon interaction are the only molecular rules that determine the outcome of the competition between different tRNA species for a codon on the ribosome. The tRNA selection problem has been viewed accordingly as a problem of geometry (Crick, 1966), which is certainly an adequate framework if the successful matching of codon and anticodon is viewed as an all-or-none event.

The natural extension of this geometric view would be to exclude the factor-dependent GTPase reactions from the accuracy problem and to restrict their influence to a catalytic function that only determines the rate of translation. This is precisely the view expounded by Spirin (1978) on the basis of a provocative series of experiments with a very special *in vitro* translation system. Spirin and his co-workers showed that it was possible to modify one of the ribosomal proteins *in situ* and thereby enhance the ability of the treated ribosomes to function in the absence of elongation factors and GTP. The apparent accuracy of translation by these modified ribosomes seems if anything better than that of normal ribosomes. However, the translation by the factor-free system is much slower than the factor-dependent system. The conclusion that followed from these observations seemed clear enough: the determination of the specificity of translation was intrinsic to the structure of the ribosome and not dependent on the factors or GTP hydrolysis.

The opposing view has a somewhat more complicated background. To begin with, it was recognized that the competition of tRNA species for a codon on the ribosome is a kinetic phenomenon and, therefore, that the relevant determinants of the error frequencies are the relative rates with which correct and incorrect species are processed by the system (Ninio, 1974). To this was added the conviction that the extent to which the competing species could be distinguished geometrically was limited and insufficient to account for the accuracy of translation (Pauling, 1957). Accordingly, a mechanism was required through which the intrinsic limits of a structural selection can be surpassed.

The solution to this problem is to repeat the elementary selection one or more times and thereby at each step to reduce the error frequency to the degree allowed by the intrinsic discrimination capacity of the system (Hopfield, 1974; Ninio, 1975). Such a strategy is now referred to as kinetic proofreading, and it differs from its less glamorous antecedent (Kornberg, 1969) only to the extent that kinetic proofreading strategies involve a multiple checking of the substrate before it is incorporated into a product, rather than after. Nevertheless, there was something new and important introduced in the discussions of kinetic proofreading. Thus, all proofreading selection systems have two characteristics that distinguish them from simple selection systems. One of these is the requirement for a branch point in the flow pattern so that a preferential flow forward to product for correct substrate is complemented by a preferential flow out of the system for the incorrect substrate along what we call a discard branch. It is the unequal partitioning of correct and incorrect substrate flows between these two branches that provides the proofreading effect. The second, related characteristic is the requirement for a driving force to activate the flow of incorrect substrate along the discard branch. This then is the new element introduced into the discussion of accuracy – the driving force.

In effect, Hopfield (1974) and Ninio (1975) realized that in addition to the constraint imposed on the system by its limited capacity to distinguish two related substrate structures in a single step, there existed a thermodynamic constraint that must be broken in order to repeat this elementary step. Thus, the tendency to establish an equilibrium between substrate bound to the selection site and the substrate ejected over the discard branch must be opposed by some force in order to obtain a net flow outward along this branch. For the ribosome, the energy source to drive the proofreading branch was identified with the GTP consumed during protein synthesis. The proofreading selection on the ribosome was envisioned by Hopfield (1974) as being dependent on the binding of the ternary complex to the ribosomal A-site leading to the formation with GTP of a 'high energy' intermediate, the decay of which would drive the proofreading branch. Evidence to support this view was offered by Thompson and Stone (1977) who observed *in vitro* an excessive hydrolysis of GTP accompanying the ribosome binding reaction of ternary complexes containing tRNA species not properly matched with the codon at the A-site.

In summary, two completely opposite views of the role of factor-dependent GTP hydrolysis in maintaining the accuracy of translation were circulating in the mid-seventies. What is more, both views appeared to be supported by mutually exclusive, but individually convincing, experimental data. One way to resolve this dilemma was to question the relevance to the problem of both sets of experiments (Kurland, 1978). The experiments of Spirin and his group as well as those of Thompson and Stone (1977) shared with the vast majority of *in vitro* translation experiments a characteristic weakness that makes their interpretation difficult. Thus, the missense error frequencies of translation *in vivo* seem to be in the 10^{-4} range whereas the corresponding error frequencies obtained with most *in vitro* systems have been at least two orders of magnitude higher. Accordingly, if the partial reactions of GTP do not seem to be necessary to maintain the accuracy of translation at the 10^{-1} level, does it follow that these partial reactions have no role to play at the 10^{-4} error level? Similarly, if there is an excess GTP hydrolysis associated with the binding of incorrectly matched tRNA to the ribosome in the absence of G factor, does it follow that such an excess hydrolysis occurs as well when the system is operating in the presence of G factor in the steady state?

Quite obviously an *in vitro* system that duplicates the accuracy of translation *in vivo* would facilitate making an experimentally meaningful decision between the opposing views of the role of the factor-dependent GTPase reactions. Such a system has been developed (Jelenc and Kurland, 1979, Wagner *et al.*, 1982a) and its characteristics suggest that the GTP reactions are involved in the aminoacyl-tRNA selection (Jelenc and Kurland, 1979; Pettersson and Kurland, 1980; Wagner and Kurland, 1980). However, before the nature of this system and the interpretation of the results obtained with it can be described, we must consider a more general relationship between the driving forces for enzymic reactions and the accuracy of substrate selections.

Although the original descriptions of proofreading associated the driving force for the discard reaction with a GTP function, they did not discuss the driving force along the substrate-product path in the same way. This, together with the obscurity of the notion of the 'high energy' intermediate, prompted a reformulation of the relationship between the driving forces along all the branches of a selection system and the accuracy of product formation (Kurland, 1978).

This more general formulation has been refined and applied to the optimization of proofreading systems (Ehrenberg and Blomberg, 1980; Blomberg and Ehrenberg, 1981; Blomberg, Ehrenberg and Kurland, 1980) and is discussed in detail in Chapter 11. The basic ideas are as follows. The standard free energy changes for the formation of biopolymers from their monomers, the nucleotides and amino acids, are to a very good approximation independent of the monomer sequences in the polymers. Therefore, if the biosynthesis of proteins and nucleic acids were to take place near chemical equilibrium, it would be impossible to carry out the accurate synthesis of genetically determined sequences. Similarly, at chemical equilibrium the flows of the relevant substrates into the polymers would be zero. Accordingly, all of the biosynthetic systems must be displaced far from equilibrium in order to obtain accurate sequence copies at appreciable rates.

The extent of displacement from chemical equilibrium for the substrates and products of a biosynthetic reaction is determined by the ratio of the concentrations of products and substrates during the reaction. When this ratio is small compared to what the ratio would be at equilibrium (i.e. the equilibrium constant), there will be a net chemical potential to drive the reaction towards product. In other words, the force that drives the chemical reaction is the tendency of the reactants to approach their equilibrium concentrations and this force is proportional to the degree of displacement from the equilibrium concentrations.

If we assume that the biosynthetic enzymes are balanced, i.e. do not have destructive kinetic interference built into them, the relationships between the displacement from equilibrium and the accuracy as well as the speed of the polymer synthesis are rather straightforward. Thus, an increase of the displacement from chemical equilibrium leads monotonically to an increase of the rate of biosynthesis until a maximum flux is obtained. This is paralleled by an increase of the accuracy of the substrate selection until a minimum error is obtained. In other words, for a balanced enzyme, the speed and accuracy of its catalysis are positive functions of the driving force. This conclusion is equally as valid for enzymes that perform their functions according to conventional (unbranched) selection schemes as for those that use proofreading (branched) selection schemes.

In the case of protein biosynthesis on the ribosomes there are two substrate species – the aminoacyl-tRNA and GTP. Furthermore, the fluxes of both these species over the ribosome are coupled to each other (Lipmann, 1969; Kaziro, 1978). This means that the displacements from equilibrium of both substrates should in general influence the rate and accuracy of translation. Furthermore, the displacement of the guanine nucleotides from equilibrium can in principle influence the accuracy of aminoacyl-tRNA selection whether or not there is proofreading taking place.

6.4 Translation *in vitro*

Although they may appear to be somewhat abstract, the arguments forwarded in the previous section have had rather concrete and useful consequences. For example, the requirement for large driving forces to obtain accurate, rapid translation means that *in vitro* systems must be supplemented with regeneration systems that can maintain the aminoacyl-tRNAs and GTP adequately displaced from equilibrium during the course of the experiments. The regeneration of both the aminoacyl-tRNAs as well as GTP can be coupled conveniently to an ATP regeneration system. In this way it could be shown directly that the accuracy of poly U translation *in vitro* is dependent on the displacement from equilibrium of the nucleoside tri-phosphates (Jelenc and Kurland, 1979).

Such results indicate that the dynamic state of the system, as defined by the displacements from equilibrium, determines how well the discrimination capacity of the system is expressed. It is worth emphasizing that this is equivalent to saying that the structures of the relevant components do not by themselves determine the accuracy of translation; they only determine the attainable limits of the accuracy.

The system used for these experiments has the important property of translating a synthetic messenger, poly U, with an error frequency close to that of *in vivo* translation, i.e. $2–6 \times 10^{-4}$. In addition to the careful tuning of the substrate regeneration components, this system is distinguishable from its predecessors by the complexity of its ionic composition, which mimics that of bacteria (Jelenc and Kurland, 1979). Thus, the polymix system contains, in addition to Mg^{++} other polycations such as Ca^{++}, and putrescine as well as spermidine. That this complex mixture of ions is important to the performance characteristics of the ribosomes could be demonstrated by comparing the activities of ribosomes in the polymix system with those in conventional buffers (Pettersson and Kurland, 1980).

The most significant conclusion that can be drawn from the properties of the polymix system is so obvious that it easily escapes notice. This is that it is possible to translate *in vitro* the UUU codon with an accuracy which matches that estimated *in vivo* (see Section 6.2). Therefore, we can argue for the first time that the apparent accuracy of flagellin translation *in vivo* reflects the precision of synthetase as well as ribosome function and is not significantly influenced by post-translational editing. This conclusion has been verified by using an *in vitro* system consisting of purified ribosomes, factors, aminoacyl-tRNA synthetases, etc. (Wagner *et al.*, 1982a).

This purified system not only supports accurate translation of poly(U); it also supports rates of translation that are close to those of the ribosomes *in vivo* (reproducibly within a factor of two). It is also worth noting that the performance characteristics of the ribosomes in this system are the same for the translation of poly(Cys Val) from poly(UG) as for poly(Phe) from poly(U); this suggests that the optimization is not specific for poly(U) translation (Andersson *et al.*, 1984). Since error rates of the order of 10^{-4} to 10^{-3} are observed at the three codons tested so far, it seems likely that the missense rates observed *in vivo* directly reflect the errors of the translation systems. Once again, the agreement of the *in vivo* and *in vitro* data suggests that post-translational editing does not play a great role in controlling the accuracy of translation.

Many parameters seem to influence the speed and accuracy of translation *in vitro* and they do so in a systematic way: in general, there is a positive correlation between these two performance characteristics of the ribosome in polymix (Jelenc and Kurland, 1979; Pettersson and Kurland, 1980; Wagner and Kurland, 1980; Wagner *et al.*, 1982a). One of these parameters, the charging level of the tRNAs, is of particular relevance to the physiological response of bacteria to limitation for amino acids.

We would expect that a dramatic decrease in the internal concentration of an amino acid would lead to a correspondingly dramatic increase of the missense error frequency at all of the codons cognate to the limiting amino acid. Such an effect would follow from the attendant change in the ratio of noncognate to cognate substrates, which in turn raises the relative chance for a noncognate substrate to be selected. Indeed, this effect is seen in the polymix *in vitro* system (Wagner and Kurland, 1980). Remarkably, wild-type *E. coli* do not respond to deprivation of a required amino acid in this way, but mutants (*relA*⁻) with a defect in the so-called stringent factor do show a significantly elevated missense error frequency when starved of an amino acid (Hall and Gallant, 1972; Edelmann and Gallant, 1977; O'Farrell, 1978; Parker, Johnston and Borgia, 1980). Clearly, wild-type *E. coli* have a strategy that can compensate the translation system for the limitation of an amino acid, and stringent factor plays some role in this strategy.

It is known that the accumulation of deacylated tRNA cognate to a particular codon leads to complex formation on the ribosome along with stringent factor, ATP and GTP (summarized in Gallant, 1979). This complex then enzymically converts the GTP to one of two hyperphosphorylated guanine nucleotides, pppGpp or ppGpp, the so-called magic spots.

The limited abundance of stringent factor in *E. coli* as well as suggestive results obtained *in vitro* prompted Gallant and Foley (1980) to suggest that it is the magic spots and not the stringent factor *per se* that protects the translation system from making frequent missense errors during amino acid deprivation. This conclusion has been verified in detail with the polymix system (Wagner and Kurland, 1980).

It was then possible to test the suggestion of O'Farrell (1978) that the magic spots exert their influence on the accuracy of translation by slowing down the flow of amino acids over the ribosome and thereby permitting the limiting amino acid to accumulate in the form of a complex with its cognate tRNA. This would have the effect of redressing the cognate/noncognate aminoacyl-tRNA ratio and, thus, maintaining the accuracy of translation. Direct measurements of the charging levels revealed the expected enhancement of the charging level for the tRNA species corresponding to the limiting amino acid when the magic spots are present *in vitro* (Wagner and Kurland, 1980).

The mechanism of this effect of magic spots is currently under study. It is now evident that both pppGpp and ppGpp can separately inhibit the functions of both EF-Tu and EF-G in protein synthesis (Wagner, Ehrenberg and Kurland, 1982b; Rojas *et al.*, 1984). However, under those conditions that mimic the *in vivo* situation most closely, the inhibition of EF-Tu by ppGpp seems to be the dominant effect.

The problem has been that the *in vivo* data do not support the O'Farrell interpretation in complete detail. On the one hand, *spoT⁻* mutants, which degrade ppGpp at a reduced rate, resume protein synthesis very slowly upon restoration of a missing amino acid, and this recovery parallels slow decline of the intracellular ppGpp level (Laffler and Gallant, 1974; O'Farrell, 1978). This supports the view that the stringent response limits the rate of protein synthesis through the inhibitory effects of ppGpp rather than through the availability of aminoacyl-tRNA. On the other hand, several studies report that the pools of aminoacyl-tRNA are not significantly different in starved *RelA⁻* and *RelA⁺* bacteria (Böck, Faiman and Neidhardt, 1966; Yegian and Stent, 1969; Piepersberg *et al.*, 1979). In other words, these data seem to contradict the expectation that the presence of ppGpp *in vivo* will raise the levels of the limiting aminoacyl-tRNA species by inhibiting their flow through the ribosome. Happily, there are alternative explanations for these data.

Initially, these observations suggested that magic spot enhances accuracy by lowering preferentially the concentration of ternary complex corresponding to the unstarved aminoacyl-tRNA species while leaving the total charging level of the limiting tRNAs virtually unchanged (Wagner *et al.*, 1982b). It may now be shown that such preferential effects can be obtained only under very specific circumstances, as, for example, when the competing tRNA species associate in ternary complexes with different stabilities (Rojas *et al.*, 1984).

A more radical and entirely general explanation of these data would postulate a direct effect of ppGpp on the accuracy of translation during amino acid limitation. According to this notion, the charging levels of aminoacyl-tRNA could remain virtually unchanged by ppGpp, if the accuracy of translation is improved by the interaction of the nucleotide analogue either with the ribosome or with EF-Tu. The most recent experiments suggest that ppGpp does not enhance the intrinsic discrimination capacity of the ribosome. Instead, they support the interpretation that an EF-Tu ppGpp complex stimulates the relative rate with which mismatched aminoacyl-tRNAs are discarded during translation (Rojas *et al.*, 1984; Kurland and Ehrenberg, 1984). These data will be discussed below.

In summary, the stringent response can be viewed as a metabolic strategy to contain the errors of translation under conditions of limiting aminoacyl-tRNA. Here, the guanine nucleotides normally required for ribosome functions are converted into analogues that selectively affect the kinetics of translation by modifying the functions of the elongation factors. Clearly, were the nucleoside triphosphates required for both tRNA charging and ribosome functions the same ones, such a metabolic strategy would not be effective. The ubiquity of the stringent response among prokaryotes is thus bound up with the use of GTP, rather than ATP, for energy dependent ribosome steps (Wagner and Kurland, 1980).

6.5 Curious consequences of proofreading

The search for a specific mechanism to explain the influence of ppGpp on the accuracy of translation obviously would be aided by detailed information about the normal mechanisms of aminoacyl-tRNA selection by codon-programmed ribosomes. Happily, recent experiments with the high performance *in vitro* system described above have revealed some of the relevant details of the selection during steady state translation under conditions comparable to those *in vivo*.

One telltale characteristic of a proofreading mechanism for the selection of ternary complexes by the ribosome would be that the efficiency of peptide bond formation is much less for ternary complexes containing noncognate tRNAs than for those containing cognate tRNAs. That is to say, proofreading should reveal itself by a dissipation of ternary complexes without concomitant peptide bond formation, and this dissipation should be much more pronounced with noncognate than with cognate tRNAs. In a sense, the enhanced accuracy is paid by the enhanced dissipation in a proofreading selection.

Accordingly, we have developed methods to measure the number of ternary complexes consumed per peptide bond formed with cognate aminoacyl-tRNA (f_c) and noncognate species (f_w) (Ruusala *et al.*, 1982b; Bohman *et al.*, 1984). The ratio f_w/f_c is a measure of the degree of enhancement of the accuracy due to the proofreading; we call this ratio the proofreading factor (F). The results of these measurements of the 'f numbers' are discussed in more detail in Chapter 11. Here, we note the following: the data indicate that there is aggressive accuracy enhancement via proofreading. The amount of proofreading varies with the identity of the aminoacyl-tRNA isoacceptor species (Ruusala *et al.*, 1982b). The proofreading flows and thereby the accuracy can be reduced by antibiotics (Jelenc and Kurland, 1984; Ruusala and Kurland, 1984). Ram mutations leading to alterations of the ribosomal protein S4 lower the accuracy by specifically reducing the proofreading (Andersson and Kurland, 1983). In contrast, mutations that effect ribosomal protein S12 and which create the restrictive streptomycin resistance (SmR) phenotype, the streptomycin dependence (SmD) or the pseudodependent phenotype (SmP) are all characterized by hypertrophied proofreading activities that reduce the errors of translation quite markedly *in vivo* as well as *in vitro* (Bohman *et al.*, 1984; Ruusala *et al.*, 1984). This variable pattern for the proofreading flows shows that external agents can modulate the accuracy through these flows and that in the case of the wild type, there is a reserve of proofreading capacity that can in principle be mobilized in an emergency. Both of these conclusions may have considerable relevance to the mechanism of ppGpp function.

Thus, it has been suggested that ppGpp could modulate the proofreading flows in such a way that during the stringent response the efficiency of ternary complex function is reduced but that the accuracy of peptide bond formation is increased (Kurland and Ehrenberg, 1984; Rojas *et al.*, 1984). In particular, it was suggested that the binding of EF-Tu.ppGpp complex to ribosomes would retard the formation of peptide bonds and, thereby, provide more time for the selective dissociation of noncognate aminoacyl-tRNA species from the ribosome. It remains to be seen if this model will survive the experimental test to which it is currently being exposed.

Before moving on to another aspect of the accuracy problem, we comment on the consequences of such a proofreading mechanism for the stability of translation (see also Chapter 2). As mentioned earlier, Orgel (1963) was concerned about the influence of errors of construction in devices such as RNA polymerase and ribosomes on the accuracy of their function. Orgel, as well as many of his followers, tacitly assumed that an error in ribosome construction must influence accuracy in a negative way. We know, however, that this is not always the case since, as we have mentioned above, there exist mutants which have ribosomes capable of abnormally aggressive proof-reading and which are significantly more accurate than the wild type. This means that an error in ribosome construction can, in principle, increase the accuracy of its function by increasing the magnitude of its proofreading flow (Kurland and Ehrenberg, 1984; Ehrenberg and Kurland, 1984). A constraint on the evolution of ribosome structure could therefore be that a sizeable proportion of random errors should have the effect of making the ribosome more, rather than less, accurate.

It may at first puzzle the reader that the wild type ribosome is not the most accurate sort of ribosome because it would seem that mutational 'mistakes' are making the altered ribosome 'better' than the wild type. However, 'more accurate' is not necessarily the same as 'better'. Thus, when the accuracy is dependent on proofreading functions, more accurate translation must be viewed as less efficient translation. Therefore, the best ribosome might be more appropriately viewed as that which optimizes the efficiency of trans-lation with its accuracy (Kurland *et al.*, 1984).

Indeed, there are at least three distinguishable components of translational efficiency that are influenced by proofreading. It has been suggested that a trade-off between, on the one hand, these efficiency parameters and, on the other hand, the inefficiency of translation reflected in error rates is determining the optimal construction of the system (Kurland and Ehrenberg, 1984; Ehrenberg and Kurland, 1984). Thus, changes in the proofreading flows will change the stoichiometric ratio of GTPs consumed per peptide bond, EF-Tu cycles dissipated per peptide bond and the saturation levels of the ribosomes. The point is that just as production of proteins that are inefficient because they contain errors will slow down the growth of the cell so too will limitations of GTP production, EF-Tu turnover, or ribosome function slow down growth. Accordingly, an optimal trade off between an acceptable error level at an acceptable efficiency cost for translation will support the maximum growth rate.

This hypothesis concerning the optimization of accuracy is supported by a number of experimental observations. First, it is possible to identify the efficiency loss in translation associated with proofreading accuracy with the parameter f_c, which paradoxically describes the number of ternary complexes dissipated per cognate peptide bond. The reason that this term dominates the dissipative losses is simply that the error level is normally so low that the contribution of the analogous term for noncognate species, f_w, is negligible when the total ternary complex dissipation for all peptide bonds is at issue. Now, the point is that the dissipative losses summarized in f_c are significantly enhanced in all the different mutants with increased accuracy of translation. Furthermore, there is an inverse correlation between f_c and the growth rates of these mutant bacteria (Bohman *et al.*, 1984; Ruusala *et al.*, 1984).

A more dramatic demonstration of this optimization principle is provided by studying the effects of Sm on the activities of dependent (SmD) and pseudodependent (SmP) mutants. This antibiotic has been shown to be a relatively specific inhibitor of ribosomal proofreading flows (Ruusala and Kurland, 1984). Accordingly, it was possible to show that the stimulation of the growth rate in SmP as well as SmD bacteria is associated with, on the one hand, an increase of the efficiency of translation and, on the other hand, a decrease of the accuracy of translation, both *in vivo* and *in vitro* (Ruusala, Anderson, Ehrenberg and Kurland, 1984). In effect, the SmP as well as SmD ribosomes are so much more accurate than wild type that excessive proofreading activity inhibits the growth of the mutant bacteria. Sm, by virtue of its effects on the dissipative losses associated with proofreading, stimulates the growth by raising the error frequencies, which is an excellent illustration of our optimization principle.

Accordingly, the consequences of errors of construction in ribosomes include, in addition to error changes, perturbations of the efficiency of translation. Hence, what was previously envisioned as an 'error loop' should be enlarged into what may be thought of as an 'efficiency loop', only one component of which is the error frequency of biosynthesis. A clear advantage of such a formulation is that it naturally leads into concern about variation in the optimal balance between efficiency and accuracy under different growth conditions (Ehrenberg and Kurland, 1984). Indeed, it can be shown formally that optimization of the accuracy of translation is different for bacteria growing in different media. Therefore, it is possible that a major function of the nucleotide analogue such as ppGpp may be to modulate the accuracy level of translation so that it approaches the different optimal level required by different growth states. Here, the stringent response could be a dramatic illustration of a more general and subtle regulatory function.

6.6 Error coupling

So far we have considered only one part of the ribosomal accuracy problem. The other relevant domain of problems concerns the maintenance of the reading frame during translation of the mRNA. This domain has been less well characterized than that of the missense errors. Nevertheless, there are strong indications that these clearly distinguishable error events are not independent; more specifically, it seems that the occurrence of a missense error can raise the probability of a subsequent reading frame error. We refer to this compounding of mistakes as error coupling. Here, we explore the mechanistic sources of error coupling.

A good starting point for this discussion is the observation of Riddle and Carbon (1973). They sequenced a mutant form of tRNA$_{GGG}^{Gly}$ that had been shown previously to be a phenotypic frameshift suppressor. The suppressor tRNA$_{sufD}^{Gly}$ was found to contain an anticodon loop with an additional C adjacent to the CCC sequence of the wild type anticodon. Here, the presumed mechanism of action of the *sufD* product was to generate a reading frame error that compensates for a frameshift mutation (nucleotide insertion) in the original mutant. Accordingly, this sequence determination showed that the structure of the tRNA, in particular that of the anticodon loop, influences the accuracy of mRNA movement.

A reasonable, specific interpretation of these results in light of the earlier Woese (1970) hypothesis was that the *sufD* suppressor, by virtue of its additional C in the anticodon loop, is able to form a four base pair interaction with a run of four Gs in the mRNA. In this way it might mediate a four base long movement of the mutant mRNA that corrects the reading frame and returns it to the proper phase. In other words, the length of the mRNA movement could be determined in this case by the number of base pairs that interact with the anticodon loop of the tRNA accepted in the A-site.

On the other hand, a counter example to this inferred rule was found in another mutant tRNA. In this case a mutant form of tRNA^{trp} was shown to translate the UGA stop codon (Sambrook, Fan and Brenner, 1967; Hirsh, 1973) as well as the Cys codon UGU (Buckingham and Kurland, 1977). Since the mutant tRNA contains the wild type anticodon CCA and is altered only in its D-stem (Hirsh, 1973), it was suggested that this mutant species responds to any codon with the structure UGN (Kurland *et al.*, 1975). In other words it would appear that the UGA suppressor tRNA^{trp} reads only the first two nucleotides in the codon. Nevertheless, it is a highly effective suppressor *in vivo* (Sambrook, Fan and Brenner, 1967), which suggests that when it translates a UGA codon, the subsequent mRNA movement is usually normal, i.e. three bases long.

For this reason an alternative interpretation of the mode of action for the *sufD* frameshift suppressor was suggested (Kurland, 1979). Here it was assumed that the ribosomal A-site is so constructed that it tends not to permit more than three base pairs to form between codon and anticodon. However, there is an ambiguity for a mutant tRNA such as the *sufD* species in that either one of the two overlapping C triplets in the mutant anticodon loop might interact with a GGG codon. If at least one of the triplet interactions of the *sufD* species with a GGG codon creates a strained or nonstandard geometry on the ribosome, it was postulated that the subsequent movement of the tRNA–mRNA complex from the A- to P-site of the ribosome could be perturbed sufficiently to cause a reading frame error.

A corollary of this interpretation is that any nonstandard codon–anticodon interaction at the A-site might lead to reading frame error during the transition to the P-site. Since a missense event results from a nonstandard codon–anticodon interaction, the expectation would be that missense events would lead, at least some of the time, to reading frame errors. Indeed, antibiotics and ribosome mutations that raise or lower the missense error frequency seem also to have parallel effects on the spontaneous frameshift suppression frequency (Atkins *et al.*, 1972). Similarly, changes in the relative concentrations of certain isoacceptor tRNA species which should increase the missense error frequencies (see previous section), lead *in vitro* to very specific reading frame errors in the translation of MS2 RNA (Atkins *et al.*, 1979).

In summary, there does seem to be a reasonable case to be made for the existence of an error coupling mechanism on the ribosome. We defer a discussion of our more recent analysis of this mechanism to the next section. Before this, we will attempt to justify the interpretation that such a coupling mechanism provides the critical element of an error containment strategy.

We note that the bacterial ribosome and its helper factors contain sequences of more than ten thousand amino acids. If the missense error frequencies for these proteins are comparable to the estimates we reviewed in Section 6.2, then the 'average' ribosome will contain three missense errors and the number of 'standard' ribosomes in a bacterium will be small (e^{-3} or 5%). Even if most of these missense errors are innocuous, we must still reckon with a significant fraction of truly defective ribosomes being generated during normal growth. Since certain amino acid substitutions of ribosomal proteins lead to error-prone ribosomes (see above), it follows that some of the normally occurring defective ribosomes will function with an unusually high error frequency, and accordingly they also may accelerate the accumulation of additional defective ribosomes (see Orgel, 1963).

In short, the normal population of ribosomes is likely to be highly heterogeneous and within this population there will be a class of ribosomes that have the potential of generating mischief at a rate much greater than the average. It would obviously be an advantage to the system if ribosomes were constructed in such a way that the error-prone ribosome could be neutralized selectively. It is here that the error coupling seems relevant.

As pointed out previously, a ribosome that makes a reading frame error is likely to encounter shortly thereafter an out-of-phase termination codon and this will result in the release of the defective polypeptide as well as its eventual enzymic degradation (Kurland, 1979). Out-of-phase termination codons are not rare; for example, there are thirteen such codons within the 130 codons long sequence that codes for the coat protein of R17. There are thirteen out-of-phase terminators in the 70 codon long sequence of the early part of the $r_{II}B$ message (Pribnow et al., 1981). At random, with three out of 64 terminator codons, there ought to be only ten rather than twenty-six out-of-phase terminators in these two sequences coding for 200 amino acids. Thus, one element in the nonrandom pattern of codon utilization (see Chapter 5) might be a positive selective value associated with sequences containing out-of-phase terminators. Accordingly, if there is a significant probability that a missense event leads to a frameshift event, the effect of the out-of-phase termination codons will be to edit away a corresponding fraction of the missense errors.

In addition, Menninger (1978) has described another relevant mechanism. He has observed the dissociation in vivo of peptidyl-tRNA from elongating ribosomes. The dissociated peptidyl-tRNA is thought then to be cleaved by a specific hydrolase and here too the released polypeptide would become a suitable substrate for degrading enzymes.

Measurements of the rates of both sorts of abortive termination events indicate that they occur in vivo at roughly the same frequencies as missense errors (Manley, 1979; Menninger, 1978). This means that the editing function of these termination events can at most reduce the average missense error frequencies by a factor of two to three. In other words, the putative editing events do not seem to contribute much to the average accuracy of ribosome function under normal conditions. Nevertheless, such termination events may function to reduce selectively the missense errors that are generated by error-prone ribosomes.

In order to illustrate this strategy we assume that the normally low mRNA movement error frequency increases very greatly when a missense error occurs: for example, the probability of a reading frame shift plus that for dissociation of peptidyl-tRNA may be as high as 0.7 when a mismatched aminoacyl-tRNA is accepted at the ribosomal A-site. The consequence of this error coupling for the normal ribosome will be marginal, but the probability of such an editing event occurring will increase exponentially with the frequency of the missense events. As a consequence, the error-prone ribosome will be selectively edited and the errors of its function selectively neutralized.

6.7 Suppression of frameshift mutations

A few years ago, the authors realized that the existence of the putative error coupling mechanism might be verified in a very striking way. A forced increase in the frequency of missense errors should also, in certain instances, phenotypically suppress frameshift mutations through compensatory ribosome frameshifts at the positions of noncognate aminoacyl-tRNA binding; in other words, a phenotypic analogue of the classic Cambridge experiments on the genetic code (Crick *et al.*, 1961).

Missense errors at any one codon family could be forced upward by limiting the supply of the amino acid or aminoacyl-tRNA and compensatory

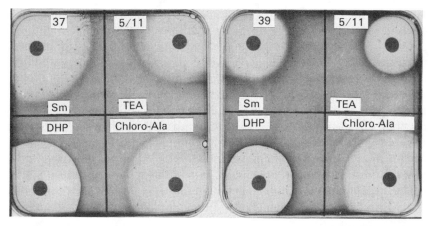

Figure 6.1 Petri plate assay for frameshift suppression induced by aminoacyl-tRNA limitation.

Frameshift mutants of *lacZ* (Newton, 1970) are plated in soft agar (0.6%) on minimal medium plates with 0.2% glucose and appropriate nutritional supplements. The top agar contains 2 mM IPTG and 5 mM cAMP to induce the *lac* operon, and an appropriate concentration (from 40 to 480 μg ml^{-1}) of [5-bromo-4-chloro-3-indolyl-]-D-galactoside (XG), the widely used histochemical stain for *lacZ* enzymatic activity. After the top agar hardens, filter discs containing appropriate quantities of amino acid analogues are placed on the plate, which are incubated at 37°C for one to three days. The analogues inhibit either the synthesis or the activation of their corresponding amino acids, and are listed below. The method provides a continuous gradient of inhibitor concentration with distance from the disc, and thus produces a circular zone of complete growth inhibition surrounded by a zone of partial growth inhibition. In the latter zone, phenotypic suppression is signalled by the blue colour of the hydrolysed and precipitated indigo dye. The isogenic strains used in the illustration carry *lacZ* frameshift mutation 38 (Newton, 1970). The strain on the plate on the left in Fig. 6.1, labelled 37, is *relA*$^+$; the strain on the plate on the right, labelled 39, is *relA*$^-$. The blue ring surrounding the disc of TEA (thienylalanine) in the case of strain 39 corresponds phenotypic frameshifting could be detected through increased leakiness of frameshift mutants, particularly in $relA^-$ bacteria (see Section 6.4 above). It remained to devise a screening programme of adequate scope, as we supposed that such phenotypic suppression might only be detected in rare instances.

The screening programme, undertaken in J. Gallant's laboratory, initially made use of a rapid Petri plate assay which permitted the testing of seventeen different *lacZ* frameshift alleles for suppression during limitation for each of fourteen different amino acids. Strong phenotypic suppression was detected in twenty-four of these 238 tests. Figure 6.1 illustrates the assay method.

Fig. 6.2 shows the map locations of suppressible and non-suppressible alleles. The typical positive result recorded visually in Fig. 6.1 corresponds to about an order of magnitude increase in leakiness, by assay of *lacZ* enzymic activity (Gallant and Foley, 1980).

Subsequently, Weiss (1983) extended this analysis to the $r_{II}B$ gene of phage T4, and discovered fourteen cases of strong phenotypic suppression amongst 107 tests. As predicted, the effect was specific to the $relA^-$ genotype (Weiss and Gallant, 1983), just as it was in the case of *lacZ* frameshift suppression (Gallant and Foley, 1980).

The highly developed genetics of the $r_{II}B$ gene enabled Weiss and Gallant (1983) to demonstrate that phenotypic suppression of a (+) frameshift allele during tryptophan limitation depended on the presence of a specific tryptophan (UGG) codon a short distance upstream. This dependence almost certainly means that suppression reflects a phenotypic shift into the (–) frame at the 'hungry' tryptophan codon, and thus supports our hypothesis in detail.

One alternative interpretation is that frameshifting at hungry codons is not

---

to a twentyfold increase in leakiness, by conventional β-galactosidase assay of cells cultivated in liquid medium (Gallant and Foley, 1980).

The inhibitors used in the experiment illustrated in Fig. 6.1 and their amino acid specificities, were as follows (those which produced phenotypic suppression with one or more mutants are underlined): O-methylthreonine (Ile); 3-fluorotyrosine (Try); 3,4-dehydroproline (Pro); 6-fluorotryptophan or indoleacrylic acid (Trp); lysine-hydroxamate (Lys); β-2-thienylalanine (Phe); aminoethylphosphonic acid (Ala); serine-hydroxamate (Ser); 1, 2, 4-triazole (Cys); azaleucine (Leu); methionine-hydroxamate (Met); α-methylhistidine (His); α-aminobutyric acid (Val); β-hydroxy-norvaline (Thr). For each inhibitor which elicited strong phenotypic suppression of one or more mutant, the amino acid specificity was confirmed by showing that the appropriate normal amino acid competed out the effect. In a few cases, the specificity was checked further by testing more than one inhibitor specific for the same amino acid: mutants which responded on 6-fluorotryptophan (Trp) also responded to 5-methyltryptophan (Trp); mutants which responded to O-methylthreonine (Ile) also responded to valine (Ile). One of the inhibitors which produced no positive responses was also checked in this way: hydroxyleucine (Leu) proved negative for all the *lacZ* mutants, and so did azaleucine (Leu).

Frameshift (positive)	18 29	55 65 85		38	36 36–16		61		24						
Deletion segments of lacZ	1	2	3	4	5	6	7	8	9	10	11	12	13	14	15	16
Frameshift (negative) 91		28 68 10					21			40			17			

Figure 6.2 Map locations of suppressible and non-suppressible frameshift alleles of *lacZ*. The central line is the series of segments of the *lacZ* gene defined by standard overlapping deletions; each frameshift allele is listed in the segment to which it maps (Newton, 1970). Those above the line in large letters exhibited strong phenotypic suppression during limitation for one or another aminoacyl-tRNA species. Those below the line in smaller characters exhibited either no or much weaker phenotypic suppression during limitation for all of the 14 amino acids (or aminoacyl-tRNAs) surveyed. One mutant (36–14) gave an intermediate response and is not listed in the figure.

a consequence of noncognate aminoacyl-tRNA binding, as we postulate, but rather of the increased time the ribosome is stalled with an empty A-site. Were this the mechanism, then tetracycline antibiotics, which inhibit ternary complex binding to the A-site, would also provoke phenotypic suppression of responsive frameshift alleles. However, partially inhibitory levels of chlortetracycline produced no suppression of any of the *lacZ* frameshift alleles which did respond to partial aminoacyl-tRNA limitation (Gallant, unpublished experiments).

Moreover, Atkins *et al.* (1979) showed that direct addition of certain purified tRNA species stimulated frameshifting during translation of phage MS2 RNA *in vitro*. Here, the pool bias which increases the frequency of noncognate aminoacyl-tRNA binding is achieved by increasing the concentration of the noncognate species, rather than by limiting that of the cognate species, and thus ribosome stalling is not involved.

6.8 Modalities of error coupling

A deeper understanding of the molecular basis of error coupling demands identification of 'shifty' sites at which noncognate aminoacyl-tRNAs provoke a frameshift error, and identification of the noncognate species responsible at each site. In the light of the aforementioned study by Weiss and Gallant (1983), sites of frameshifting induced by aminoacyl-tRNA limitation are directly identified as codons calling for the limiting type. To identify the shifter tRNA, Weiss and Gallant sought to redress the pool imbalance between noncognate and cognate species by double amino acid (or aminoacyl-tRNA) limitation. They found that trp-limited phenotypic suppression at one site was abolished by simultaneous limitation for Leu-tRNA, but was essentially unaffected by secondary limitation for each of seven other species. The specificity of this result strongly suggests that a leucyl-tRNA is the culprit in this instance.

The aforementioned experiments of Atkins *et al.* (1979), which depend on tRNA addition, directly identify shifter tRNAs. To identify the shifty sites, Atkins and co-workers also sought to redress the pool imbalance, in this case by systematically *adding* a second purified tRNA. They found that frame-shifting on message for MS2 coat protein induced by a particular Ser-tRNA was specifically antagonized by a particular Ala-tRNA; and that frameshifting on MS2 replicase induced by a particular Thr-tRNA was antagonized by a particular Pro-tRNA. Atkins and co-workers drew the reasonable inference that the competing species were those cognate to the positions of induced frameshifting, and thus the codon specificities of the competitors identified the codons at which the shifts occurred.

The methods of aminoacyl-tRNA limitation and tRNA addition are both to some degree inferential. However, the two methods have been cross-checked in the MS2 system, and have been found to confirm one another in each of four cases (Gallant *et al.*, 1985). It is therefore reasonably safe to conclude that either method correctly identifies shifty sites and shifter tRNAs. Reassured by this concordance, Gallant *et al.* (1985) have recently used the double limitation method to identify two more putative pairs of shifty sites and their counterpart shifter tRNAs.

The results of all these investigations are summarized in Table 6.1. Two aspects of the table are noteworthy. In each case, the first base of the shifty codon is identical to the second base normally recognized by the shifter tRNA. This correspondence is surely no coincidence: the probability that it would occur by chance in seven cases is $(1/4)^7 = 6.1 \times 10^{-5}$. Moreover, the first two bases of each shifty codon might be identical to the second two bases recognized by the shifter tRNA (there is some ambiguity due to third position degeneracy).

These regularities were recognized by Weiss, after the first three cases listed in Table 6.1 were documented, and led him to propose a speculative molecular model of error coupling (Weiss, 1984). We might note that the next four cases listed in Table 6.1 were established afterwards, and therefore should be viewed as confirmations of the model's general predictions. The tenets of the Weiss model are as follows. First, each identified shifty codon

Table 6.1 Shifty codons and shifter tRNAs. Frameshifting in translation of $r_{II}B$ was assessed by phenotypic suppression of nearby frameshift mutant alleles; the amino acid specificities of the shifty sites and putative shifter tRNAs were assessed by aminoacyl-tRNA limitation *in vivo* (Weiss and Gallant, 1983; Gallant *et al.*, 1985). Frame-shifting in translation of MS2 *in vitro* was assessed by formation of the elongated, frameshift variants of coat and replicase described by Atkins *et al.* (1979). Shifter tRNAs were identified by direct addition of the purified tRNA, as described in the latter report. The shifty sites were identified by competition with purified tRNAs which are presumably cognate to the codon at which the shifter tRNA acts, as described in the text, and this inference was confirmed in each case by amino acid limitation *in vitro*.

Case	Shifty codon	Shifter tRNA	Bases translocated	Direction of frameshift	Gene	Reference
A	UGG	Leu (CUX, UUA/G)	4	(−)	T4 r_{II}B	Weiss and Gallant, 1983
B	GCA	Ser$_3$(AGU/C)	2	(+)	MS2 Coat	Atkins *et al.*, 1979
C	CCC/U	Thr$_{maj}$(ACU/C)	2	(+)	MS2 Synthetase	Atkins *et al.*, 1979;
					MS2 Coat	Gallant *et al.*, 1985
D	CCC/U	Thr$_{maj}$(ACU/C)	2	(+)	MS2 Coat	Gallant *et al.*, 1985
E	UCA	Leu$_3$(CUU/C)	2	(+)	MS2 Coat	Gallant *et al.*, 1985
F	AAA/G	Gln (CAA/G)	4	(−)	T4 r_{II}B	Gallant *et al.*, 1985
G	AAA/G	Glu (GAA/G)	2	(+)	T4 r_{II}B	Gallant *et al.*, 1985

can associate with its counterpart shifter tRNA by means of an offset pairing shifted by one base to the 5' side of the normal anticodon at nucleotides 34, 35 and 36 of the tRNA. In this offset pairing configuration, nucleotides 33, 34 and 35 comprise the anticodon; the universal U at position 33 becomes the wobble position of the offset anticodon.

An earlier analysis of frameshift suppressor tRNAs which have an extra base in the anticodon suggested the idea that some lateral play is possible in the triplet which functions as the anticodon (Kurland, 1979). Weiss extended this idea to normal tRNAs, paired in the postulated offset fashion, and, more importantly, specified the details of its possible effect on the translocation step.

The conventional picture of translocation is that messenger RNA is dragged along to a new position through its association with the tRNA in the movement from the ribosome's A-site to the P-site. If the mRNA is paired one base up the 5' side of the anticodon loop, as in the offset pairing model, then it will be swung to an abnormal position, supplying an obvious explanation of the abnormal translocation which underlies phenotypic frameshifting.

This picture of the translocation process seemingly predicts only a single mode of abnormal translocation as a consequence of offset pairing, one which corresponds to a (+) frameshift. However, the cases of phenotypic frameshifting we have discovered include shifts in both the (+) and (−) direction.

The solution to this dilemma may lie in the 'reciprocating ratchet' model of translocation (Woese, 1970). In Woese's heterodox model – long ignored by the ribosome confraternity – the tRNA as a whole does not move relative to the ribosome in the transition from the A- to the P-state. The only movement is a stacking transition within the anticodon loop, which has the effect of inverting the codon:anticodon pair with respect to the helical axis of the anticodon stem. Weiss analysed space-filling models of this form of translocation and discovered that offset pairing could lead to either of two aberrations of the stacking transition, the geometry of which correspond plausibly to a (+) frameshift in one case or a (−) frameshift in the other.

From this point of view, our hypothesis of error coupling and Woese's unconventional view of translocation turn out to be intimately connected by way of Weiss's model of ribosome frameshifting. Critical tests of this entire picture, through construction of tRNA molecules with specific alterations of the anticodon loop, are now within the realm of possibility.

Two other puzzles remain to be explained. First, amino acid limitation or tRNA addition provokes frameshifting at only a few of many sites which have been tested. There are evidently strong context constraints governing both the occurrence of error coupling and, perhaps surprisingly, the direction of the frameshift error. Analysis of UGG codons in the $r_{II}B$ gene indicates that the identity of the 3' neighbouring base plays an important role in determining stiffness (Weiss *et al.*, 1984; Weiss and Gallant). This is reminiscent of the strong effect of 3' context on nonsense suppressibility (Bossi, 1983; Miller and Albertini, 1983), and the basic mechanism is equally mysterious.

Second, aminoacyl-tRNA limitation also leads to phenotypic suppression of the certain nonsense alleles, and this phenomenon, like frameshift suppression, is specific to the *relA⁻* genotype (Gallant *et al.*, 1982). Such nonsense suppression events could be accounted for in terms of a multiplicity of reading frameshifts that lead the ribosome through the nonsense codon out-of-phase and then re-establish the correct reading frame. However, we would expect such multiple events to occur at much lower frequencies than the single frameshift suppression events. They do not and this is not easily explained.

One possible explanation is that reading frameshifts normally occur at a particularly high frequency within (or adjacent to?) nonsense codons. Indeed, the classic monograph on frameshift mutants (Barnett *et al.*, 1967) includes precisely this suggestion. The authors were led to this hypothesis by certain anomalies in the properties of double frameshift mutant combinations which geneticists have wisely concealed from a generation of students. Whatever the ultimate account turns out to be, the phenotypic suppression of nonsense codons under conditions of aminoacyl-tRNA limitation provides a vivid illustration of our central thesis, that different kinds of translation errors are coupled to each other.

6.9 Concluding remarks

It will be evident to the reader that we have devoted relatively little space in our discourse to the 'nuts and bolts' of translation. Instead we have focused attention on those aspects of the system behaviour of the translation apparatus that influence the accuracy of its function. This choice of emphasis was dictated by two considerations. One is the conviction that much of the accepted lore concerning the structural basis of ribosome function is ripe for experimental revision. This is so because most of the relevant *in vitro* studies have been carried out in the past without any regard for the dynamic character of protein synthesis and under conditions that demonstrably obliterate the accuracy of the process.

A second motivation for our oblique approach is more positive. We wish to demonstrate that the translation apparatus is not designed simply to carry out the accurate production of proteins under stable, optimal conditions. Instead, we hope to have illustrated the idea that the dynamic organization of the system permits it to solve a variety of problems created by what we may loosely call the uncertainties of life. Thus, the accuracy of translation must be maintained within a background of accumulated errors as well as under transient conditions that limit the availability of substrates.

These and other boundary conditions are likely to have influenced the particular ways that the translation system has evolved. In other words, we cannot hope to understand the principles of ribosome construction until we understand the variety of problems it solves.

References

Andersson, D. and Kurland, C. G. (1983) Ram ribosomes are defective proofreaders. *Mol. Gen. Genet.*, **191**, 378–381.

Andersson, S. G. E., Buckingham, R. H. and Kurland, C. G. (1984) Does codon-composition influence ribosome function? *EMBO J.*, **3**, 91–94.

Atkins, J. F., Elseviers, D. and Gorini, L. (1972) Low activity galactosidase in frameshift mutants of *Escherichia coli. Proc. Natl Acad. Sci. USA*, **69**, 1192–1195.

Atkins, J. F., Gesteland, R. F., Ried, B. R. and Anderson, C. W. (1979) Normal tRNAs promote ribosomal frameshifting. *Cell*, **18**, 1119–1131.

Barnett, L., Brenner, S., Crick, F. H. C., Shulman, R. G. and Watts-Tobin, R. J. (1967) Phase-shift and other mutants in the first part of the $r_{II}B$ cistron of bacteriophage T4. *Phil. Trans. Royal Soc. B*, **252**, 387–560.

Blomberg, C. and Ehrenberg, M. (1981) Energy considerations for kinetic proofreading in biosynthesis. *J. Theor. Biol.*, **88**, 631–670.

Blomberg, C., Ehrenberg, M. and Kurland, C. G. (1980) Free-energy dissipation constraints of the accuracy of enzymatic selections. *Quart. Rev. Biophys.*, **13**, 231–254.

Böck, A., Fairman, L. E. and Neidhardt, F. C. (1966) Biochemical and genetic characterization of a mutant of *Escherichia coli* with a temperature-sensitive valyl ribonucleic acid synthetase. *J. Bacteriol.*, **92**, 1076–1082.

Bohman, K. T., Ruusala, T., Jelenc, P. C. and Kurland, C. G. (1984) Hyper-accurate ribosomes inhibit growth. *Mol. Gen. Genet.*, **198**, 90–99.

Bossi, L. (1983) Context effects: translation of UAG codon by suppressor tRNA is affected by the sequence following UAG in the message. *J. Mol. Biol.*, **164**, 73–87.

Bossi, L. and Roth, J. R. (1980) The influence of codon contest on genetic code translation. *Nature*, **286**, 123–127.

Bouadloun, F., Donner, D. and Kurland, C. G. (1983) Codon-specific missense errors *in vivo. EMBO J.*, **2**, 1351–1356.

Buckingham, R. H. and Kurland, C. G. (1977) Codon specificity of UGA suppressor tryptophan tRNA from *E. coli. Proc. Natl Acad. Sci. USA*, **74**, 5496–5499.

Crick, F. H. C. (1958) On protein synthesis. *Symp. Soc. Exp. Biol. XII*, 138–163.

Crick, F. H. C. (1966) Codon-anticodon pairing: the wobble hypothesis. *J. Mol. Biol.*, **19**, 548–555.

Crick, F. H. C., Barnett, L., Brenner, S. and Watts-Tobin, R. J. (1961) The general nature of the genetic code for proteins. *Nature*, **192**, 1227–1232.

De Wilde, M., Cabezon, T., Herzog, A. and Bollen, A. (1977) Apport de la génétique à la connaissance du ribosome bactérien. *Biochimie*, **59**, 125–140.

Edelmann, P. and Gallant, J. (1977) Mistranslation in *E. coli. Cell*, **10**, 131–137.

Ehrenberg, M. and Blomberg, C. (1980) Thermodynamic constraints on kinetic proofreading in biosynthetic pathways. *Biophys. J.*, **31**, 333–358.

Ehrenberg, M. and Kurland, C. G. (1984) Costs of accuracy determined by a maximal growth rate constraint. *Quart. Rev. Biophys.*, **17**, 45–82.

Ellis, N. and Gallant, J. (1982) An estimate of the global error frequency in translation. *Mol. Gen. Genet.*, **188**, 169–172.

Fluck, M. M., Salser, W. and Epstein, R. H. (1977) The influence of the reading context upon the suppression of nonsense codons. *Mol. Gen. Genet.*, **151**, 137–149.

Gallant, J. (1979) Stringent control in *E. coli. Ann. Rev. Genetics*, **13**, 393–415.

Gallant, J. and Foley, D. (1980) On the causes and prevention of mistranslation. In *Ribosomes, Structure and Function, and Genetics* (eds G. Chambliss, G. R. Craven, J. Davies, K. Davis, L. Kahan and M. Nomura), University Park Press, Baltimore, pp. 615–638.

Gallant, J., Erlich, H., Weiss, R., Palmer, L., and Nyari, L. (1982) Nonsense suppression in aminoacyl-tRNA limited cells. *Mol. Gen. Genet.*, **186**, 221–226.

Gallant, J. and Palmer, L. (1979). Error propagation in viable cells. *Mech. Ageing Dev.*, **10**, 27–38.

Gallant, J., Weiss, R., Murphy, J. and Brown, M. (1985) Some puzzles of translational accuracy. In *The Molecular Biology for Bacterial Growth* (eds M. Schaechter, F. Neidhardt and J. Ingraham), Barlett and Jones, Boston, Mass.

Garen, A. and Siddiqi, O. (1962) Suppression of mutations in the alkaline phosphatase structure cistron of *E. coli. Proc. Natl Acad. Sci. USA*, **48**, 1121–1126.

Gorini, L. (1974) Streptomycin and misreading of the genetic code. In *Ribosome* (eds M. Nomura, A. Tissieres, and P. Lengy) Cold Spring Harbor Laboratory, New York, pp. 791–804.

Hall, B. and Gallant, J. (1972) Defective translation in RC⁻ cells. *Nature New Biol.*, **237**, 131–135.

Hirsh, D. (1973) Tryptophan transfer RNA as the UGA suppressor. *J. Mol. Biol.*, **58**, 439–458.

Hopfield, J. J. (1974) Kinetic proofreading. *Proc. Natl Acad. Sci. USA*, **71**, 4135–4141.

Jelenc, P. C. and Kurland, C. G. (1979) Nucleoside triphosphate regeneration decreases the frequency of translation errors. *Proc. Natl Acad. Sci. USA*, **76**, 3174–3178.

Jelenc, P. C. and Kurland, C. G. (1984) Multiple effects of kanamycin on translational accuracy. *Mol. Gen. Genet.*, **194**, 195–199.

Kaziro, Y. (1978) The role of guanosine 5′-triphosphate in polypeptide chain elongation. *Biochim. Biophys. Acta*, **505**, 95–112.

Kornberg, A. (1969) Active Center of DNA polymerase. *Science*, **163**: 1410–1414.

Kuhlberger, R., Piepersberg, W., Petzet, A., Buckel, P. and Böck, A. (1979) Alteration of ribosomal protein L6 in gentamycin resistant strains *Escherichia coli*. Effects on fidelity of protein synthesis. *Biochemistry*, **18**, 187–193.

Kurland, C. G. (1978) Role of guanine nucleotide in protein biosynthesis. *Biophys. J.*, **22**, 373–392.

Kurland, C. G. (1979) Reading frame errors on ribosomes. In *Nonsense Mutations and tRNA Suppressors* (eds J. E. Celis and J. D. Smith), Academic Press, London, pp. 97–108.

Kurland, C. G. (1983) Accuracy strategies. *Biochem. Soc. Symp.*, **47**, 1–9.

Kurland, C. G. and Ehrenberg, M. (1984). Optimization of translational accuracy. *Progr. Nucl. Acid Res. Mol. Biol.*, **31**, 191–219.

Kurland, C. G., Andersson, D. I., Andersson, S. G. E., Bohman, K., Bouadloun, F., Ehrenberg, M., Jelenc, P. C. and Ruusala, T. (1984) Translational accuracy and bacterial growth. In *Alfred Benzon Symposium 19, Gene expression* (eds B. F. C. Clark and H. V. Peterson), Munksgaard, Copenhagen, pp. 193–207.

Kurland, C. G., Rigler, R., Ehrenberg, M. and Blomberg, C. (1975) Allosteric mechanism for codon-dependent tRNA selection on ribosomes. *Proc. Natl Acad. Sci. USA*, **72**, 4248–4251.

Laffler, T. and Gallant, J. (1974) Stringent control of protein synthesis in *E. coli. Cell*, **3**, 47–49.

Langridge, J. (1968) Thermal responses of mutant enzymes and temperature limits to growth. *Mol. Gen. Genet.*, **103**, 116–126.

Lipmann, F. (1969) Polypeptide chain elongation in protein biosynthesis. *Science*, **164**, 1024–1031.

Loftfield, R. and Vanderjagt, D. (1972) The frequency of errors in protein synthesis. *Biochem. J.*, **128**, 1353–1356.

Manley, J. L. (1979) Synthesis and degradation of termination and premature termination fragments of β-galactosidase *in vitro. J. Mol. Biol.*, **125**, 407–432.

Menninger, J. (1978) The accumulation as peptidyl-transfer RNA of isoaccepting transfer RNA families in *E. coli* with temperature-sensitive peptidyl transfer RNA hydrolase. *J. Biol. Chem.*, **253**, 6808–6813.

Miller, J. H. (1979) Genetic studies of the lac repressor: XI On aspects of the *lac* repressor structure suggestion of genetic experiments. *J. Mol. Biol.*, **131**, 249–258.

Miller, J. H. and Albertini, A. M. (1983) Effects of surrounding sequence on the suppression of nonsense codons. *J. Mol. Biol.*, **164**, 59–71.

Newton, A. (1970) Isolation and characterization of frameshift mutations in the *lac* operon. *J. Mol. Biol.*, **49**, 589–601.

Ninio, J. (1974) A semiquantitative treatment of missense and no sense suppression in the *str* A and *ram* ribosomal mutants of *E. coli. J. Mol. Biol.*, **84**, 297–313.

Ninio, J. (1975) Kinetic amplification of enzyme discrimination. *Biochimie*, **57**, 487–595.

O'Farrell, P. H. (1978) The suppression of defective translation by ppGpp and its role in the stringent response. *Cell*, **14**, 545–557.

Orgel, L. E. (1963) The maintenance of the accuracy of protein synthesis and its relevance to ageing. *Proc. Natl Acad. Sci. USA*, **49**, 517–521.

Parker, J., Johnston, T. C. and Borgia, P. T. (1980) Mistranslation in cells infected with bacteriophage MS2: direct evidence of Lys for Asn substitution. *Mol. Gen. Genet.*, **180**, 275–281.

Parker, J., Flanagan, J., Murphy, J. and Gallant, J. (1981) On the accuracy of protein synthesis in *Drosphila melanogaster. Mech. Ageing Dev.*, **16**, 127–139.

Pauling, L. (1957) *In Festschrift Arthur Stoll*, Birkhauser Verlag, Basel, pp. 597–602.

Pettersson, I. and Kurland, C. G. (1980) Ribosomal protein L7/L12 is required for optimal translation. *Proc. Natl Acad. Sci. USA*, **77**, 4007–4010.

Picard-Bennoun, M. (1976) Genetic evidence for ribosomal anti-suppressions in *Podospora anserina. Mol. Gen. Genet.*, **147**, 299–306.

Piepersberg, W., Geyl, D., Buclel, P. and Böck, A. (1979) Studies on the coordination of tRNA charging and polypeptide synthesis activity in *E. coli*. In *Regulation of Macromolecular Synthesis by Low Molecular Weight Mediators* (eds G. Koch and D. Richter), Academic Press, New York.

Pribnow, D., Sigurdson, D. C., Gold, L., Singer, B. S., Napoli, C., Brosius, J., Dull T. J. and Noller, H. F. (1981) r_{II} cistrons of bacteriophage T4: DNA sequence around the intercistronic divide and positions of genetic landmarks. *J. Mol. Biol.*, **149**, 337–376.

Riddle, D. and Carbon, J. (1973) Frameshift suppression: A nucleotide addition in the anticodon of a glycine transfer RNA. *Nature New Biol.*, **242**, 230–234.

Rojas, A.-M., Ehrenberg, M., Andersson, S. G. E. and Kurland, C. G. (1984) ppGpp inhibition of elongation factors Tu, G and Ts during polypeptide synthesis. *Mol. Gen. Genet.*, **197**, 36–45.

Ruusala, T., Ehrenberg, M. and Kurland, C. G. (1982a) Catalytic effects of elongation factor Ts on polypeptide synthesis. *EMBO J.*, **1**, 75–78.

Ruusala, T., Ehrenberg, M. and Kurland, C. G. (1982b) Is there proofreading during polypeptide synthesis? *EMBO J.*, **1**, 741–745.

Ruusala, T., Andersson, D., Ehrenberg, M. and Kurland, C. G. (1984) Hyper-accurate ribosomes inhibit growth. *EMBO J.*, **3**, 2575–2580.

Ruusala, T. and Kurland, C. G. (1984) Streptomycin perturbs preferentially ribosomal proofreading. *Mol. Gen. Genet.*, **198**, 100–104.

Sambrook, J. F., Fan, D. P. and Brenner, S. (1967) A strong suppressor specific for UGA. *Nature*, **214**, 452–453.

Spirin, A. S. (1978) Energetics of the ribosome. *Prog. Nucl. Acids Res. Mol. Biol.*, **21**, 39–63.

Thompson, R. C. and Stone, P. J. (1977) Proofreading of the codon-anticodon interaction on ribosomes. *Proc. Natl Acad. Sci. USA*, **74**, 198–202.

Thompson, R. C., Dix, D. B., Gerson, R. B. and Karin, A. M. (1981) A GTPase reaction accompanying the ejection of Leu-tRNA by UUU-programmed ribosomes. *J. Biol. Chem.*, **256**, 81–86.

Wagner, E. G. H. and Kurland, C. G. (1980) Translational accuracy enhanced *in vitro* by (p)ppGpp. *Mol. Gen. Genet.*, **180**, 139–145.

Wagner, E. G. H., Jelenc, P. C., Ehrenberg, M. and Kurland, C. G. (1982a) Rate of elongation of polyphenylalanine *in vitro*. *Eur. J. Biochem.*, **122**, 193–197.

Wagner, E. G. H., Ehrenberg, M. and Kurland, C. G. (1982b) Kinetic suppression of translational errors by (p)ppGpp. *Mol. Gen. Genet.*, **185**, 269–274.

Weiss, R. and Gallant, J. (1983) Mechanism of ribosome frameshifting during translation of the genetic code. *Nature*, **302**, 389–393.

Weiss, R. (1983) Mechanism of ribosome frameshifting during translation of the genetic code. PhD Thesis, University of Washington.

Weiss, R. (1984) Molecular model of ribosome frameshifting. *Proc. Natl Acad. Sci. USA*, **81**, 5797–5801.

Weiss, R., Murphy, J., Wagner, G. and Gallant, J. (1984) The ribosome's frame of mind. In *Alfred Benzon Symposium 19: Gene Expression, the Translational Step and its Control* (eds B. F. C. clark and H. U. Peterson), Munksgaard, Copenhagen, pp 208–220.

Woese, C. (1970) Molecular mechanism of translation: a reciprocating ratchet mechanism. *Nature*, **226**, 817–820.

Yegian, C. D. and Stent, G. S. (1969) An unusual condition of leucine transfer RNA appearing during leucine starvation of *Escherichia coli*. *J. Mol. Biol.*, **39**, 45–58.

7 The accuracy of RNA synthesis

R. P. ANDERSON and J. R. MENNINGER

7.1 Introduction

For the normal metabolism of cells it is essential to transfer the genetic information contained in a DNA sequence into an RNA molecule. In order for the information to be faithfully expressed, the DNA sequence must be accurately transcribed and the primary RNA product molecule must be correctly processed and modified in order to produce a tRNA, mRNA, rRNA or other mature RNA.

This chapter will review some of the research bearing on accuracy during initiation, elongation and termination of RNA synthesis and during post-transcriptional processing. The discussion of accuracy during the polymerization of RNA will include estimates of the fidelity of transcription *in vitro* and *in vivo*. The context of bases in the template DNA, the structure of RNA polymerase, and various cofactors have been shown to influence accuracy during the polymerization of RNA; these will also be discussed. The last part of our review will focus on how accuracy is maintained during post-polymerization processing. For the purpose of analysing the principles involved two examples of how RNA is processed will be presented: (1) mRNA splicing; (2) maturation of the 3' terminus of mRNA.

7.2 Accuracy during RNA polymerization

Errors during the synthesis of RNA, although not as permanent and damaging as errors during DNA replication, can have significant physiological consequences in the cell. An error during the polymerization of RNA will alter its primary sequence and could also alter the secondary or tertiary structure of the resulting macromolecule. An error in the primary structure of an mRNA could alter several sequence-dependent functions, including a variety of post-transcriptional RNA processing events, the initiation or termination of protein synthesis, or the amino acid sequence of a completed protein. Errors that change its secondary or tertiary structure can affect the function of an RNA. For example, tRNAs require an appropriate conformation both to be properly processed to a mature form (Altman, 1981) and for specific bases within the tRNA to be properly modified (Agris and Söll, 1977).

The frequency of misincorporating ribonucleotides into RNA during polymerization must be at least as low as the observed frequency of amino acids misincorporated into completed proteins. Therefore, it is possible to estimate the maximal error frequency for the polymerization of RNA if one assumes that all the errors observed in completed proteins are the result of failures to transcribe faithfully the DNA, rather than failures of the translational process. The observed frequency of erroneous amino acids in completed proteins should be equal to the product of the probability per elongated base of an erroneous ribonucleoside triphosphate being incorporated into a nascent RNA, the number of bases required to code for an amino acid, and the probability that such a transcriptional error will sufficiently alter the bases so they code for a new amino acid. The frequency of errors taken as typical for completed proteins is $20\text{--}40 \times 10^{-5}$ per amino acid position (Loftfield and Vanderjagt, 1972; Edelmann and Gallant, 1977; Ellis and Gallant, 1982). If one assumes that all transcriptional errors occur at the same frequency and that only the first two nucleotides of a codon are responsible for determining an amino acid, the maximum expected transcriptional error frequency would be $10\text{--}20 \times 10^{-5}$. Even though this value is based on several simplifying assumptions and even though the observed frequency of translational errors can vary significantly with the type or context of the error analysed (Bouadloun, Donner and Kurland, 1983; Parker *et al.*, 1983; Johnston, Borgia and Parker, 1984), this estimate of the expected transcriptional error frequency provides a useful framework in which to evaluate the experimental data.

7.2.1 ACCURACY DURING POLYMERIZATION OF RNA *in vitro*

Springgate and Loeb (1975) analysed the *E. coli* DNA-dependent RNA polymerase for its accuracy when incorporating ribonucleoside triphosphates (rNTPs) *in vitro* into RNA. The fidelity of the process was determined by comparing the incorporation of complementary and non-complementary nucleotides, using poly $[d(A\text{--}T) \cdot d(A\text{--}T)]$ and poly $[d(C) \cdot d(G)]$ templates. The probability of erring was estimated as the molar ratio of non-complementary to complementary precursors incorporated into the polyribonucleotide product. Erroneous base pairing during transcription was found to be infrequent. The highest error ratios observed for CTP and GTP incorporated into RNA synthesized by *E. coli* RNA polymerase transcribing a poly $[d(A\text{--}T) \cdot d(A\text{--}T)]$ template were 42×10^{-5} and 2.4×10^{-5}, respectively. An analysis of the nearest neighbour frequency in the RNA product transcribed from the poly $[d(A\text{--}T) \cdot d(A\text{--}T)]$ template revealed that the incorporation of CMP in place of UMP was the more frequent error. For ATP and UTP incorporated into RNA during the transcription of a poly $[d(C) \cdot d(G)]$ template the highest error ratios observed were 11×10^{-5} and 5.0×10^{-5}, respectively.

E. coli DNA polymerase I was used to synthesize the templates, under conditions that should have minimized contamination of the product DNA by bases complementary to the incorrect rNTP to be used in the fidelity assay. The infidelity of DNA polymerase I has been observed to be less than 1×10^{-5} for the incorporation of non-complementary dGTP (Radding and Kornberg, 1962). This suggests that the misincorporation of rNTPs occurred during elongation of the RNA chain.

The reported data on the accuracy of base selection during transcription by *E. coli* DNA-dependent RNA polymerase vary substantially. The error ratios determined for the incorporation of GTP into a polyribonucleotide product transcribed from a poly $[d(A-T) \cdot d(A-T)]$ template include 120×10^{-5} (Ozoline, Oganesjan and Kamzolova, 1980), 100×10^{-5} (Bick, 1975) and 2.4×10^{-5} (Springgate and Loeb, 1975). When poly (dT) was used as a template, the error ratios determined for the non-complementary incorporation of GTP included 3300×10^{-5} (Strniste, Smith and Hayes, 1973), and 59×10^{-5} (Bass and Polonsky, 1974). Several factors might account for these varying results. There were differences in the methods used to prepare RNA polymerase, the methods used to remove unincorporated nucleotides from the reaction mixture, the temperatures of incubation, and the concentrations of monovalent and divalent cations. The more useful values for error rates might be those of Ozoline *et al.* (1980), Bick (1975) and Springgate and Loeb (1975), since conditions in their assays most accurately reflected physiological conditions. Only the values reported by Springgate and Loeb (1975), however, are less than the maximum expected from the observed error rates in proteins.

7.2.2 ACCURACY DURING POLYMERIZATION OF RNA *in vivo*

Estimates of transcriptional error rates *in vivo* are needed as a frame of reference for the studies of transcription fidelity *in vitro*. Experiments using *E. coli* cells have recently measured the transcriptional error rate at a specific codon in the *lac* operon (Rosenberger and Foskett, 1981). Transcriptional error frequencies were measured using an extremely polar *lacZ* mutant which contains a nonsense codon at amino acid position 23 of the β-galactosidase gene. Complete transcripts of *lacZ* mRNA are produced at very low levels in extremely polar mutants since ribosomes that accurately translate the nonsense codon dissociate, which leads to premature termination of transcription. Only a fragment of the *lac* mRNA is transcribed and that is probably degraded rapidly.

Control experiments suggested that translational errors make an insignificant contribution to the residual β-galactosidase activity. It was assumed that any observed enzyme levels in the strains with the extremely polar *lacZ* mutation could only be the result of transcriptional errors, which result in changing the nonsense codon to one that codes for an amino acid. This would prevent premature translation termination and allow the synthesis of a complete mRNA. Several independent experiments demonstrated that amino acid substitutions within the first twenty-seven residues of β-galactosidase do not greatly reduce its activity. Rosenberger and Foskett made the assumption that amino acid substitution errors, due to transcriptional erring at the nonsense codon, would produce β-galactosidase molecules that function equivalently to those produced by the wild type strain. Based on these assumptions, the transcriptional error rate at the nonsense codon was estimated to be 14×10^{-5}. This value is approximately equivalent to that of the most common transcriptional error measured *in vitro* by Springgate and Loeb (1975), the substitution of C for U at a rate of 42×10^{-5}.

Using a similar method, Rosenberger and Hilton (1983) were able to estimate the transcriptional error rate at two additional strongly polar *lacZ* mutations. By assaying the kinetics of β-galactosidase induction, they were also able to demonstrate that the residual enzyme activity in both strongly polar mutants and in one out of three non-polar *lacZ* nonsense mutants was primarily the result of transcriptional errors. Assuming that the residual β-galactosidase activity is completely the result of transcriptional errors and that errors at these sites do not affect the activity of the error-containing molecules, the frequencies of transcriptional errors at the polar alleles U118[UAA] and U118[UAG] were 0.85×10^{-5} and 8.8×10^{-5}, respectively. For the non-polar allele 625[UAG] the frequency of transcriptional errors was 25×10^{-5}.

The frequencies of errors made by *E. coli* RNA polymerase during transcription *in vivo* at a nonsense codon are close to the error rates determined *in vivo* for the reversion frequency of an extracistronic mutant of the RNA phage Qβ (Batschelet, Domingo and Weissmann, 1976). A Qβ phage with an A-to-G mutation near the 3′ terminus was serially propagated for ten or more infection cycles, either in a pure culture or a culture mixed with an equal number of plaque-forming units of wild type phage, to analyse the effect of intracellular competition between the wild type and mutant phage (Domingo, Flavell and Weissmann, 1976). Batschelet *et al.* (1976) used these results, along with a mathematical model which describes the competition of a mixed phage population as a function of the mutation and growth rates of the wild type and mutant, to estimate the error rate of Qβ replicase. By determining the probability of site-specific reversion of the mutant sequence to the wild type sequence the site-specific mutation rate *in vivo* was estimated as 10×10^{-5} per phage doubling.

7.2.3 STRUCTURAL FACTORS THAT INFLUENCE THE ACCURACY OF RNA SYNTHESIS DURING ELONGATION

Both correct and incorrect rNTPs interact with a site on the RNA polymerization complex that consists of the RNA polymerase, the DNA template, and associated cofactors. Specificity during the selection process is a consequence of the free energy difference between the binding of an incorrect rNTP and the binding of a correct rNTP. Correct rNTPs form additional and/or stronger contacts with the polymerization complex and are, therefore, used preferentially for elongation. Such specific interactions are primarily the result of the hydrogen bonds formed between the correct rNTP and the complementary base within the template. Any perturbation that alters either the specific or non-specific interactions between the rNTPs and the polymerization complex would be expected to affect the accuracy of RNA polymerization.

(a) Template context effects
Modifying the bases in the DNA template induces transcriptional errors. These have been used to analyse the importance of the hydrogen bonds between the precursors and the template. Kröger and Singer (1979) constructed template polynucleotides which contained approximately 10% of a base with a modification that would block essential hydrogen bonding sites. Nearest neighbour analysis of the RNA product indicated there was little or no specificity of incorporation exhibited at the position of the modified bases. From this and similar studies, Singer and Spengler (1981) concluded the RNA polymerase itself requires only that the appropriate stacking forces exist between the ribonucleotide precursor and the template base in order for elongation to occur. It is the correct length and number of hydrogen bonds between the RNA polymerase-bound rNTP and the template base that mainly control the specificity of base selection.

The precise effects of template context on accuracy *in vivo* cannot presently be evaluated as sufficient data are not available. One would expect that the accuracy of adding any base to the nascent RNA would be influenced by the adjacent bases because it is known that the accuracy both of DNA replication (involving DNA–DNA interactions) and of nonsense mutant suppression (involving RNA–RNA interactions) are affected by the local context (Fluck, Salser and Epstein, 1977; Fluck and Epstein, 1980). The stability of binding an rNTP to the polymerization complex would be expected to be influenced by stacking forces resulting from the alignment of adjacent bases. Strong stacking forces will reduce the difference in total free energy between the correct and incorrect rNTPs and lead to higher error rates.

It would seem possible to approach the influence of context by assaying its effects on the transcriptional suppression of nonsense mutations, using assays like those of Rosenberger and Foskett (1981). Values of the free energy changes for addition of a specific base pair adjacent to any other base pair within an RNA double helix have been calculated (Tinoco *et al.*, 1973; Salser, 1977). They could be used as a guide to predict the effects on transcriptional accuracy of site-specific mutations immediately preceding a *lacZ* nonsense codon. Provided the most common error was that due to mistranscription at the first residue of the nonsense codon, transcriptional suppression levels would be expected to vary systematically with the context-dependent change in the stability of binding an rNTP at that position.

(b) RNA polymerase effects
The structure of RNA polymerase is also involved in the ribonucleotide selection process during transcription. Mutants of *E. coli* that are rifampicin-resistant have an altered β subunit in their RNA polymerases. These mutant enzymes incorporate non-complementary rNTPs into poly (A–U), synthesized using a poly [d(A–T)·d(A–T)] template, more frequently than does the RNA polymerase isolated from a wild type strain of *E. coli* (Ozoline, Oganesjan and Kamzolova, 1980). The error ratio observed for the incorporation of non-complementary GTP by rifampicin-resistant RNA polymerase was 540×10^{-5}, while the error ratio observed using wild type RNA polymerase was 120×10^{-5}. An independently isolated rifampicin-resistant mutant has been reported to be more prone to transcriptional errors, both *in vitro* and *in vivo* (J. Gallant, personal communication). This mutant RNA polymerase increased the misincorporation of GTP into RNA, directed by a poly [d(A–T)·d(A–T)] template, and increased the residual level of expression of several early polar *lacZ* nonsense mutants. These observations indicate that at least the β subunit of *E. coli* RNA polymerase can influence the accuracy of base selection during transcription.

Springgate and Loeb (1975) analysed the effects of several cofactors of RNA polymerase on the transcriptional error frequency *in vitro*. They observed that the substitution of Mn^{2+} for Mg^{2+} increased the fidelity of RNA synthesis by about 50%. The addition of the sigma factor to the RNA polymerase core enzyme, however, had no significant effect on accuracy.

7.2.4 IS THERE ERROR CORRECTION DURING RNA POLYMERIZATION?

In addition to the structure of RNA polymerase, the rNTPs and the bases in the DNA template, all of which are known to influence the accuracy of RNA polymerization, there is also the possibility that a mechanism might exist for the correction of transcriptional errors. Volloch, Rits and Tumerman (1979) isolated a nucleoside triphosphate phosphohydrolase (NTPase) activity from a preparation of *E. coli* RNA polymerase. They hypothesized that NTPase, acting in conjunction with RNA polymerase, corrects errors during transcription by removing non-complementary rNTPs from the elongation site of RNA polymerase, hydrolysing them to ribonucleoside diphosphates. Since

RNA polymerase must have nucleoside triphosphates in order to elongate a nascent RNA chain, hydrolysis of an NTP should be sufficient to prevent it from participating in elongation. Volloch and coworkers observed that when an RNA polymerase preparation containing the NTPase activity was used in their fidelity assay, non-complementary rNTPs of the same heterocyclic class as the complementary precursor – such as a non-complementary purine substituting for a complementary purine – were hydrolysed to nucleoside diphosphates. When the NTPase activity was absent, similar but incorrect NTPs were not hydrolysed and there was a thirty- to fiftyfold increase in the incorporation of the non-complementary NTP of the same heterocyclic class as the appropriate precursor.

Volloch, Rits and Tumerman (1979) proposed that selection of NTPs could include two tests of appropriateness. The first would evaluate a randomly chosen precursor, via weak interactions, at an entry site common both to correct and incorrect NTPs. An appropriate NTP, one which is complementary to the next template base to be copied, would be selected by its inducing a conformational change in the RNA polymerase, causing a subsequent high affinity binding of the appropriate NTP to the functional elongation site (Chamberlain and Berg, 1964). In the elongation site, the appropriateness of the NTP would be tested a second time, by the NTPase activity. Non-complementary NTPs that occasionally bind tightly to the elongation site of the RNA polymerase would be hydrolysed to ribonucleoside diphosphates, thus preventing their incorporation into the nascent RNA. Those inappropriate NTPs that infrequently become bound tightly to the elongation site and are not hydrolysed to ribonucleoside diphosphates by the NTPase activity would be misincorporated in the RNA product.

The proposed model includes two discrimination steps, one during precursor selection and one that edits erroneous precursors before incorporation into the RNA product. Superficially this is similar to the proofreading models proposed by Hopfield (1974) and Ninio (1975). The proposed model does not, however, meet the formal kinetic requirements for a proofreading mechanism. Volloch and coworkers do not separate their two proposed discrimination steps by a process that is essentially irreversible. In a kinetic proofreading mechanism it is this irreversibility that allows the proofreading step to be independent of the preceding discrimination and thereby gives an overall error frequency that is the product of the error frequencies of the two steps. Volloch and coworkers did not analyse the kinetics of their proposed error correction mechanism.

7.3 Accuracy during initiation of RNA synthesis

This phase of transcription requires binding of RNA polymerase to a special sequence of DNA (called the promoter), 'melting in' by the enzyme to form a stable 'open' complex with separated template strands, binding of two triphosphate precursors, and the formation of the first phosphodiester bond. Only an erroneous 5' terminal dinucleotide (pppXpY) is detectable as a loss of accuracy during initiation, although failures are possible in the DNA sequence specificity of any of these three processes. Later processing of the 5' end may obscure errors in the initiation step(s).

Initiation does not appear to be very accurate, either in prokaryotes or in eukaryotes. Carpousis, Stefano and Gralla (1982) have provided the most careful report of heterogeneity in the 5' terminus; they studied *lac* mRNA transcribed by *E. coli* RNA polymerase (Fig. 7.1). As templates they used DNAs containing mutant promoters that do not require cAMP nor the cAMP activator protein for efficient transcription. As precursors they used γ-^{32}P-labelled rNTPs. The product RNA was subjected to partial sequence analysis, using various specific nucleases. Under reaction conditions in which each added promoter was saturated with RNA polymerase, they detected two major and two minor 5' termini. If the promoter DNA sequence contained different mutations the distribution of transcripts varied among the four

Promoter	Percent starts at				Total RNA (pmol)
	G1	A1	G2	A2	
UV5	29	55	9	7	0.82
L305 x UV5	23	52	19	6	1.01
p^s	12	51	31	6	0.86
L305 x p^s	4	26	66	4	0.64

(a)

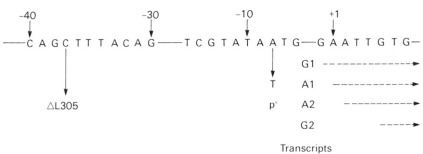

(b)

Figure 7.1 The distribution of start sites in *lac* mRNA for different promoters. (a) The percent starts detected at the four initiation sites by Carpousis, Stefano and Gralla (1982), and the total RNA synthesized when directed by certain lac promoters. (b) The DNA sequence in the sense strand of the *lac* UV5 mutant promoter region. The positions of the L305 deletion and the p^s transversion are also shown, as well as the start sites near position + 1 that are referred to in the text. (Used by permission.)

detected start sites, although the total synthesis of *lac* mRNA was not much affected (Fig. 7.1).

Carpousis *et al.* (1982) interpreted their data to mean that the heterogeneous initiations from each DNA sequence were all from the same promoter, as opposed to being from a set of overlapping promoters. This interpretation was strengthened by their demonstration that varying the ratio of precursors caused variation in the distribution of start sites transcribed from a single DNA sequence. Since open complexes of template and RNA polymerase are stable, a single promoter must be able to be recognized ambiguously when the precursor triphosphates are added.

Previous studies by Maizels (1973) and Majors (1975) had shown that the *lac*UV5 and wild type *lac* promoters programmed inaccurate initiations. Maizels observed only the two major species (G1:A1 = 15:85) but also found that the relative frequency of initiation at these two start sites could be perturbed by starving for a precursor triphosphate. Majors had to add cAMP and the cAMP activator protein to the reaction mixture since he was measuring transcription from the wild type promoter. Unlike Maizels' earlier work, he observed an equal probability for initiating at the G1 and A1 sites, although the reaction conditions were asserted to perturb this ratio. Using dinucleotides as primers, Majors found that GpA would stimulate three times as much transcription (presumably at the A1 site) as GpG (presumably at the G1 site), and many times as much as ApA (probably at A2). Dinucleotide UpG did not stimulate detectable initiation.

With the distributions of initiation sites reported above, it is, as a practical matter, not simple to obtain a measure of accuracy. The essence of the definition of error is a clear definition of truth and we have from these data at best a rather flabby idea of the 'normal' initiation site. If one takes the mode of the distribution of start sites as normal then the error rates reported in Fig. 7.1 are no lower than 34% (for the case of L305 × p^s promoter). Whether this high level of apparent erring is of physiological import is equally difficult to assess. The initiation region for translation of the message into a protein is located downstream from the transcription start site and no coding errors (in the sense of one amino acid being substituted for another) will result from heterogeneity of the 5′ terminus of the mRNA. Nevertheless, we should not assume too much about the lack of a functional role for the end of the mRNA. Molecules that modify the RNA polymerase, either proteins or small metabolites, might also perturb the distribution of 5′ termini, and regulation of translation or of other cellular functions may be affected by the terminal triphosphate.

The literature describing accuracy during initiation by eukaryotic RNA polymerases is mainly concerned with transcription from alternative promoters. It seems likely that the choice among promoters will be strongly influenced by the physiological state of the cell. Until we know more about the

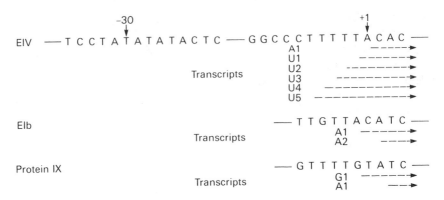

Figure 7.2 Start sites for adenovirus 2 mRNAs. The DNA sequence around the mRNA start sites of three adenovirus 2 genes (EIV, EIb, protein IX) are tabulated, as well as part of the promoter sequence for EIV. (From Baker and Ziff (1981); used with permission.)

error-free expectation for a particular physiological state we cannot estimate the accuracy of promoter choice.

There are some data which show heterogeneity of the cap sites of viral RNAs (Baker and Ziff, 1981). The different start sites are all adjacent or very close to one another and it seems reasonable to propose these cases as due to errors in initiation. Baker and Ziff infected HeLa cells with adenovirus 2 and then isolated ^{32}P-labelled RNA on poly U sepharose. Adenovirus mRNAs were identified by hybridization to specific viral DNA sequences and the 5' termini were sequenced after removing the cap structure.

Examination of the EIV mRNA revealed six start sites (Fig. 7.2). The fraction starting at each site was not reported by the authors but their published autoradiograms suggest they occurred in the order:

$$A1 > U5 > U4 > U3 \gg U2 > U1.$$

Other ambiguities were observed for other transcription initiations: for EIb, $A1 \gg A2$; for protein IX, $A1 \gg G1$ (Fig. 7.2). On the other hand, the EIa mRNA had only a single start site, as did the major late mRNA.

The experiments of Baker and Ziff have the advantage of having been done *in vivo*. The observed heterogeneity can therefore be presumed actually to be physiologically relevant and, by perturbing the cells, in the future it may be possible to demonstrate physiological responses in the distribution of 5' termini. Unfortunately, the terminal structures measured by such experiments do not necessarily reflect only events during the initiation of RNA synthesis. The same distribution of 'start sites' of the EIV mRNA, for example, could be generated by a single initiation (upstream from or at the U5 site) followed by exonucleolytic cleavage to varying extents before completing the synthesis of the cap structure. It is possible to prove the cap sites are the actual initiation sites of transcription, by labelling with β-^{32}P-labelled NTPs. It has been shown that the cap structure includes the β-phosphate of the 5′ terminal nucleotide for SV40 mRNAs (Contreras and Fiers, 1981; Gidoni *et al.*, 1981) and for mouse β-globin and adenovirus 2 major late mRNAs (Hagenbüchle and Schibler, 1981). It would be helpful if future observations of 5′ terminal heterogeneity were accompanied by experiments demonstrating each start sequence can be labelled by the appropriate β-^{32}P-NTP.

7.4 Accuracy during termination of RNA synthesis

There are several ways the 3′ end of an RNA may be constructed inaccurately. In this section we shall consider only the process of termination – the formation in the primary transcript of the last phosphodiester bond and dissociation of the RNA polymerase complex. Terminating 'prematurely' and failure to terminate may also be erroneous, but a discussion of those cases should wait until we understand more about the physiological signals that define the 'true' termination process.

One of the best described systems for examining termination is the attenuator of the tryptophan operon in *E. coli*, studied in great detail by Yanofsky and his colleagues. In the presence of excess tryptophan there is maximal termination of the leader transcript, which prevents synthesis of the distal mRNA that specifies the enzymes for tryptophan biosynthesis. Bertrand *et al.* (1977) labelled *E. coli* under these conditions with ^{32}PO$_4$, isolated the leader RNA by hybridization to a DNA probe containing the *trp* leader, followed by gel electrophoresis, and then determined the 3′ sequence. They observed five different termination sites, clustered at a group of consecutive U residues (Fig. 7.3). The efficiency of termination was also variable, ranging from 85% down to 25%. Errors in termination have also been observed during transcription of the *trp* leader by purified RNA

—G C C C G C C T A A T G A G C G G G C T T T T T T T T G A A—

	Transcripts			
		U1	———————	0.07
		U2	—————————	0.45
		U3	————————————	0.31
		U4	————————————	0.12
		U5	——————————————	0.05

Figure 7.3 Termination sites for the *trp* attenuator. *E. coli trpR* cells were labelled with ^{32}PO$_4$ in the presence of excess tryptophan. The labelled RNA was purified by hybridization to $\lambda \phi 80$ *trpE1* (leader) DNA and gel electrophoresis, then sequenced. The figure shows the sequence of the DNA near the end of the leader and the terminal sequences of the observed transcripts. The column on the right gives the relative molar yields of the various 3′ termini. (From Bertrand *et al.* (1977); used with permission.)

polymerase holoenzyme (Lee *et al.*, 1976). Termination sites U4 and U5 were reported to occur at about equal frequency (although the published auto-radiograms appear to show U4 > U5) and the efficiency of termination was asserted to be 95%.

The 3' sequence of the tryptophan attenuator of *E. coli* is an example of a 'strong terminator', one associated with a GC-rich region that can form a stem/loop conformation, followed by a string of consecutive T residues in the DNA coding strand (Fig. 7.3). The stem/loop slows the RNA polymerase and it dissociates from the cluster of As in the template strand. The consecutive Us should tend to weaken the association of the nascent RNA with the DNA template since hybrid $(rU \cdot dA)_n$ duplexes are especially unstable (Martin and Tinoco, 1980). This is because $oligo(dA)_n$ sequences tend to have the B conformation, which is not compatible with the RNA–DNA duplex. Ryan and Chamberlain (1983) showed that heteroduplex DNA, mismatched in the stem/loop region, would still act as a proper template for transcription termination, provided the sequence in the non-coding strand was correct. This implies that RNA–RNA interactions in the stem/loop determine termination. Termination also is influenced by RNA polymerase mutations (Farnham and Platt, 1980).

The strong terminator structures are able to stop polymerization *in vitro* without participation of the termination factor rho, although it may alter the efficiency of termination at these sites *in vivo*. There are other stop sequences, much less well defined, that do not terminate with high efficiency unless rho factor is present (Holmes, Platt and Rosenberg, 1983). The participation of rho, and anti-termination factors, is probably under physiological control and there are too few available data to discuss the accuracy of these processes.

There is a substantial literature concerning variability of the site of poly (A) addition in eukaryotic mRNAs. This site is almost certainly generated, however, by processing a much longer primary transcript. Hofer, Hofer-Warbinek and Darnell (1982) found that the primary transcript of the major β-globin gene terminated in a region about 1400 ± 100 nucleotides beyond the poly (A) addition site. Their experiments were done with isolated nuclei from mouse erythroleukaemia cells, treated to induce globin synthesis. The resolution of the sequence analysis, however, was not sufficient to answer questions about the accuracy of termination. They also presented evidence suggesting instability of the sequences beyond the poly (A) addition site. Rapid processing of the primary transcript *in vivo* or in isolated nuclei will probably confound most fidelity experiments in such preparations.

Primary transcripts that extend beyond the 3' terminus of the mature mRNA have also been observed for viral systems: adenovirus 2 (Fraser and Hsu, 1980) and SV40 (Ford and Hsu, 1978). The former authors showed by electron microscopy that many transcripts of late adenovirus genes extend at least 2000 nucleotides beyond the poly (A) addition site.

RNA polymerase III in eukaryotic cells transcribes the genes for transfer RNAs and for the 5S ribosomal RNA. Heterogeneity is seen when cloned 5S DNA or 5S chromatin is transcribed by purified RNA polymerase III (Cozzarelli *et al.*, 1983). The 90% homogeneous enzyme could not properly initiate 5S rRNA transcripts on purified genes (initiation does occur non-specifically on such templates, in the absence of additional protein factors) but it could be used to evaluate termination from cloned templates containing the 3' two-thirds of the 5S gene and the 3' spacer. There are three terminator sequences in the 5S rRNA gene, two weak ones at approximately 120 and 129 nucleotides from the start site, and a strong one at 139 nucleotides, each with a series of Ts in the DNA coding strand. Two or three stop sites for each of three clusters of termination sequences were detected by Cozzarelli and coworkers but the sites were not sequenced.

The longest transcripts of 5S rRNA from RNA polymerase III are similar to those found after labelling whole oöcytes of *Xenopus laevis* (Denis and Wegnez, 1973; Gurdon and Brown, 1978). Bogenhagen and Brown (1981) warn that to see this heterogeneity in the intact 5S rRNA requires the use of thin gels and a brief autoradiographic exposure. (That advice might be instructive for anyone seeking an understanding of the fidelity of primary transcript length.) The transcription was shown by its sensitivity to α-amanitin (50% inhibited by 50 μg ml^{-1}) to have been performed by RNA polymerase III (Gurdon and Brown, 1978). Newly synthesized 5S rRNAs of *Xenopus laevis* oöcytes have a 3' terminal sequence of two to four Us with 80% having two Us (–CUU–OH, 120 nucleotides total) and 20% being longer. Processing probably reduces the length of RNA polymerase III transcripts terminating in the distal sites to the equivalent of termination in the proximal region. When injected into oöcytes, however, a longer transcript of 5S DNA plus 3' spacer – synthesized by the RNA polymerase of *E. coli* – is not matured properly (Gurdon and Brown, 1978).

7.5 Accuracy during mRNA splicing

An analysis of accuracy during the polymerization of RNA is insufficient to evaluate the overall accuracy of RNA synthesis since the nucleotide sequence of a mature mRNA is also altered by processing reactions. Unfortunately, a thorough discussion of all possible post-polymerization processing reactions involved in the maturation of tRNA, rRNA, mRNA and other RNA molecules is beyond the scope of this review. Instead we shall focus on accuracy during RNA splicing and during the cleavage and polyadenylation of the 3' terminus of mRNA precursors.

The process known as splicing involves the removal of intervening sequences (termed introns) which do not appear in the mature RNA while at the same time joining together in the correct order the expressed sequences (termed exons) which are part of the functional RNA. In order for an intron to be accurately removed from a precursor mRNA (pre-mRNA) the splicing system must recognize and cleave at the exact sites which define the beginning and end of the intron, and then ligate the two ends of the adjoining exons. One expects the enzyme(s) involved in splicing to be quite accurate since an error made during the splicing reaction would be very costly to the cell. If the splice junction were displaced erroneously by a single base, the reading frame of all coding regions downstream from the splice site would be changed. Failure to cleave the pre-mRNA would result in adding an entire intron to the coding region of the mRNA; failure to ligate the exons would result in the production of only a fragment of the mature RNA. Splicing pre-mRNA molecules that contain multiple introns could remove exons as well as introns if the 3' end of one exon was mistakenly joined to the 5' end of a second exon farther downstream. When a pre-mRNA is incorrectly spliced, the cell wastes not only the time, energy and resources required to synthesize the mRNA, but also those necessary to translate the erroneous mRNA repeatedly into a non-functional protein, a process which lasts until the aberrantly spliced mRNA is degraded.

Many of the specific details of the splicing machinery are incompletely understood and are presently being investigated. These include character-ization of the enzyme(s) and cofactors involved in splicing, the process of selection of one potential splicing site over another, and the mechanisms of the cleavage and splicing reactions. However several important aspects of the process, involving alignment of the ends of a single intron so that accurate cleavage at the splice junction can occur, are more fully explored. The model that accounts for this and the experimental evidence to support it will serve as background to several detailed genetic and biochemical analyses of the influence of the nucleotide sequence at the exon–intron borders in deter-mining the accuracy of the splicing reaction. Models will then be presented which attempt to explain how multiple introns can be accurately processed from a pre-mRNA without deleting any of the coding sequences.

7.5.1 A MODEL FOR INTRON AND EXON ALIGNMENT AT A SPLICE JUNCTION

Several similar models have been offered to explain how the cell is able to recognize the two cleavage sites at the borders of an intron and bring the ends of the exons into intimate contact so they can be spliced together (Murray and Holliday, 1979a, b; Davidson and Britten, 1979; Lerner et al., 1980; Rogers and Wall, 1980). Basically, the models propose that another RNA molecule is hydrogen bonded to the 5' splice site (5' SS; the nucleotide sequence adjacent to the 5' boundary of an intron) and to the 3' splice site (3' SS; the nucleotide sequence adjacent to the 3' boundary of an intron). The resulting double-stranded RNA molecule holds the splice junction in the appropriate position for the cleavage and ligation reactions, catalysed by the rest of the splicing system. The structure of the intermediate, with the intron looped out, is similar to that proposed to occur in DNA recombination (the cruciform or Holliday structure). Because of the observed complementarity between the 5' end of the U1 small nuclear RNA (snRNA) and a conserved sequence at the splice sites, U1 snRNA was nominated as such a splicing jig (Lerner et al., 1980; Rogers and Wall, 1980).

Recent observations by a number of investigators have largely supported this original idea. Over a hundred different splice junctions have been sequenced and the consensus sequence derived from 139 different 5' SSs and 130 different 3' SSs (Mount, 1982) is predicted to form a very stable complex with the 5' end of U1 snRNA (Fig. 7.4). In the consensus sequence only two bases at the 5' end and two bases at the 3' end of the intervening sequence are present at all the splice junctions. Considerable variability can exist among the other bases present.

After the addition of antibodies which precipitate U1-containing small ribonuclear proteins (snRNPs), it was observed that the splicing of adenoviral RNA molecules synthesized in HeLa cell nuclei was inhibited up to 95% (Yang et al., 1981). The first eight nucleotides of U1 snRNA can be removed by site-directed hydrolysis with ribonuclease H, in the presence of a synthetic

Figure 7.4   The consensus sequence of the splice sites derived from the DNA sequences of 139 different 5' SSs and 130 different 3' SSs. The arrows indicate the points of cleavage. Numbers below the bases are the percent occurrence of that nucleotide in the population of splice sites. The number below the run of eleven consecutive conserved pyrimidines in the 3' SS represents the mean percent occurrence of a C or T (Mount, 1982). Below is shown the 5' end of U1 snRNA and the splice site consensus sequence (based on suggestions by Lerner et al., 1980 and Rogers and Wall, 1980). (N represents any nucleotide.)

complementary oligonucleotide. When this treated U1 snRNA is included in HeLa cell nuclear extracts, splicing of the major late adenovirus pre-mRNA is undetectable (Krämer *et al.*, 1984). Tatei *et al.* (1984) observed that U1-containing snRNPs from HeLa cells or rat liver cells could preferentially bind small RNAs whose sequence included either a 5' or 3' SS and that the affinity of the 5' or 3' SSs for the U1-containing snRNPs could be significantly reduced when normally highly conserved nucleotides located at the intron borders were changed.

MacCumber and Ornstein (1984) have analysed the original model for mRNA splicing in terms of its stereochemical and thermodynamic plausibility. The resulting molecular model of the mRNA splice region consists of the 5' and 3' SSs hydrogen bonded in an A-form helical RNA to a complementary RNA molecule which serves to hold the splice sites in a favourable position for cleavage. (This model may have to be reworked as potential intermediates of the splicing reaction become characterized further. For example, Ruskin *et al.* (1984) and Padgett *et al.* (1984) have isolated 'lariat' structures in which the 5' end of the intron is bound by a 5'-to-2' linkage to an adenosine residue near the 3' cleavage site.)

7.5.2 FACTORS THAT INFLUENCE THE ACCURACY OF RNA SPLICING

The model proposed by Lerner *et al.* (1980) and Rogers and Wall (1980) and its supporting data explain in large part how a single intron with well-defined splice sites can be processed out of a pre-mRNA. Among the questions which remain to be evaluated are what factors influence the accuracy of splicing and how accurate is the splicing process. In order to answer these questions it is necessary to develop assays in which aberrant splicing can be detected. Aberrant splicing has been most extensively studied in systems where the normal pattern has been altered through mutation, with the result that more than one stable mRNA is produced from the processing of one pre-mRNA. This review will consider only cases where multiple mRNAs are produced from a single pre-mRNA, apparently as the result of failure of the splicing system to process accurately, and not those cases that are the result of differentiation or developmental regulation.

The accuracy of RNA splicing can be influenced by altering the nucleotide sequence within the 5' or 3' SS which is proposed to bind to U1 snRNA. Wieringa *et al.* (1983) systematically analysed the accuracy and relative efficiency of expression of spliced mRNAs produced from a rabbit β-globin gene carrying one of six point mutations within the 5' SS of the second intron. Mutant and normal β-globin genes, carried on a plasmid vector along with a mouse β-globin gene, were introduced into HeLa cells and the expression of the resulting rabbit globin mRNAs was measured relative to the reference mouse globin mRNA. Amounts, cleavage sites and sequences of the mRNAs were assayed by mapping S1-protected DNA probes hybridized to total cellular RNA, and by primer extension of DNA probes hybridized to total poly $(A)^+$ RNA.

Two purine transition mutations, at the third and fourth nucleotides downstream from the 5' splice site, had no observable effect on the accuracy or efficiency of splicing (Table 7.1, lines 2 and 3). Although both these mutations were located in the sequence that is expected to pair with the 5' end of U1 snRNA, neither mutation would be expected to decrease substantially the stability of the complex formed. Only one of the six mutations examined affected RNA splicing, a G-to-A transition located at the first position upstream of the 5' cleavage site (Table 7.1, line 4). This mutation changes the

Table 7.1 Normal and mutant 5' splice sites. The position and type of mutations within the 5' splice sites of the genes of several mRNAs are tabulated. The consequences of these mutations on the accuracy and efficiency of splicing at these sites are discussed in the text. The arrows indicate the site of cleavage during splicing. Coding sequences lie to the 5' side of the cleavage site. Sequences on the 3' side are expected to pair with the 5' end of U1 snRNA.

Source	Mutation	Sequence	Reference
Rabbit β–globin Intron 2	Normal SS	A G G↓G T G A G T	Wieringa et al., 1983
	G → A	A G G G T A A G T	
	A → G	A G G G T G G G T	
	G → A	A G G A T G A G T	
Human β-globin Intron 1	Normal SS	C A G↓G T T G G T	Treisman, Orkin and Maniatis, 1983
	G → C	C A G G T T G C T	
	T → C	C A G G T T G G C	
Human β-globin Exon 1	Cryptic 5'SS	G T G↓G T G A G G	Goldsmith et al., 1983
	T → A	G A G G T G A G G	
Human α2-globin Intron 1	Normal SS	G A G↓G T G A G G	Felber, Orkin and Hamer, 1982
Exon 1	Normal SS	G G G↓G T A A G G	
Consensus sequence		$^{C}_{A}$ A G G T $^{A}_{G}$ A G T	Mount, 1982

invariant GU dinucleotide in the normal 5′ SS consensus sequence to an AU. Normal splicing at this site was reduced to undetectable levels and aberrant splicing at three previously undetectable 5′ SSs was observed.

Treisman, Orkin and Maniatis (1983) observed that mutations in the 5′ SS in a human β-globin gene, at positions 5 or 6 bases upstream from the normal cleavage site between exon 1 and intron 1, reduced the amount of mRNA produced from splicing at the normal site and induced splicing at three cryptic 5′ SSs (Table 7.1, lines 6 and 7). Both mutations would be expected to reduce the stability of the U1 snRNA·5′ SS complex but no quantitative data were reported on the relative amounts of normally processed and aberrantly spliced mRNAs.

In the study by Wieringa *et al.* (1983), where quantitative data are available, a correlation exists between the relative level of expression of the aberrantly spliced mRNAs and the stability of the complementary complex that can be formed between the sequences at the cryptic 5′ SSs and the 5′ end of U1 snRNA. Determination of the stability of the complexes in which the invariant GU dinucleotide of the consensus 5′ SS was conserved revealed that the more thermodynamically stable the complex, the greater the relative level of expression of the spliced mRNA.

These results are in agreement with those of Avvedimento *et al.* (1980), who studied the processing of a fragment of the chicken α2 collagen gene. They observed three additional 5′ SSs within a single intron. They also found a correlation between the stability of these U1·SS complexes and the detected amounts of various fragments that would be expected from splicing at the internal sites. It was discovered in both studies that aberrant splicing occurred at a 5′ SS in which the invariant GU dinucleotide was not conserved. In both instances, however, relatively stable base-paired complexes with U1 snRNA could still be formed if a single base within the 5′ SS were looped out. Assuming that the relative levels of spliced mRNAs found *in vivo* are an accurate reflection of the amount of splicing occurring at a particular site, and not the result of differential rates of degradation of the variously spliced mRNAs, such observations indicate that the stability of the double-stranded RNA complex at the splice junction influences the accuracy and efficiency of splicing.

In other systems, mutations that alter sites affecting the proposed U1 snRNA binding also alter the accuracy of splicing, but in many cases the results cannot be explained by the stability of the double-stranded RNA at the splice junction. Using methods similar to those of Wieringa *et al.* (1983), Felber, Orkin and Hamer (1982) analysed a mutant human α2-globin gene which contained a pentanucleotide deletion of the second to sixth bases of intron 1. The mutation destroyed the invariant GU dinucleotide of the 5′ SS and thus eliminated the complementarity between this region and U1 snRNA. Nevertheless the mutant α2-globin gene produced a pre-mRNA that was processed, by utilizing a cryptic 5′ SS found within the first exon (Fig. 7.5, line 11). The cryptic 5′ SS actually can form a more stable complex with U1 snRNA than can the normal 5′ SS. An analysis of the expression of the normal α2-globin gene revealed, however, that although splicing at the cryptic 5′ SS was clearly observable, it occurred at a level of <1% of that at the normal site. A similar result was obtained by Fukumaki *et al.* (1982). A mutation within intron 1 of the human β-globin gene activated a cryptic 3′ SS that was used in preference to the normal 3′ SS, even though the mutant 3′ SS was less complementary to U1 snRNA. Eighty to 90% of the total β-globin mRNA was abnormally spliced in monkey kidney cells.

The accuracy of splice site selection has also been affected by altering bases which are near the U1 snRNA binding site but do not directly interact with it. For example, Goldsmith *et al.* (1983) observed that a cryptic 5′ SS within exon 1 of human β-globin could be activated by a mutation two bases downstream from the cryptic cleavage site (Table 7.1, line 9). The abnormally spliced mRNA was expressed at 30% of the level of normally spliced β-globin mRNA.

These observations indicate that although the stability of the splice junction may influence the accuracy of splice site selection, a number of other factors must also play a role. The splicing apparatus has not yet been isolated and its specific interaction with the nucleotides at the splice junction are not yet known. One would expect, however, that the splicing apparatus would have to be in direct contact with a number of nucleotides at the splice junction in order to maintain the bases in the proper orientation and configuration to ensure accurate cleavage and ligation. Mutant nucleotides at the 5′ and 3′ SSs that change the accuracy of splice site selection without increasing the stability of the double-stranded RNA complex formed with U1 snRNA may be the nucleotides involved in direct interaction with the splicing apparatus. They could therefore increase the overall affinity of the splicing apparatus for a cryptic splice site, making possible a more effective competition with the normal splice site.

The accuracy of splice site selection could also be affected by the secondary and tertiary structure of the pre-mRNA. Wieringa *et al.* (1983) noted a potential 5′ SS from which no spliced mRNA was detected, despite its being located between two cryptic 5′ SSs that were both utilized. They hypothesized that failure to use such potential splice sites could result from a secondary structure of the mRNA that masks the splice site or an unfavourable tertiary structure that positions a 5′ or 3′ SS in such a way that it cannot be paired with other splice sites. It is important to note, however, that a specific tertiary structure is probably not required for accurate splicing to occur. The tertiary structure of an RNA can be changed by large deletions of nucleotides within an intron and accurate splicing still occurs (Wieringa, Hofer and Weissmann, 1984).

7.5.3 MEASUREMENTS OF THE ACCURACY OF RNA SPLICING

The most relevant measures of the accuracy of splicing are those for which the frequency of aberrantly spliced mRNA is assayed relative to a normally spliced mRNA in a system where splicing at the two sites is not affected by mutation or by developmental or tissue-specific regulation. Aberrant splicing in such systems has been detected. As mentioned earlier, Felber, Orkin and Hamer (1982) detected mRNAs spliced at a cryptic 5' SS during the expression of a wild type α2-globin gene in transfected monkey cells. The amount of aberrantly spliced mRNA was <1% that of the normally spliced mRNA. King and Piatigorsky (1984) have observed the alternate use of both a 5' and 3' SS in a pre-mRNA which codes for the αA2 rat lens protein. The mRNAs produced from splicing at the alternate sites are found at levels 10 to 20% that of the normally spliced mRNA. There appears to be no development-specific or tissue-specific regulation of the expression of this gene and the polypeptide produced from the aberrantly spliced mRNA is functionally equivalent to the normal αA2 polypeptide.

The accuracy of splicing *in vivo* is difficult to evaluate unambiguously. The level of expression of any mRNA product is the result of the steady state that exists between its rate of synthesis and its rate of degradation. It would be advantageous to the cell if an aberrantly spliced mRNA were less stable than the normally spliced mRNA. In this way a cell might prevent the further waste of resources in the synthesis of abnormal proteins. However if such differential rates of degradation of aberrantly spliced mRNAs did occur *in vivo*, measurements of their fractional contribution to the total would over-estimate the accuracy of the splicing apparatus. An *in vitro* system where splicing can occur independently of RNA degradation would be expected to yield the most reliable estimates of the accuracy of RNA splicing.

Krainer *et al.* (1984) have developed an *in vitro* splicing system which can accurately process human β-globin mRNA precursors in HeLa cell nuclear extracts. They were able to analyse *in vitro* several β-globin mutants previously characterized *in vivo* by Treisman, Orkin and Maniatis (1983). In this system, Krainer and coworkers observed many more aberrantly cleaved pre-mRNAs than were observed *in vivo*. It is not yet clear whether these represent normal intermediates of the splicing process or erroneously cleaved mRNAs. Even though the same kinds of aberrantly spliced mRNAs were observed, significant quantitative differences existed between the results of the two studies. The total amount of mRNA spliced from the mutant genes *in vivo* is much less than that observed when the wild type gene is used. The levels *in vitro* are not significantly different. Also the relative levels of normally and aberrantly spliced mRNAs produced from the mutants *in vivo* are different from the relative levels *in vitro*. These results may be explained if aberrantly spliced mRNAs are preferentially degraded *in vivo*.

7.5.4 ACCURACY DURING THE REMOVAL OF MULTIPLE INTRONS

The problem of removing multiple introns present in a single transcript should be especially apparent during the processing of the chick α2-collagen gene, which contains 51 introns. If each intron were bordered by an identical 5' and 3' SS, it would be possible to combine any splice sites during processing and some exons could be eliminated during processing of the pre-mRNA (Sharp, 1981). The simplest way to explain how multiple introns can be accurately removed is a 'processive scanning model', in which the splicing apparatus attaches to one end of the RNA and moves unidirectionally until one end of an intron is encountered. Scanning continues until the other end of the intron is recognized, then is followed by removal of the intron. The scanning process would continue until all the introns are removed from the messenger (Lewin, 1980; Sharp, 1981).

Various aspects of the model have been evaluated by analysing the spliced mRNA products of recombinant genes that have been constructed with tandem duplications of either the 5' SS or the 3' SS. Lang and Spritz (1983) examined spliced products from a human G_γ-globin gene that contained duplications of either the 5' SS or the 3' SS of intron 2. Consistent with a 5'-to-3' scanning model, they observed that mRNAs were spliced only at the most upstream 5' and 3' SS. However, Kühne et al. (1983) analysed a rabbit β-globin gene that contained either duplicated 5' or 3' SSs bordering intron 2 and observed that only the splice sites distal to the intron were used. This result is inconsistent with the unidirectional scanning model. It is also inconsistent with a bidirectional scanning model, in which the splicing apparatus binds to the intron and diffuses along the RNA to locate the 5' and 3' SSs.

Kühne and coworkers proposed two main influences on the selection of appropriate splice sites in pre-mRNAs that contain multiple introns: (1) the nucleotide sequence of the splice regions, and (2) the relative position of a pair of 5' and 3' SSs. The nucleotide sequence would determine the binding affinity of the splicing apparatus to the splice sites. But secondary and tertiary structures could mask these sites or modify their binding affinity for the splicing apparatus. The relative position of the 5' and 3' SSs is proposed to determine the probability of splicing between any pair of splice sites. At any particular time in a pre-mRNA with multiple introns, only one pair of 5' and 3' SSs (or at the most a small number of easily distinguishable 5' and 3' SSs) may be unmasked by the secondary and tertiary structure and positioned appropriately, relative to one another, for splicing to occur. Removal of one intron would then alter the structure of the pre-mRNA so that other introns could be accurately removed. Structural unmasking of groups of introns would not be expected to result in a precisely defined order of intron removal, but could result in a preferential sequence.

In support of this model, several investigators have observed that efficient removal of one intron is dependent on the efficient removal of another (Busslinger, Moschonas and Flavell, 1981; Treisman *et al.*, 1982). This model may also explain why in some cases introns contain internal 5' and 3' SSs (Avvedimento *et al.*, 1980; Donaldson *et al.*, 1982). The stepwise removal of pieces of a single intron may be required to form an appropriate structure to insure accurate removal of other introns.

The preceding discussion has focused on how an individual splice site is accurately chosen, rather than on the exact specificity of the cleavage event itself. No studies have addressed this question, even though an error of one or two bases either added or deleted at the splice junction would have disastrous effects on the reading frame of the mRNA downstream. It would, unfortunately, be difficult to detect a very low level of misspliced mRNAs by S1 nuclease mapping or by labelled primer extension analysis, especially when the aberrant product may be only one or two bases longer or shorter than the normally spliced mRNA.

7.6 Accuracy during maturation of the 3' terminus of an mRNA

In many mRNA transcripts from a variety of sources it appears that a nucleolytic cleavage event is required to generate the mature 3' terminus. Transcription termination occurs downstream from the polyadenylation sites in the mouse β-globin gene (Hofer and Darnell, 1981) and in several viral genes from SV40 (Ford and Hsu, 1978) and adenovirus type 2 (Fraser *et al.*, 1979; Nevins, Blanchard and Darnell, 1980). Other studies also indicate that post-transcriptional processing of the 3' end occurs in mRNAs for chicken histone H2B (Krieg and Melton, 1984) and *Drosophila* histone H3 (Price and Parker, 1984). It may be premature to assume a general requirement for 3' processing of an mRNA transcript but the current data for mammalian cells are not inconsistent with this idea.

Maturation of the mRNA 3' terminus requires that the enzyme(s) involved be able to recognize the appropriate site, cleave the mRNA and, for most eukaryotic cell mRNAs, add a poly (A) tail. An error in processing can arise from failure to cleave the RNA or cleavage at an incorrect site. The latter might result in only a fragment of the mRNA being produced while the former could result in a subsequent failure to polyadenylate the message. Lack of a poly (A) tail can greatly decrease the stability of an mRNA in the cytoplasm (reviewed by Nevins, 1983).

Recognition of a poly (A) addition site is a sequence-dependent event. Proudfoot and Brownlee (1976) observed a highly conserved hexanucleotide sequence, 5'-AAUAAA-3', that is located in the 3' non-coding region of six different mRNAs. Recently Berget (1984) analysed the sequences adjacent to the poly (A) sites of a selected sample of 61 different mRNAs (Fig. 7.5).

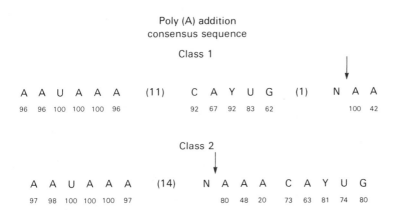

Figure 7.5 The consensus sequences for two classes of poly (A) addition sites. These were derived from an analysis of 61 different examples from chicken, mouse, rabbit, rat, and man. The arrows above the sequences designate the proposed site of cleavage and poly (A) addition. Numbers below the bases indicate the percent occurrence of an individual nucleotide. Numbers in parentheses represent the positions of unconserved nucleotides (from Berget, 1984). (N is any nucleotide and Y is either pyrimidine.)

Berget observed that the highly conserved AAUAAA sequence was located from about ten to thirty (with an average of sixteen) nucleotides upstream from the poly (A) addition site. In addition, a pentanucleotide sequence, CAYUG, was conserved in most of the mRNAs examined (where Y represents either pyrimidine). Consensus sequences were determined for two classes of mRNA: Class 1 mRNA has the pentanucleotide sequence located upstream and Class 2 downstream from the poly (A) addition site.

Several lines of evidence point to the AAUAAA sequence being necessary for accurate cleavage and polyadenylation. When Fitzgerald and Shenk (1981) constructed a deletion mutant in the late SV40 gene that lacked the AATAAA sequence, cleavage and polyadenylation did not occur at the normal site but were shifted to a site slightly downstream from another AAUAAA. Montell *et al.* (1983) constructed mutant adenovirus type 2 with a single T-to-G transversion in the AATAAA site of the early IA gene. This mutation reduced both the efficiency and accuracy of cleavage at the poly (A) site. The extent of cleavage was reduced more than 95%, but without a significant effect on the efficiency of polyadenylation of those transcripts that were cleaved. Also, the specificity of cleavage was altered so at least one additional site was used, located two bases upstream from the normal cleavage site. Higgs *et al.* (1983) characterized a cloned mutant α2-globin gene in which the normal AATAAA sequence was changed to AATAAG. When the gene was expressed in HeLa cells, the majority of the transcripts produced were not cleaved at the normal site but were presumably terminated farther along, in sequences derived from the cloning vector DNA.

The accuracy and efficiency of cleavage have also been altered by changing the sequences adjacent to the 3' side of the AAUAAA site. Small deletions of three to fourteen nucleotides immediately downstream from the AATAAA site constructed in SV40 by Fitzgerald and Shenk (1981), resulted in polyadenylation occurring between eleven and nineteen nucleotides from the 3' boundary at the AAUAAA sequence, immediately following a CA dinucleotide. The normal site of poly (A) addition in SV40 is twelve nucleotides downstream from the AAUAAA site.

McDevitt et al. (1984) analysed the efficiency and accuracy of poly (A) addition of an adenovirus E2A transcription unit in human 293 cells. A series of mutants was constructed in which portions of the gene downstream from the polyadenylation site were deleted. McDevitt and coworkers observed that a mutant containing the thirty-five nucleotides to the 3' side of the AAUAAA site produced mRNA that was normally cleaved and polyadenylated. The extent of cleavage was 90% that found in the transcript from a wild type gene. Deletion mutants that contained only twelve or twenty of the downstream nucleotides reduced the amount of cleavage to about 10% of normal. These results indicate that deletion of sequences between twenty and thirty-five nucleotides downstream of the AATAAA site significantly affects the processing of the 3' end of the transcript. The sequence within this region is complementary to a preceding sequence that includes the AATAAA site. The formation of a stem and loop structure by pairing of these two sites in the RNA may function as a substrate or a recognition site for the appropriate endonuclease. The requirement for a particular secondary structure of the mRNA near the site of cleavage and polyadenylation could explain why not all potential AAUAAA sequences are used.

Quantitative measurements of the accuracy of the cleavage and polyadenylation of the 3' terminus of mRNAs have not yet been made. Currently available data indicate that deletion and alteration of the AAUAAA site can change the site of polyadenylation or result in small variations in the exact site of cleavage. As the *in vitro* polyadenylation systems are further refined, reliable quantitative evaluations of the accuracy of this process, free from the complications of intermediates and degradation of aberrant products, should be obtainable. Further characterization of the influence on accuracy of 3' processing of the mRNA secondary and tertiary structures, and snRNPs and/or snRNAs, will provide valuable information on the question of how differential polyadenylation of a single transcript is regulated during development and differentiation.

An interesting addition to the factors which help to determine the site of cleavage and polyadenylation is the possible involvement of snRNA U4-containing snRNPs. The consensus sequence for poly (A) addition sites determined by Berget (1984) is also complementary to a number of sites in U4 snRNA. U4 RNA contains three closely spaced sequences complementary to the AAUAAA sequence and five different sequences within its first seventy 5' terminal nucleotides that are complementary to the consensus CAYUG sequence. Moore and Sharp (1984) reported evidence that U4 is important for normal transcription: they demonstrated that accurate polyadenylation in HeLa cell extracts of L3 transcripts from adenovirus is inhibited by antisera that precipitate RNPs containing a variety of snRNAs, including U1 and U4. Despite these indications, the role of snRNAs such as U4 should be considered speculative until direct evidence of their involvement can be obtained.

7.7 Conclusion

From the standpoint of an individual cell, there are advantages to the error-free transfer of information from the genome to other DNA, RNA and protein molecules. It would minimize the cost of the time, energy and material resources necessary to synthesize potentially non-functional products. The second law of thermodynamics, on the other hand, guarantees that errors will occur in any information transfer process that uses material objects. Only by investing a sufficiently large amount of energy can errors be minimized to vanishingly small levels. Thus, cellular metabolism must maintain a balance between the energy cost of errors and the energy cost of maintaining very high levels of accuracy (see Chapter 12).

The overall accuracy of any one stage of information transfer can be no better than that of the stage immediately preceding. It would be wasteful, for example, to devote energy to maintaining a high level of accuracy during transcription if the primary source of errors in completed RNAs were the result of failures during DNA replication. The fidelity of purified DNA polymerase can vary with its type and source (reviewed by Loeb and Kunkel, 1982). The error rate determined for the misincorporation of dGTP into a polydeoxynucleotide product, using poly $(dA \cdot dT)$ as template, is 1.3×10^{-5} for E. coli DNA polymerase I (Agarwal, Dube and Loeb, 1979) and 1.0×10^{-5} for DNA polymerase α from regenerating rat liver (Salisbury, O'Conner and Saffhill, 1978). This is not very much better than the error ratio determined for the misincorporation of GTP into a polyribonucleotide product synthesized from a poly $[d(A-T) \cdot d(A-T)]$ template of 2.4×10^{-5} (Springgate and Loeb, 1975). The overall accuracy of DNA replication in prokaryotic cells, however, also reflects post-polymerization events, primarily the $3' \rightarrow 5'$ exonuclease activities involved in proofreading. This additional step allows the overall accuracy of DNA replication to rise to very high levels.

Whether there are post-polymerization events that can improve the accuracy of transcription is still an unanswered question. The NTPase activity described by Volloch, Rits and Tumerman (1979) may play such a role. There is also the possibility that other proteins and/or small metabolites (e.g., ppGpp and pppGpp) might affect the fidelity of RNA polymerases. More data are needed on the error frequencies of transcription *in vivo* to establish whether its overall accuracy is better than indicated from the behaviour of purified enzymes. The range of error frequencies reported for transcription *in vitro* likewise suggests a need for better designed experiments. Most useful would be measures of errors during the transcription of natural templates, rather than the artificial repetitive templates used thus far.

RNA synthesis in eukaryotic cells is even less well characterized. The most critical need is for purified enzyme complexes capable of initiating, elongating, and terminating natural substrates. The analysis of errors during initiation of capped mRNAs should be approachable with present techniques. Using β-^{32}P-labelled rNTPs, it can be established whether the heterogeneity observed in cap sites is due to heterogeneity in the initiating nucleotide. Some progress might be made in the analysis of termination errors by experiments designed to address that issue directly. There are techniques available to characterize the 3' terminus of RNA, but they have only rarely been used to ask questions about heterogeneity. Experiments in nuclear or cell extracts would be improved if it were possible to block the processing enzymes that make the 3' sequences difficult to preserve.

It is probably too early in the study of the function of promoters, terminators, poly (A) addition signals, and the like, to make sense of the question of accuracy. Nevertheless it would be helpful if experiments devoted to these questions were to be designed with the possible functional consequences of errors in mind.

References

Agarwal, S. S., Dube, D. K. and Loeb, L. A. (1979) On the fidelity of DNA replication: accuracy of *Escherichia coli* DNA polymerase I. *J. Biol. Chem.*, **254**, 101–106.

Agris, P. F. and Söll, D. (1977) The modified nucleosides in transfer RNA. In *Nucleic Acid–Protein Recognition* (ed. H. J. Vogel), Academic Press, New York, pp. 321–344.

Altman, S. (1981) Transfer RNA processing enzymes. *Cell*, **23**, 3–4.

Avvedimento, U. E., Vogeli, G., Yamada, Y., Maizel, J. V., Pasten, I. and deCombrugghe, B. (1980) Correlation between splicing sites within an intron and their sequence complementarity with U1 RNA. *Cell*, **21**, 689–696.

Baker, C. C. and Ziff, E. B. (1981) Promoters and heterogeneous 5' termini of the messenger RNAs of adenovirus serotype 2. *J. Mol. Biol.*, **149**, 189–221.

Bass, I. A. and Polonsky, Jr, S. (1974) On the fidelity of *in vitro* polynucleotide synthesis by *E. coli* RNA polymerase. *FEBS Lett.*, **48**, 306–309.

Batschelet, E., Domingo, E. and Weissmann, C. (1976) The proportion of revertant and mutant phage in a growing population, as a function of mutation and growth rate. *Gene*, **1**, 27–32.

Berget, S. M. (1984) Are U4 small nuclear ribonucleoproteins involved in poly-adenylation? *Nature*, **309**, 179–181.
Bertrand, K., Korn, L. J., Lee, F. and Yanofsky, C. (1977) The attenuator of the tryptophan operon of *Escherichia coli*. Heterogeneous 3'-OH termini *in vivo* and deletion mapping of functions. *J. Mol. Biol.*, **117**, 227–247.
Bick, M. D. (1975) Misincorporation of GTP during transcription of poly dAT–dAT and poly dABU–dABU. *Nucl. Acids Res.*, **2**, 1513–1523.
Bogenhagen, D. F. and Brown, D. D. (1981) Nucleotide sequences in *Xenopus* 5S DNA required for transcription termination. *Cell*, **24**, 261–270.
Bouadloun, F., Donner, D. and Kurland, C. G. (1983) Codon-specific missense errors *in vivo*. *EMBO J.*, **2**, 1351–1356.
Busslinger, M., Moschonas, N. and Flavell, R. A. (1981) β^+ Thalassemia: Aberrant splicing results from a single point mutation in an intron. *Cell*, **27**, 289–298.
Carpousis, A. J., Stefano, J. E. and Gralla, J. D. (1982) 5' nucleotide heterogeneity and altered initiation of transcription at mutant *lac* promoters. *J. Mol. Biol.*, **157**, 619–633.
Chamberlain, M. and Berg, P. (1964) Mechanism of RNA polymerase action: Characterization of the DNA-dependent synthesis of polyadenylic acid. *J. Mol. Biol.*, **8**, 708–726.
Contreras, R. and Fiers, W. (1981) Initiation of transcription by RNA polymerase II in permeable, SV40-infected or noninfected, CV1 cells; evidence for multiple promoters of SV40 late transcription. *Nucl. Acids Res.*, **9**, 215–236.
Cozzarelli, N. R., Gerrard, S. P., Schlissel, M., Brown, D. D. and Bogenhagen, D. F. (1983) Purified RNA polymerase III accurately and efficiently terminates transcription of 5S RNA genes. *Cell*, **34**, 829–835.
Davidson, E. H. and Britten, R. J. (1979) Regulation of gene expression: Possible role of repetitive sequences. *Science*, **204**, 1052–1059.
Denis, H. and Wegnez, M. (1973) Recherche biochimiques sur l'oogenèse. 7. Synthèse et maturation du RNA 5S dans les petit oocytes de *Xenopus laevis*. (Biochemical research on oogenesis. 7. Synthesis and maturation of 5S RNA in the small oöcytes of *Xenopus laevis*.) *Biochimie*, **55**, 1137–1151.
Domingo, E., Flavell, R. A. and Weissmann, C. (1976) *In vitro* site-directed mutagenesis: Generation and properties of and infectious extracistronic mutant of bacteriophage Qβ. *Gene*, **1**, 3–25.
Donaldson, D. S., McNab, A. R., Rovera, G. and Curtis, P. J. (1982) Nuclear precursor molecules of the two β-globin mRNAs in Friend erythroleukemia cells. *J. Biol. Chem.*, **257**, 8655–8660.
Edelmann, P. and Gallant, J. (1977) Mistranslation in *E. coli*. *Cell*, **10**, 131–137.
Ellis, N. and Gallant, J. (1982) An estimate of the global error frequency in translation. *Mol. Gen. Genet.*, **188**, 169–172.
Farnham, P. J. and Platt, T. (1980) A model for transcription termination suggested by studies on the *trp* attenuator *in vitro* using base analogs. *Cell*, **20**, 739–748.
Felber, B. K., Orkin, S. H. and Hamer, D. H. (1982) Abnormal RNA splicing causes one form of α thalassemia. *Cell*, **29**, 895–902.
Fitzgerald, M. and Shenk, T. (1981) The sequence 5'-AAUAAA-3' forms part of the recognition site for polyadenylation of late SV40 mRNAs. *Cell*, **24**, 251–260.
Fluck, M. M. and Epstein, R. H. (1980) Isolation and characterization of context

mutations affecting the suppressibility of nonsense mutations. *Mol. Gen. Genet.*, **177**, 615–627.

Fluck, M. M., Salser, W. and Epstein, R. H. (1977) The influence of the reading context upon the suppression of nonsense codons. *Mol. Gen. Genet.*, **151**, 137–149.

Ford, J. P. and Hsu, M. T. (1978) Transcription pattern of *in vivo* labeled late SV40 RNA. *J. Virol.*, **28**, 795–801.

Fraser, N. W. and Hsu, M.-T. (1980) Mapping of the 3′ terminus of the large late Ad-2 transcript by electron microscopy. *Virology*, **103**, 514–516.

Fraser, N. W., Nevins, J. R., Ziff, E. and Darnell, J. E. (1979) The major late adenovirus type-2 transcription unit: Termination is downstream from the last poly (A) site. *J. Mol. Biol.*, **129**, 643–656.

Fukumaki, Y., Ghosh, P. K., Benz, E. J., Reddy, V. B., Lebowitz, P., Forget, B. G. and Weissman, S. M. (1982). Abnormally spliced messenger RNA in erythroid cells from patients with β^+ thalassemia and monkey cells expressing a cloned β^+-thalassemic gene. *Cell*, **28**, 585–593.

Gidoni, D., Kahana, C., Canaani, D. and Groner, Y. (1981) Specific *in vitro* initiation of transcription of simian virus 40 early and late genes occurs at the various cap nucleotides including cytidine. *Proc. Natl Acad. Sci. USA*, **78**, 2174–2178.

Goldsmith, M. E., Humphries, R. K., Ley, T., Cline, A., Kantor, J. A. and Nienhuis, A. W. (1983) 'Silent' nucleotide substitution in a β^+-thalassemia globin gene activates splice site in coding sequence RNA. *Proc. Natl Acad. Sci. USA*, **80**, 2318–2322.

Gurdon, J. B. and Brown, D. D. (1978) The transcription of 5S DNA injected into *Xenopus laevis* oocytes. *Develop. Biol.*, **67**, 346–356.

Hagenbüchle, O. and Schibler, U. (1981) Mouse β-globin and adenovirus-2 major late transcripts are initiated at the cap site *in vitro*. *Proc. Natl Acad. Sci. USA*, **78**, 2283–2286.

Higgs, D. R., Goodbourn, S. E. Y., Lamb, J., Clegg, J. D. and Weatherall, D. J. (1983) α-Thalassaemia caused by a polyadenylation signal mutation. *Nature*, **306**, 398–400.

Hofer, E. and Darnell, J. E. (1981) The primary transcription unit of the mouse β-major globin gene. *Cell*, **23**, 585–593.

Hofer, E., Hofer-Warbinek, R. and Darnell, J. E. (1982) Globin RNA transcription: a possible termination site and demonstration of transcriptional control correlated with altered chromatin structure. *Cell*, **29**, 887–893.

Holmes, W. M., Platt, T. and Rosenberg, M. (1983) Termination of transcription in E. coli. *Cell*, **32**, 1029–1032.

Hopfield, J. J. (1974) Kinetic proofreading: A new mechanism for reducing errors in biosynthetic processes requiring high specificity. *Proc. Natl Acad. Sci. USA*, **71**, 4135–4139.

Johnston, T. C., Borgia, P. T. and Parker, J. (1984) Codon specificity of starvation induced misreading. *Mol. Gen. Genet.*, **195**, 459–465.

King, C. R. and Piatigorsky, J. (1984) Alternative splicing of αA-crystallin RNA: structural and quantitative analyses of the mRNAs for the αA$_2$- and αA^{ins}-crystallin polypeptides. *J. Biol. Chem.*, **259**, 1822–1826.

Krainer, A. R., Maniatis, T., Ruskin, B. and Green, M. R. (1984) Normal and mutant human β-globin pre-mRNAs are faithfully and efficiently spliced *in vitro*. *Cell*, **36**,

993–1005.

Krämer, A., Keller, W., Appel, B. and Lührmann, R. (1984) The 5' terminus of the RNA moiety of U1 small nuclear ribonucleoprotein particles is required for the splicing of messenger RNA precursors. *Cell*, **38**, 299–307.

Krieg, P. A. and Melton, D. A. (1984) Formation of the 3' end of histone mRNA by post-transcriptional processing. *Nature*, **308**, 203–206.

Kröger, M. and Singer, B. (1979) Ambiguity and transcriptional errors as a result of methylation of N-1 of purines and N-3 of pyrimidines. *Biochemistry*, **18**, 3493–3500.

Kühne, T., Wieringa, B., Reiser, J. and Weissmann, C. (1983) Evidence against a scanning model of RNA splicing. *EMBO J.*, **2**, 727–733.

Lang, K. M. and Spritz, R. A. (1983) RNA splice site selection: Evidence for a 5' → 3' scanning model. *Science*, **220**, 1351–1355.

Lee, F., Squires, C. L., Squires, C. and Yanofsky, C. (1976) Termination of transcription *in vitro* in the *Escherichia coli* tryptophan operon leader region. *J. Mol. Biol.*, **103**, 383–393.

Lerner, M. R., Boyle, J. A., Mount, S. M., Wolin, S. L. and Steitz, J. A. (1980) Are snRNPs involved in splicing? *Nature*, **283**, 220–224.

Lewin, B. (1980) Alternatives for splicing: Recognizing the ends of introns. *Cell*, **22**, 324–326.

Loeb, L. A. and Kunkel, T. A. (1982) Fidelity of DNA synthesis. *Ann. Rev. Biochem.*, **52**, 429–457.

Loftfield, R. and Vanderjagt, D. (1972) The frequency of errors in protein synthesis. *Biochem. J.*, **128**, 1353–1356.

MacCumber, M. and Ornstein, R. L. (1984) Molecular model for messenger RNA splicing. *Science*, **224**, 402–405.

Maizels, N. M. (1973) The nucleotide sequence of the lactose messenger ribonucleic acid transcribed from the UV5 promoter mutant of *Escherichia coli*. *Proc. Natl Acad. Sci. USA*, **70**, 3585–3589.

Majors, J. (1975) Initiation of an *in vitro* mRNA synthesis from the wild-type *lac* promoter. *Proc. Natl Acad. Sci. USA*, **72**, 4394–4398.

Martin, F. and Tinoco, I. (1980) DNA–RNA hybrid duplexes containing (dA:rU) sequences are exceptionally unstable and may facilitate termination of transcription. *Nucl. Acids Res.*, **8**, 2295–3000.

McDevitt, M. A., Imperiale, M. J., Ali, H. and Nevins, J. R. (1984) Requirement of a downstream sequence for generation of a poly (A) addition site. *Cell*, **37**, 993–999.

Montell, C., Fisher, E. F., Caruthers, M. H. and Berk, A. J. (1983) Inhibition of RNA cleavage but not polyadenylation by a point mutation in mRNA 3' consensus sequence AAUAAA. *Nature*, **305**, 600–605.

Moore, C. L. and Sharp, P. A. (1984) Site-specific polyadenylation in a cell-free reaction. *Cell*, **36**, 581–591.

Mount, S. M. (1982) A catalogue of splice junction sequences. *Nucl. Acids Res.*, **10**, 459–472.

Murray, V. and Holliday, R. (1979a) Mechanism for RNA splicing of gene transcripts. *FEBS Letts*, **106**, 5–7.

Murray, V. and Holliday, R. (1979b) A mechanism for RNA–RNA splicing and a model for the control of gene expression. *Genet. Res.*, **34**, 173–188.

Nevins, J. R. (1983) The pathway of eukaryotic mRNA formation. *Ann. Rev.*

Biochem., **52**, 441–466.

Nevins, J. R., Blanchard, J. M. and Darnell, J. E. (1980) Transcription units of adenovirus type 2: Termination of transcription beyond the poly (A) addition site in early regions 2 and 4. *J. Mol. Biol.*, **144**, 377–386.

Ninio, J. (1975) Kinetic amplification of enzyme discrimination. *Biochimie*, **57**, 587–595.

Ozoline, O. N., Oganesjan, M. G. and Kamzolova, S. G. (1980) On the fidelity of transcription of *Escherichia coli* RNA polymerase. *FEBS Letts*, **110**, 123–125.

Padgett, R. A., Konarska, M. M., Grabowski, P. J., Hardy, S. F. and Sharp, P. A. (1984). Lariat RNAs as intermediates and products in the splicing of messenger RNA precursors. *Science*, **225**, 898–903.

Parker, J., Johnston, T. C., Borgia, P. T., Holtz, G., Remaut, E. and Fiers, W. (1983) Codon usage and mistranslation: *In vivo* basal level misreading of the MS2 coat protein message. *J. Biol. Chem.*, **258**, 10 007–10 012.

Price, D. H. and Parker, C. S. (1984) The 3' end of *Drosophila* histone H3 mRNA is produced by a processing activity *in vitro*. *Cell*, **38**, 423–429.

Proudfoot, N. J. and Brownlee, G. G. (1976) 3' Non-coding region sequences in eukaryotic messenger RNA. *Nature*, **263**, 211–214.

Radding, C. M. and Kornberg, A. (1962) Enzymatic synthesis of deoxyribonucleic acid: XIII. Kinetics of primed and *de novo* synthesis of deoxynucleotide polymers. *J Biol. Chem.*, **237**, 2877–2882.

Rogers, J. and Wall, R. (1980) A mechanism for RNA splicing. *Proc. Natl Acad. Sci USA*, **77**, 1877–1879.

Rosenberger, R. F. and Foskett, G. (1981) An estimate of the frequency of *in vivo* transcriptional errors at a nonsense codon in *Escherichia coli*. *Mol. Gen. Genet.* **183**, 561–563.

Rosenberger, R. F. and Hilton, J. (1983) The frequency of transcriptional and translational errors at nonsense codons in the *lacZ* gene of *Escherichia coli*. *Mol. Gen. Genet.*, **191**, 207–212.

Ruskin, B., Krainer, A. R., Maniatis, T. and Green, M. R. (1984) Excision of an intact intron as a novel lariat structure during pre-mRNA splicing *in vitro*. *Cell*, **38**, 317–331.

Ryan, T. and Chamberlain, M. J. (1983) Transcriptional analyses with heteroduplex *trp* attenuator templates indicate that the transcript stem and loop structure serves as the termination signal. *J. Biol. Chem.*, **258**, 4690–4693.

Salisbury, J. G., O'Conner, P. J. and Saffhill, R. (1978) Molecular size and fidelity of DNA polymerase α from the regenerating liver of the rat. *Biochim. Biophys. Acta*, **517**, 181–185.

Salser, W. (1977) Globin mRNA sequences: Analysis of base pairing and evolutionary implications. *Cold Spring Harbor Symp. Quant. Biol.*, **42**, 985–1002.

Sharp, P. A. (1981) Speculations on RNA splicing. *Cell*, **23**, 643–646.

Singer, B. and Spengler, S. (1981) Ambiguity and transcriptional errors as a result of modification of exocyclic amino groups of cytidine, guanosine, and adenosine. *Biochemistry*, **20**, 1127–1132.

Springgate, C. F. and Loeb, L. A. (1975) On the fidelity of transcription by *Escherichia coli* ribonucleic acid polymerase. *J. Mol. Biol.*, **97**, 577–591.

Strniste, G. F., Smith, D. A. and Hayes, F. N. (1973) X-ray inactivation of the

Escherichia coli deoxyribonucleic acid dependent ribonucleic acid polymerase in aqueous solution. II. Studies on initiation and fidelity of transcription. *Biochemistry*, **12**, 603–608.

Tatei, K., Takemura, K., Mayeda, A., Fujiwara, Y., Tanaka, H., Ishihama, A. and Ohshima, Y. (1984) U1 RNA-protein complex preferentially binds to both 5' and 3' splice junction sequences in RNA or single-stranded DNA. *Proc. Natl Acad. Sci. USA*, **81**, 6281–6285.

Tinoco, I., Borer, P. N., Dengler, B., Levine, M. D. and Oblenbeck, O. C. (1973) Improved estimation of secondary structure in ribonucleic acids. *Nature*, **246**, 40–41.

Treisman, R., Orkin, S. H. and Maniatis, T. (1983) Specific transcription and RNA splicing defects in five cloned β-thalassaemia genes. *Nature*, **302**, 501–596.

Treisman, R., Proudfoot, N. J., Shander, M. and Maniatis, T. (1982) A single-base change at a splice site in a β^0-thalassemic gene causes abnormal RNA splicing. *Cell*, **29**, 903–911.

Volloch, V. Z., Rits, S. and Tumerman, L. (1979) A possible mechanism responsible for the correction of transcription errors. *Nucl. Acids Res.*, **6**, 1535–1546.

Wieringa, B., Hofer, E. and Weissmann, C. (1984) A minimal length but no specific internal sequence is required for splicing the large rabbit β-globin intron. *Cell*, **37**, 915–925.

Wieringa, B., Meyer, F., Reiser, J. and Weissmann, C. (1983) Unusual splice sites revealed by mutagenic inactivation of an authentic splice site of the rabbit β-globin gene. *Nature*, **301**, 38–43.

Yang, V. W., Lerner, M. R., Steitz, J. A. and Flint, S. J. (1981) A small nuclear ribonucleoprotein is required for splicing of adenoviral early RNA sequences. *Proc. Natl Acad. Sci. USA*, **78**, 1371–1375.

8 DNA replication fidelity and base mispairing mutagenesis

M. F. GOODMAN and E. W. BRANSCOMB

8.1 Introduction

Central to understanding evolution and to unravelling the complex relationships connecting mutagenesis to ageing and cancer is an appreciation of processes designed to maintain the integrity of genetic information. It is reasonable to begin addressing these processes by directing one's attention initially to a simple class of mutations. It will be our main purpose in this chapter to focus upon the events leading to the formation of single base-pair purine–pyrimidine mismatches and their subsequent partial elimination from DNA. Single base mismatches in DNA can occur spontaneously and by exposure of a cell to certain groups of mutagenic agents.

Purine–pyrimidine mispairs, e.g., A–C or G–T base-pairs, were the first to be considered on a molecular level. In fact, in the landmark papers by Watson and Crick (1953a, b) on the structure of DNA, there is a clearly stated model explaining how spontaneous transition mutations might arise. They suggest that imino and enol tautomeric forms of the nucleotide bases could have the correct geometric configuration to base-pair 'properly' with non-Watson–Crick partners.

The idea that minor isomeric forms of the bases can be responsible for base-mispairing mutagenesis has continued to generate interest. Freese (1959) published an influential paper in which the base analogues 2-aminopurine (AP) and 5-bromouracil (BU) were proposed to undergo tautomeric transitions more readily and, as a result, to exhibit ambiguities in base-pairing properties leading to enhanced rates of induction of both $A \cdot T \rightarrow G \cdot C$ and $G \cdot C \rightarrow A \cdot T$ transitions. Topal and Fresco (1976) have suggested that anti- and syn-isomers allow for the formation of purine–purine and pyrimidine–pyrimidine transversions. Yet, despite the attractive simplicity of these models, there is currently no direct experimental evidence to support them. What can be said with some assuredness concerning minor tautomers is that they do not compete for incorporation on an equal footing with the normal bases; their equilibrium concentrations in solution are thought to be several orders of magnitude larger than spontaneous base substitution mutation rates (Katritzky and Waring, 1962; Wolfenden, 1969) which are in the range of 10^{-6}–10^{-10} per base-pair per round of replication in micro-organisms (Drake, 1969) and perhaps even less in higher organisms. Rare tautomer form models, whether ultimately shown to be right or wrong, have nevertheless served the very useful purpose of calling attention to the requirement for cellular error correction processes to achieve the fidelity levels exhibited by normal cells.

It is a common occurrence in science that avenues of experimental investigation do not always intersect at any given time with related theoretical developments. In a three-year period, 1965–1968, a series of important papers appeared (Speyer, 1965; Speyer, Karam and Lenny, 1966; Freese and Freese, 1967; Drake and Allen, 1968) which had no immediate connection to the tautomer models. These genetic studies offered conclusive evidence that bacteriophage T4 DNA polymerase must play an essential role in determining the fidelity of T4 DNA replication. The essence of the studies was that single base changes in gene 43, the gene coding for T4 DNA polymerase, can cause a significant increase or *decrease* in spontaneous and base analogue induced mutation frequencies compared to wild type T4 (43^+). For some alleles in the r_{II} region of T4 the difference between the r_{II} reversion frequencies in *ts*L56 mutator and *ts*L141 antimutator backgrounds can be as much as 10^3–10^4-fold (Speyer, 1965; Drake and Allen, 1968; Drake *et al.*, 1969); the spontaneous reversion frequencies ($A \cdot T \rightarrow G \cdot C$) at the r_{II} UV199 locus in *ts*L56, 43^+, and *ts*L141 backgrounds are about 10^{-5}, 10^{-7}, and 10^{-9}, respectively (see, for example, Goodman, Hopkins and Gore, 1977).

Brutlag and Kornberg (1972) using DNA polymerase I from *Escherichia coli* and Muzyczka, Poland and Bessman (1972) using T4 mutator, wild type, and antimutator polymerases proposed that the 3'-exonuclease activity which appears to be an integral component in all prokaryotic DNA polymerases studied to date is responsible for editing newly synthesized DNA at the 3'-OH growing point. When DNA polymerases from *E. coli* and T4 are presented with specially constructed DNA templates containing single nucleotide mismatches at the 3'-OH end, their associated 3'-exonuclease activities excise terminal mismatches (Brutlag and Kornberg, 1972; Muzyczka, Poland and Bessman, 1972) to allow synthesis to begin at properly paired termini. An essential concept emanating from Bessman's studies is the importance of the relative *rates* (as opposed to specificities) of exonuclease and polymerase activities in determining nucleotide misincorporation frequencies. An enzyme exhibiting large exonuclease 'turnover' (Muzyczka, Poland and Bessman, 1972; Bessman *et al.*, 1974) to polymerase ratio (t/P) (*ts*L141 antimutator polymerase) makes fewer errors than one with small t/P ratio (*ts*L56 mutator). The point to be emphasized is that mutator and antimutator polymerases may, under some circumstances, exhibit similar specificities for both insertion and removal of mismatched nucleotides, while substantial differences, in the neighbourhood of thirty-fold, have been measured in t/P activity ratios (Bessman *et al.*, 1974).

During this same period, however, Hershfield (1973) and Nossal and Hershfield (1973) showed that the T4 mutator polymerase, tsL88, enhances insertion of mismatched bases by comparison with 43^+ polymerase. More recently, Reha-Krantz and Bessman (1981) have discovered another mis-insertion mutator tsM19. Thus, the full range of enzymological studies using T4 mutant and wild type polymerases taken together show that both the insertion and excision functions of the DNA polymerase as well as their relative activities, have an essential role in determining fidelity.

In recent years the work from Kornberg's group (Schekman, Wickner and Kornberg, 1974) and by Wickner and Hurwitz (1976) using *E. coli* and small coliphages and from Alberts' group (Alberts *et al.*, 1975) and by Nossal (1979) using T4 clearly shows that DNA replication requires a well-defined complex of proteins. Proteins other than DNA polymerase are required, for example, to initiate Okazaki fragments and to destabilize and unwind the DNA double helix at the replication fork. Since there is now compelling evidence demonstrating a direct physical interaction between the various proteins in the replication complex, it is reasonable to suppose that replication complex proteins other than polymerase affect fidelity either directly or indirectly by causing perturbations in the properties of the polymerase.

In genetic studies similar to those with T4 mutants in gene 43, Bernstein *et al.* (1972), Watanabe and Goodman (1978), and Mufti (1979) observed the effect of mutations in T4 genes 32, 41, 45, 44, 62, which together are now known to comprise the replication complex, on reversion rates at other genetic loci, principally in the r_{II} region. Significant mutator and antimutator effects from mutants in the five T4 replication complex proteins were observed, but these effects appear to be at least an order of magnitude smaller compared to those induced by DNA polymerase mutators and antimutators. Studies from Alberts' laboratory with the T4 replication complex system *in vitro* (Liu *et al.*, 1978; Hibner and Alberts, 1980) suggest that the intact complex synthesizes DNA with measurably enhanced fidelity compared with T4 DNA polymerase alone. However, these data are still essentially consistent with the genetic studies leading to a probable conclusion that the predominant means of controlling replication fidelity in T4 is via the DNA polymerase. One possible effect of other replication proteins may be to increase the turnover/polymerase ratio of DNA polymerase, thereby lowering mutation rates.

The T4 system may prove to be atypical in its reliance on a polymerase associated exonuclease to mediate DNA synthesis fidelity. The exonuclease/polymerase activity ratios in the case of T4 polymerases are exceptionally large. It is interesting to note, for example, that the strong mutator L56 polymerase, which has a smaller t/P ratio than 43^+ polymerase, has nevertheless about an order of magnitude larger ratio than *E. coli* Pol I. Although, as mentioned earlier, Brutlag and Kornberg (1972) have shown that Pol I removes mismatched 3'-OH termini prior to nucleotide addition, and Clayton *et al.* (1979) have observed that Pol I was able to turn over a 2-aminopurine (AP) deoxynucleotide in preference to dAMP at a growing 3'-terminus, Loeb *et al.* (1981) have argued on the basis of the low turnover activity and high intrinsic error rates measured on synthetic copolymer templates *in vitro* that Pol I's 3'-exonuclease may not be playing a predominant error-correcting role *in vivo*.

It is also important to realize that DNA polymerases from animal cells have no measurable exonuclease activities – see, for example, Sedwick, Wang and Korn (1975) and Weissbach (1977). Obviously, it is still possible that error-correcting nucleases can exist independently of polymerases, but the question which currently remains open is whether it is generally required that nuclease editors act in concert with polymerases at the replication fork since errors might also be eliminated post-replicatively. It has already been shown that post-replication repair of mismatched bases occurs in *E. coli* (Nevers and Spatz, 1975; Wagner and Meselson, 1976; Rydberg, 1977, 1978), and may reduce mutation frequencies possibly on the order of 10^3-fold (Glickman and Radman, 1980). Post-replication error correction can take place selectively on a newly synthesized strand of DNA which is distinguishable from the parental DNA by its lack of methylated adenines (Glickman and Radman, 1980; Radman *et al.*, 1980). Although there is no supporting evidence as yet, it would not be surprising if post-replication removal of single base mismatches were found to be an important means of insuring fidelity in animal cells.

There are a number of additional primary effects on replication fidelity which occur more or less independently from those due to polymerases, nucleases, replication complex proteins, and post-replication repair systems. These include base context and DNA sequence configuration effects (Koch, 1971; Ronen and Rahat, 1976; Ronen, Rahat and Halevy, 1976; Bessman and Reha-Krantz, 1977; Coulondre *et al.*, 1979; Hopkins and Goodman, 1980; Topal, DiGiuseppi and Sinha, 1980), hydrogen bonding free energy differences between 'competing' correct and incorrect nucleotides at the growing point of DNA (Galas and Branscomb, 1978; Clayton *et al.*, 1979), and perturbations in deoxyribonucleoside triphosphate pool concentrations (Clayton *et al.*, 1979; Fersht, 1979; Hopkins and Goodman, 1980). It might even turn out that enzyme complexes in prokaryotes (Tomich *et al.*, 1974; Mathews, North and Reddy, 1979) and eukaryotes (Reddy and Pardee, 1980) which are believed to be involved in deoxyribonucleotide biosynthesis and compartmentalization also affect fidelity.

We have recently suggested (Goodman *et al.*, 1980) that it is now feasible to investigate quantitative models relating fidelity as measured *in vitro* to mutagenesis *in vivo*. Part of this paper will be devoted to showing how some *in vitro–in vivo* relationships have been established. We shall show that deoxyribonucleotide pool sizes, which can be controlled *in vitro* by the experimenter but which the cell for the most part controls *in vivo*, and differences in base-pairing free energies, which may be largely independent of the various replication and error correction proteins, are central to establishing these relationships.

To illustrate these ideas, we will evaluate a K_m discrimination model (Galas and Branscomb, 1978; Clayton *et al.*, 1979; Goodman *et al.*, 1980) for error correction by DNA polymerase in which it is assumed *a priori* that the polymerase has no ability to distinguish between correct and incorrect competing candidate deoxyribonucleotides. A critical evaluation of model predictions will be made using data obtained from the T4 mutator–antimutator system. The T4 system provides a vast dynamic range in genetic marker reversion frequencies (10^1–10^4-fold differences between mutators and antimutators at different marker alleles). The system is also strongly susceptible to mutagenesis by the base analogues 2-aminopurine (AP) and 5-bromouracil (BU). The dynamic range and base analogue features will prove to be particularly convenient in establishing examples where simple hydrogen bonding ideas succeed and also where they largely fail to provide an explanation for the large differences in mutator and antimutator mutation frequencies.

8.2 K_m Discrimination model

The simplest and most economical assumption we can make regarding the mechanism responsible for fidelity during DNA synthesis is that differences in base-pairing strengths account entirely for the discrimination against mispairs. Such an assumption permits stringent, quantitative predictions to be made and should therefore be the most easily disproved. For polymerases that contain an editing exonuclease, the specificity of mispair rejection for both the polymerizing reaction and the exonuclease reaction must be quantitatively accounted for in terms of the base-pairing bond strengths. We will review a model of this type (Galas and Branscomb, 1978; Clayton *et al.*, 1979) illustrated in Fig. 8.1. In this model the DNA polymerase and associated 3′-exonuclease activity, if present, are both viewed as illiterate elements in the sense that they cannot distinguish between correct and incorrect nucleotides *per se*, whether or not the other proteins of the replication complex are present. This K_m discrimination model is one possible alternative to earlier proposals suggesting a literate role for the polymerase (Koch and Miller, 1965; Speyer, Karam and Lenny, 1966). To develop the K_m

Insertion step

Misinsertion ratio

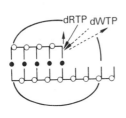

$$\frac{I\,(W)}{I\,(R)} = \frac{[dWTP]}{[dRTP]}\, e^{\frac{-\Delta G_p}{RT}} = \frac{[dWTP]}{[dRTP]}\, \frac{K_{eq}^{d}}{K_{eq}^{W}}$$

Valid for most polymerases independent of
total [dNTP], + or – replication complex

Editing ratio

Editing step

$$\frac{\text{Fraction wrong left in}}{\text{Fraction right left in}} \begin{cases} \geqslant e^{\frac{-\Delta G_E}{RT}} & \text{– strong antimutator (L141)} \\ \leqslant 1 & \text{– strong mutator (L56) or} \\ & \text{mammalian polymerases} \end{cases}$$

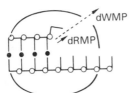

Misincorporation ratio

$$\frac{I\,(W)}{I\,(R)} \underset{\sim}{<} \frac{[dWTP]}{[[dRTP]}\, e^{\frac{-\Delta G_E}{RT}}\, e^{\frac{-G_p}{RT}}$$

Depends on total [dNTP], replication proteins

Figure 8.1 K_m discrimination model. Two substrate deoxyribonucleotides, dRTP (right) and dWTP (wrong), are shown in the upper diagram competing for insertion opposite a template site. The similar association rates of the substrates to the polymerase–DNA complex are illustrated by arrows of equal length. The unequal length arrows represent a more rapid dissociation of dWTP compared with dRTP because of reduced H-bond stability of mismatched compared with properly matched base-pairs. In the lower diagram, the unequal length arrows represent the different rates for melting out a terminal nucleotide; dWMP melts out faster than dRMP, permitting its preferential excision by a polymerase associated 3′-exonuclease. ΔG is the equilibrium free energy difference between matched and mismatched base-pairs. The misincorporation ratio, I (W)/I (R), is approximately equal to the product of the misinsertion ratio and the editing ratio. The latter ratio is constrained to be greater than or equal to the factor $\exp(-\Delta G_E/RT)$ corresponding to a maximum exploitation of the differences in melting rates.

discrimination ideas we will consider separately the two steps of the polymerization process: first insertion and then excision of a nucleotide.

8.2.1 FIDELITY AT THE INSERTION STEP

We begin by considering the ways in which enzymic reaction rates for competing substrates can be controlled differentially. Figures 8.2(a) and (b) illustrate two possible enzymic methods for 'selecting' against insertion of a wrong deoxyribonucleotide substrate in competition with a right substrate. In

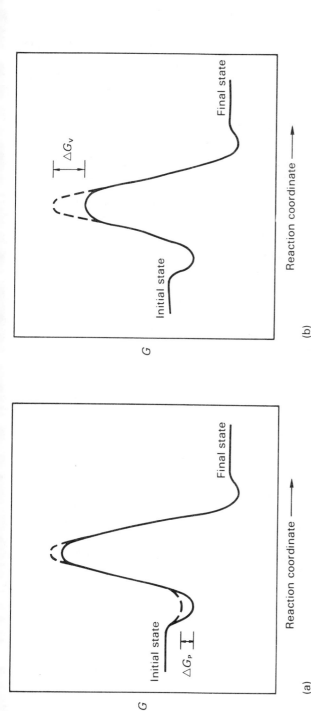

Figure 8.2 Free energy reaction path diagrams for enzyme mediated K_m and V_{max} discrimination. Free energy reaction paths for the polymerization reaction are shown as a function of a formal reaction coordinate. Reactants (the enzyme–template–primer complex and dNTPs) progress from a free initial state to an associated complex at lower energy, then through high energy activated intermediate states to the final state in which a new nucleotide has been added to the primer. The path followed by the right dNTP is indicated by a solid line, that followed by the wrong dNTP by a dashed line. In K_m discrimination, part (a), competing substrates (dNTPs) differ in the strength of the bonds formed in the associated complex (by the energy ΔG_p). Incorrect substrates associate more weakly, dissociate more rapidly, and have a correspondingly reduced probability of entering the activated intermediate state. In V_{max} discrimination, part (b), competing substrates are distinguished by the height of the activation energy barrier they must cross. Correct substrates see a lower barrier and therefore react more readily than do incorrect substrates.

Fig. 8.2(a), selection occurs at the triphosphate binding step. We will refer to this type of selection by the term 'K_m discrimination'. This is illustrated by indicating (dashed lines in the left hand well) a smaller free energy for binding the wrong substrate. The *difference* in free energies between deoxynucleotide substrates bound 'wrong' and those bound 'right', ΔG_P, is a direct measure of the DNA polymerase insertion discrimination:

$$\frac{I(W)}{I(R)} = \frac{\text{Insertion wrong base}}{\text{Insertion right base}} = \frac{[dWTP]}{[dRTP]}\exp(-\Delta G_P/RT) \qquad (8.1)$$

where [dWTP] and [dRTP] are the pool size concentrations of competing wrong and right deoxyribonucleoside triphosphates respectively.

There is an obvious physical mechanism which can provide template-dependent binding discrimination as shown in Fig. 8.2(a); we will describe this K_m discrimination mechanism in some detail since it serves as the basis for the model. Let us imagine that two deoxynucleotide substrates compete for access to the polymerase triphosphate binding site on the enzyme. The 'on'-rate (association rate constant, a in Equation (8.2)) is assumed to be diffusion limited and is therefore the *same* for both correct and incorrect substrates *when the substrates are at equal concentration*.

$$(E-\text{DNA}_n) + \text{dNTP} \underset{d}{\overset{a}{\rightleftharpoons}} (E-\text{DNA}_n-\text{dNTP}) \overset{p}{\longrightarrow} (E-\text{DNA}_{n+1}) + \text{PPi}$$

$$(8.2)$$

However, the 'off'-rates can be very different. Suppose, for example, the off-rate, or equivalently the length of time (residence time) that a substrate remains bound on a polymerase–DNA complex, is controlled by the sum of three independent binding terms: (1) nucleotide non-specific binding at the polymerase triphosphate binding site, (2) hydrogen bonds between substrate and template bases, and (3) nearest neighbour base stacking interactions. The ratio of the residence times for correct versus incorrect nucleotides is given by an exponential of the difference in total substrate binding energy between right and wrong bases and is therefore a function of the difference in base-pairing stabilities at the template site.

Assuming that binding at the triphosphate site is of roughly equal magnitude for both substrates, the stability for any particular base-pair is determined by specific Watson–Crick H-bonds and by non-specific nearest neighbour base stacking interactions. Since base stacking effects serve only to lessen the specificity inherent in Watson–Crick H-bonds, it may be that the replication complex proteins act to minimize base stacking contributions to base-pairing free energy differences. In any case, the misinsertion frequency averaged over all stacking environments should depend only on the relative stabilities of the Watson–Crick H-bonds found. If we now assume that the insertion rate for a given nucleotide is proportional to the residence time, then the ratio of the rates of insertion for wrong and right nucleotides, $I(W)/I(R)$, is given by the ratio of Michaelis constants ($K_m = (d+p)/a$), i.e.,

$$e^{-\Delta G_P/RT} = \frac{K_m^R}{K_m^W} \tag{8.3}$$

so that $\dfrac{I(W)}{I(R)} = \dfrac{[dWTP]}{[dRTP]} e(-\Delta G_P/RT) = \dfrac{[dWTP]}{[dRTP]} \dfrac{K_m^R}{K_m^W}$ $\qquad$ (8.4)

In the limit where phosphodiester bond formation is slow compared with substrate dissociation rates, the misinsertion ratio is proportional to the ratio of dissociation rate constants for right and wrong substrates.

In this simple picture, differences in base-pairing stabilities control DNA polymerase discrimination during the nucleotide insertion step. The enzyme is playing an illiterate role in the sense that it does not modify its binding affinities or its catalytic rates in response to the base presented on the template in order to select or reject nucleotides. Instead, the differences in residence times are governed by differences in H-bond stabilities between *any* two candidate nucleotides. A 'wrong' substrate simply diffuses off the enzyme–template complex at a more rapid rate than a 'right' substrate. The parameter ΔG_P (Fig. 8.2(a) and Equation (8.1)) enters as a free energy difference between correct and incorrect base-pair insertions, e.g., A·T versus A·C. Later we will consider the complications introduced into this picture by base stacking energies.

The value of ΔG_P, determined from experimental values of $I(W)$ and $I(R)$, can be used to define a quantitative free energy scale proceeding from the most stable base-pair G.C to least stable base-pairs G.T and A.C (Goodman, Watanabe and Branscomb, 1982). If the assumptions in the model are reasonable, then this parameter should provide a physical basis for comparing different pairs of nucleotides competing for insertion at a 3'-OH primer site on DNA. The misinsertion ratio $I(W)/I(R)$, which defines ΔG_P (Equation (8.1)), is obtained by measuring wrong to right insertions during DNA synthesis. We have carried out measurements of this type for the base analogue 2-aminopurine (AP) competing with A for insertion opposite a template T (Clayton *et al.*, 1979), and for T and C competing opposite templates AP and A (Watanabe and Goodman, 1981).

If the polymerase is in fact playing an illiterate role in insertion, then $I(W)/I(R)$ should depend solely on H-bond differences and should be the same for *all* normal DNA polymerases. It has been observed, for example, that several T4 mutators, wild type, and antimutator polymerases misinsert AP opposite T at similar frequencies (Bessman *et al.*, 1974; Clayton *et al.*, 1979); this is true also for calf thymus α, and *E. coli* Pol I (Clayton *et al.*, 1979). Since a wide variety of competing base-pairs can be analysed in this manner, it will be possible to observe whether different polymerases exhibit similar or different misinsertion ratios for other competing base-pairs, e.g., A versus C for insertion opposite T. Although an illiterate polymerase in general cannot improve on the insertion accuracy determined by the hydrogen binding free energies, it can degrade this accuracy. For example, any distortions of the steric constraints imposed by the enzyme on the competing base-pair could modify the base-pairing free energies and thus change, and in general degrade, the discrimination between candidate nucleotides. We have evidence to suggest that the mutagenicity of Mn^{2+} results from such effects (Goodman *et al.*, 1983), and it is possible that the insertion mutators L88 (Hershfield, 1973; Nossal and Hershfield, 1973) and M19 (Reha-Krantz and Bessman, 1981) also operate in such a way.

A DNA polymerase might, in addition, discriminate between substrates in a template controlled way by adjusting its intrinsic catalytic rate (V_{max}). Figure 8.2(b) illustrates this second type of kinetic discrimination mechanism in which $\Delta G_P = 0$, but the rate of phosphodiester bond catalysis depends on whether a right or wrong nucleotide is resident on the enzyme–template complex. The difference ΔG_V in the free energy of activation determines the difference in velocity between two competing substrates, i.e., V_{max} discrimination. Since the two substrates are assumed to have roughly similar K_ms, a lowered velocity for inserting the wrong substrate would enable it to dissociate preferentially from the enzyme–template complex. The insertion discrimination is again given by Equation (8.1) with ΔG_V replacing ΔG_P.

In contrast to K_m discrimination, we know of no simple mechanism by which a template base can directly affect the V_{max} of the reaction to favour the properly paired nucleotide. For this reason, finding a difference in the V_{max} for insertion between two corresponding nucleotides would suggest that the enzyme is playing a literate role in base selection. Such literate discrimination mechanisms could in principle achieve very great accuracies. A theoretical accuracy limit is of course imposed by the difference in free energy between the reactants and products in the overall reaction (Equation (8.2)).

8.2.2 FIDELITY AT THE EXCISION STEP

A second ΔG discrimination step can be achieved for polymerases which contain associated $3' \rightarrow 5'$ exonuclease activities, a property shared by most prokaryotic DNA polymerases studied to date (Kornberg, 1980). In a manner analogous to insertion, it is possible to describe nucleotide excision (proofreading) in terms of Michaelis–Menten enzyme kinetic mechanisms. Here, just as with the insertion step, either illiterate or literate mechanisms may be involved. For example, one can imagine that the enzyme senses the presence of a mispaired base and modifies its exonucleolytic properties (binding affinities or reaction rate) so as to increase the probability of terminal base excision. Alternatively, the physical properties of malformed base-pairs may themselves cause misinserted bases to be preferentially excised by an enzyme whose catalytic properties are unchanging.

Since it is only in this latter illiterate case that a simple quantitatively predictive model can be generated, we will proceed with the discussion of a model of this type. This illiterate picture again uses the notion that H-bonding and not enzyme (in this case exonuclease) specificity controls the removal of $3'$-nucleotides at a primer terminus. This picture can be justified by recognizing that polymerase associated $3' \rightarrow 5'$ exonucleases are known to be much more active in digesting single-stranded as compared to double-stranded DNA (Kornberg, 1974). We may therefore postulate (Galas and Branscomb, 1978) that the exonuclease active site is positioned on the enzyme so that it can act only on melted-out primer termini, not on annealed ones. Since a misinserted nucleotide is more likely to be melted-out, that is, more likely to appear single-stranded, than a properly paired one, it is to the same degree more likely to be excised. To be an efficient editor, an enzyme need only excise non-annealed termini with high probability before they serve as primer sites for additional DNA synthesis.

A free energy diagram representing K_m exonuclease editing is shown in Fig. 8.3. The parameter ΔG_E represents the difference in free energy between the melted-out states of right and wrong nucleotides at a given template site. By analogy with the assumption governing insertion competition between two nucleotides, it is assumed that the enzyme can excise nucleotides in proportion to their residence time at the exonuclease active site, i.e. K_m discrimination. Melted-out termini are potential residents while annealed termini are not. The specificity with which such a mechanism can reject wrong bases and retain right ones is limited by the differences in their probabilities for melting away from the template strand and is determined therefore by the difference ΔG_E (Fig. 8.3) in their melting and reannealing free energies.

A literate (as opposed to illiterate) exonuclease could in principle discriminate to a virtually unlimited extent by refusing (1) to excise a melted-out correct terminus and (2) to add a nucleotide to an annealed wrong terminus. This could be reflected on the free energy diagram, Fig. 8.3, by making $G_E^R \neq G_E^W$ and by having a large barrier height G_E^R. The evidence gathered to date (Bessman et al., 1974; Liu et al., 1978; Clayton et al., 1979) gives no indication that the polymerase associated exonuclease is in any way able to remove wrong nucleotides selectively. The data for the removal of dAPMP and dAMP opposite template T sites (Bessman et al., 1974; Clayton et al., 1979) is consistent with an illiterate exonuclease in which T4 mutator, wild type, and antimutator polymerase associated exonucleases exhibit

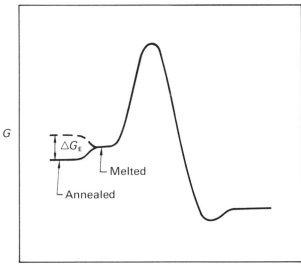

Reaction coordinate ⟶

Figure 8.3   Free energy reaction path diagrams for exonuclease discrimination. (Same conventions as in Figs 8.2 (a) and (b).) From the initial state, on the left, with the primer terminus annealed, the system passes to the state with the primer terminus melted away from the template then through an activated intermediate to the final state in which the exonuclease reaction is complete and a free monophosphate is released. For incorrectly paired termini (dashed line) the annealed state is at higher energy, by the amount ΔG_E, than it is for correctly paired termini (solid line). The former are therefore more often melted-out than are the latter, and enter the subsequent reaction path correspondingly more often.

different activities but similar specificities for removal of wrong versus right nucleotides (Goodman *et al.*, 1974; Galas and Branscomb, 1978; Clayton *et al.*, 1979).

It should become clear from our discussion in the next two sections that while a K_m model describing nucleotide insertion and excision is attractive in an 'Occam's razor' sense, and while various kinetic schemes of this type can be analysed with only a moderate amount of mathematical complication, the present model appears unable to explain some extremely large differences (10^3–10^4-fold) between T4 mutators and antimutators *in vivo*. However, the model is found to be considerably more successful in explaining fidelity data for numerous purified DNA polymerases *in vitro* and may offer, therefore, useful insight into questions of how and to what extent DNA polymerases control replication fidelity *in vivo*.

8.2.3 EXPERIMENTS DESIGNED TO TEST THE K_m DISCRIMINATION MODEL

In the report from Clayton *et al.* (1979), a direct measurement of dATP and dAPTP competing for template T sites (Fig. 8.4) showed that K_m^{AP} is six-fold larger than K_m^A (ΔG_P = 1.1 kcal/mol) while $V_{max}^{AP} \approx V_{max}^A$. These data are consistent with the K_m discrimination model in which hydrogen bonding free energy differences determine the relative frequencies of inserting dAPTP and dATP opposite T. However, the free energy difference between A·T and AP·T base-pairs is not very large, and it is certainly possible that V_{max} or perhaps a more complex base-pairing discrimination mechanism might be operating for cases involving larger ΔG mispairs, e.g., the insertion of dCTP opposite a template A and AP in competition with dTTP (this competition is discussed in Section 8.3.4).

To distinguish experimentally between K_m and V_{max} discrimination, one can compare the K_ms and V_{max}s for inserting right and wrong nucleotides as carried out in the study by Clayton *et al.* (1979). Figure 8.5 illustrates the measurement of an effective free energy difference between dTTP and dCTP competing for insertion opposite a template AP. The ratio of [^{32}P]dCMP to

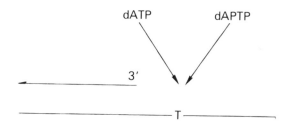

Figure 8.4 Competition between A and AP for insertion opposite T. The next base to be attached to the 3'-OH terminus of the primer strand (top) must pair with a T on the template strand (bottom); dATP and dAPTP are shown competing for association at this site.

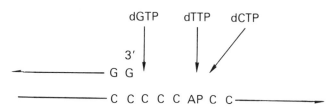

Figure 8.5 Competition between T and C for insertion opposite AP in a p(dC) template. On a p(dC) template, with interspersed APs, primed with p(dG), T and C are shown competing for insertion opposite an AP. The presence of dGTP ensures net forward synthesis.

[³H]TMP insertion (nucleotide incorporation plus turnover) into DNA is measured at known [³²P]dCTP and [³H]dTTP concentrations, and Equation (8.1) is used to determine ΔG. The experiment is carried out at saturating concentrations of unlabelled dGTP which cannot compete with dTTP or dCTP for insertion opposite AP. K_m and V_{max} measurements are then made independently with [³²P]dCTP in the absence of [³H]dTTP, and vice versa, at saturating levels of dGTP.

For K_m discrimination to hold, Equation (8.4) must be satisfied. That is, the ratio $[dCTP]K_m^T/[dTTP]K_m^C$ should be similar to the insertion ratio $I(C)/I(T)$ measured under conditions where the two substrates are competing simultaneously for insertion opposite a template AP. Further, the V_{max} of the two substrates should be very nearly identical. It is important to note that the poly (dC–dAP) template strand must be dilute in AP. If there are significant numbers of two or more nearest neighbour aminopurines on the template strand, then V_{max}^R will be *artificially* reduced by a failure in being able to add successive cytosines due to poor priming.

A simple test to insure that the correct V_{max}^R is being measured would be to vary the C/AP template ratios to show that V_{max}^R increases with C/AP ratios and finally reaches a plateau when successive template aminopurines have been reduced to the point where priming is no longer rate limiting. Note that if one finds $V_{max}^R > V_{max}^W$, then a K_m discrimination model will probably not be adequate to explain the data. If, however,

$$V_{max}^R \approx V_{max}^W, \text{ and } K_m^W/K_m^R = I(R)/I(W)$$

then the assumptions, attributing control of fidelity to base-pair H-bonding, are not violated. A more compelling test of the model is to acquire data on a number of base competitions and determine if they all appear consistent with the K_m discrimination model. Several different pairwise competitions between right and wrong base-pairs are currently being investigated experimentally by the criteria described above. In Section 8.3.4, we will show that AP·T versus AP·C base-pair competition also satisfies K_m discrimination model predictions (Watanabe and Goodman, 1981, 1982).

From the pairwise competition experiment illustrated in Figs 8.4 and 8.5, one obtains a direct measure of the different insertion rates of right and wrong nucleotides at a given template site. In the case of mammalian polymerases, the 'wrong/right' insertion ratio, Equation (8.4), is given directly by a measurement of the misincorporation ratio because exonucleolytic proofreading is absent. The observed misincorporation ratios define the intrinsic accuracy of the purified mammalian enzyme. If it should turn out to be generally true, as our data (Watanabe and Goodman, 1982) now indicate for α-polymerase from KB cells (Sedwick, Wang and Korn, 1975; Fisher, Wang and Korn, 1979), that K_m (binding) rather than V_{max} (velocity) discrimination governs the enzyme's accuracy, then there would have to be means, independent from the polymerase itself, to insure adequate replication accuracy in animal cells. The maximum attainable K_m discrimination for insertion against base transition mutations in the absence of proofreading, e.g. A·C versus G·C, is of the order of 5–6 kcal/mol (Goodman, Watanabe and Branscomb, 1982). This free energy difference would by itself yield too high a mutation frequency per base-pair, exp $(-\Delta G/RT)$, of between 2×10^{-4}–4×10^{-5} per round of replication. More reasonable values for actual mutation rates would be 10^{-8}–10^{-10} per base-pair (Drake, 1969) which is closer to exp$(-2\Delta G/RT)$. We should emphasize that for enzymes with editing exonucleases both turnover and incorporation must be measured to separate the contributions to overall accuracy from the two reactions involved. In this case according to the K_m discrimination model both reactions must show discrimination specificities not significantly greater than the relevant base-pairing free energy differences would allow. Maximum obtainable accuracies are determined by the model's detailed mechanisms (see Section 8.4) and are achieved only in the limit of 'infinite' deoxyribonucleotide turnover.

8.3 Evidence in support of a K_m discrimination model for fidelity

In the previous section we discussed in general terms physical mechanisms which can govern the fidelity of DNA replication. In Section 8.4, we will indicate how these mechanisms can be included in a specific mathematical model which then enables the prediction of error rates as a function of deoxyribonucleotide pool concentrations, base context, and base-pairing free energy differences. In this section, we analyse recent *in vitro* and *in vivo* data obtained using the T4 mutator–antimutator system. From the data, it is now possible to begin evaluating some essential model predictions. The following definitions will be used throughout:

The misincorporation frequency is:

$$P(W) = \frac{\text{nmol dWMP incorporated}}{\text{nmol dWMP incorporated} + \text{nmol dRMP incorporated}} \qquad (8.5)$$

where dWMP and dRMP represent wrong and right nucleotides respectively. The removal frequencies for wrong and right nucleotides are defined as:

$$R(W) = \frac{\text{nmol dWMP turned over}}{\text{nmol dWMP turned over} + \text{nmol dWMP incorporated}} \qquad (8.6)$$

$$R(R) = \frac{\text{nmol dRMP turned over}}{\text{nmol dRMP turned over} + \text{nmol dRMP incorporated}} \qquad (8.7)$$

(Mis)insertion frequency is defined as:

$$I(W) = \frac{\text{nmol dWMP incorporated} + \text{nmol dWMP turned over}}{\text{nmol dWMP inc.} + \text{nmol dWMP t.o.} + \text{nmol dRMP inc.} + \text{nmol dRMP t.o.}}.$$

(8.8)

Note that $I(W) + I(R) = 1$.

We will focus our interest on an experimental system which uses the base analogue 2-aminopurine as a measure of aberrant DNA synthesis (Bessman *et al.*, 1974; Clayton *et al.*, 1979). This system is exceedingly valuable in at least four important respects.

(1) AP drives $A \cdot T \rightleftharpoons G \cdot C$ transition mutations in the T4 system (see Ronen, 1979).
(2) The population of each base-pair involved in AP-induced mutagenesis $A \cdot T$, $AP \cdot T$, $AP \cdot C$, $G \cdot C$ can be quantified *in vivo* (Goodman, Hopkins and Gore, 1977; Hopkins and Goodman, 1979).
(3) Purified DNA polymerases from mutator, wild type, and antimutator gene 43 alleles have been used alone in cell free systems (Bessman *et al.*, 1974; Clayton *et al.*, 1979) or in the presence of additional T4 replication complex proteins (Liu *et al.*, 1978) to measure $AP \cdot T$ and $AP \cdot C$ base-mispairing frequencies *in vitro*.
(4) The T4 system is itself sufficiently defined biochemically and genetically so that based on properly defined models, measurements made *in vitro* can be used to predict events occurring *in vivo* and vice versa. The closely coupled *in vivo–in vitro* measurements allow us to test models of fidelity as well as to obtain a defined and predictive connection between biochemical and genetic observations.

8.3.1 AP·T BASE-MISPAIRS

The strongest data presently available in support of the idea that H-bonding free energy differences between competing base-pairs determine the error frequency results from an *in vitro* analysis of the frequency of forming $AP \cdot T$ base-pairs by mutator, wild type, and antimutator DNA polymerases (see Fig. 8.4 and Clayton *et al.*, 1979). As stated earlier, a K_m discrimination model makes the prediction that the insertion specificity for a given pair of competing nucleotides should be the same for all 'normal' polymerases. It has been shown for the case of five polymerases studied to date (T4 mutator (L56), wild type (43^+), and antimutator (L141), *E. coli* Pol I, and calf thymus α) that the specificity of AP insertion is the same for all of the above enzymes in the range of 13–16% for equimolar concentrations of deoxyribonucleotide substrates (Bessman *et al.*, 1974; Clayton *et al.*, 1979). A similar AP insertion frequency (13%) has recently been observed in a HeLa cell nuclei system

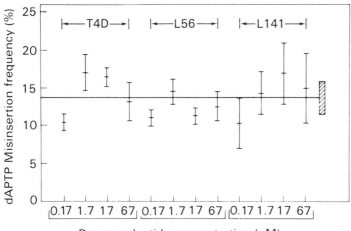

Figure 8.6 Aminopurine misinsertion frequency as a function of dNTP concentration for T4D, L56, and L141 DNA polymerases. The misinsertion frequency is defined as the sum of incorporation and turnover of dAPMP divided by the sum of incorporation and turnover of dAPMP and dAMP (Equation (8.8)). Incorporation and turnover assays were performed as described in Clayton *et al.* (1979). The vertical bars at each data point represent 1 S.E. The solid horizontal line at 13.9% is the average of all data points. The cross-hatched vertical bar indicates 1 S.E. for the average of the data points for the three enzymes taken together.

(Wang, Stellwagen and Goodman, 1981). This gives $\Delta G = 1.1$ kcal/mol for the differences in H-bond free energy between $A \cdot T$ and $AP \cdot T$ base-pairs. When all substrate concentrations are set equal to each other, this specificity is constant over a 400-fold variation in absolute dNTP concentrations (Fig. 8.6). It has recently been shown (Liu *et al.*, 1978), that the insertion specificity for AP by the 43^{+} enzyme is not altered in the presence of the replication complex composed of 44/62, 45, and 32 proteins.

As mentioned before, for K_m discrimination mechanisms the insertion specificity must be given by the ratio of K_ms

$$I(AP)/I(A) = [dAPTP]K_m^A/[dATP]K_m^{AP}.$$

Here we find that the K_m ratios for A and AP are similar for the mutant and wild type T4 polymerases and equal to about 13–16%, consistent with the model. Note, however, that these two measurements are completely independent. The $I(AP)/I(A)$ determination is obtained by allowing dAPTP and dATP to compete simultaneously for insertion opposite a template T. However, the determinations for K_m are carried out for each nucleotide in the *absence* of its competitor. One obvious additional test is deliberately to bias the ratio of dAPTP/dATP pools by the ratio K_m^A/K_m^{AP} in order to equalize the residence times of the competing triphosphates at the polymerase active site. When [dATP]/[dAPTP] is set equal to 14% and the two substrates are added simultaneously, we measure, as predicted, an I(AP)/I(A) ratio of one (Clayton *et al.*, 1979). Furthermore the V_{max}s for A and AP appear to differ by less than 30% (Clayton *et al.*, 1979). It thus appears that K_m discrimination with an effective free energy of 1.1 kcal/mol, governs the insertion of AP versus A opposite T.

In the case of mammalian DNA polymerases having no measurable nucleotide turnover activity, (mis)incorporation equals (mis)insertion. However, in the case of prokaryotic polymerases, and in the model of the illiterate exonuclease outlined in Section 8.2.2, excision of the terminal nucleotide is possible when the 3'-terminus of the primer strand is melted out. A competition therefore exists between chain elongation, for which the terminal nucleotide must be annealed, and excision (turnover) of dNMP (both wrong and right) – see Section 8.4. At saturating deoxyribonucleoside triphosphate concentrations, the mutator L56 polymerase which has a relatively weak 3'-exonuclease activity (low turnover/polymerase ratio) will be successful in removing only a small fraction of melted-out termini before these termini transiently reanneal to become proper substrates for elongation. However, an antimutator polymerase (L141 has about a ten-fold higher turnover/polymerase activity ratio on AP substrates than mutator L56) can remove most melted-out termini before they reanneal even in the presence of maximal forward synthesis rates (Bessman *et al.*, 1974; Clayton *et al.*, 1979). In so doing, many correctly paired but transiently melted A·T base-pairs are also excised. This turnover of a significant fraction of proper base-pairs may impose an unacceptably large cost for only modest gains in accuracy (Galas and Branscomb, 1978; Clayton *et al.*, 1979) and is perhaps one reason why antimutator phage have not replaced the wild type. Wild type replication error rates can be estimated at about 10^{-6} or roughly one per ten T4 genomes. A meaningful contribution to viability by preventing genetic death would not result from a further reduction in mutation rate. The antimutator's rate of one mutation per 1000 genomes must therefore represent a considerable expense in deoxyribonucleotide turnover in this organism whose chief metabolic aim is DNA synthesis.

An important point is that it is not necessary to require that differences exist among mutant and wild type 3'-exonucleases' intrinsic abilities to distinguish right from wrong nucleotides to account for the observed differences in misincorporation frequencies. The differences in AP misincorporation frequency observed in the T4 system *in vitro* are accounted for in part by large differences in the relative activities of the polymerase-associated 3'-exonucleases, not by differences in the exonuclease specificity. The specificity for removal of AP compared with A is about five to one for the two mutant

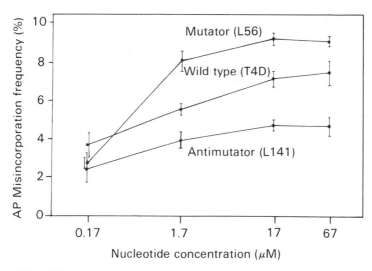

Figure 8.7 Misincorporation of aminopurine as a function of dNTP concentration for T4D (wild type), L56 (mutator), and L141 (antimutator) DNA polymerases. Polymerase assays were carried out at four dNTP concentrations as described by Clayton *et al.* (1979). The misincorporation frequency is expressed as P(AP) which is the nanomoles of dAPMP incorporated divided by the sum of nanomoles of dAPMP and dAMP incorporated (Equation (8.5)). Vertical error bars at each data point indicate 1 S.E.M.

and wild type T4 polymerases and, as was true also for the insertion step, reflects the difference in free energies for an AP·T versus A·T base-pair. A misinserted AP is about five times more likely to be melted-out than A, and the active L141 exonuclease has an 80% chance of removing AP under conditions of maximal DNA synthesis (saturating dNTP concentrations), while the inactive L56 exonuclease is able to remove AP with an efficiency of only 10–20% (Bessman *et al.*, 1974; Clayton *et al.*, 1979).

A strong prediction of the K_m discrimination model is that if it were possible to slow down the forward polymerization rate to allow the inactive mutator 3'-exonuclease more time to excise melted primer termini, then the mutator polymerase should exhibit an enhanced DNA synthesis fidelity. Slowing the polymerization rate for antimutator, however, should show little or no effect since the antimutator exonuclease is already excising most APs along with a substantial fraction of As. A controlled reduction in polymerization rate can be achieved by reducing all the dNTP concentrations coordinately. When this was done (Fig. 8.7), it was observed that both mutator and wild type polymerase error frequencies converged to the measured antimutator fidelity while the misinsertion (Fig. 8.6) and excision specificities remained constant over a 400-fold co-ordinate variation in dNTP substrate concentration (Clayton *et al.*, 1979).

In the K_m discrimination model, there are two limiting fidelity bounds corresponding approximately to one-step and two-step error correction limits. The one-step limit applies to polymerases, such as those from mammalian cells, having no associated 3'-exonuclease activity. Here discrimination occurs only at the insertion step and is due to the relative differences in dissociation rates for two competing dNTP substrates. In two-step discrimination a second step is added involving a maximally active 3'-exonuclease which excises melted-out nucleotides at 3'-primer termini. The absolute accuracy limit for the ideal two-step case does not result in a reduction of errors to zero (Goodman et al., 1974; Galas and Branscomb, 1978); rather the rate of AP to A incorporation is bounded by the dAPTP/dATP pool size ratio multiplied by $\exp(-(\Delta G_P + \Delta G_E)/RT)$ where ΔG_P and ΔG_E are the H-bond free energy differences between AP·T and A·T base-pairs at the polymerase and nuclease steps, respectively. Although we have obtained a good fit to the data (Clayton et al., 1979) by assuming $\Delta G_P = \Delta G_E$ for AP·T versus A·T, the two free energy differences need not necessarily be the same at polymerase and nuclease sites. The steric constraints imposed on the base-pairing bonds by the enzyme–template–primer complex may be different at the two sites (see Section 8.4).

Within the theoretical constraints of our model, it should not be possible to reduce the ratio of AP/A incorporation to much below $\exp(-2.2/0.6) = 2.6\%$ at equimolar ratios of competing deoxyribonucleotide substrates at 30°. It is interesting, and we believe non-coincidental, that wild type T4 polymerase assayed in the presence of 44/62, 44, and 32 proteins has an AP/A incorporation ratio of 3.2% (Liu et al., 1978), which is similar to the L141 polymerase by itself (Bessman et al., 1974; Clayton et al., 1979). An analysis of AP and A nucleotide turnover data (Liu et al., 1978) showed that in the presence of the replication complex proteins, the 43^+ enzyme removes with greater efficiency both melted AP·T and A·T primer termini. It would be interesting to determine experimentally whether the predicted AP/A lower bound of 2.6% is violated by the antimutator polymerase in the presence of the other four proteins comprising the replication complex.

8.3.2 In vitro PREDICTIONS – in vivo MEASUREMENTS: DEOXYRIBO-NUCLEOTIDE POOL SIZE EFFECTS ON FIDELITY

If, as stipulated by the K_m discrimination model, the relative formation of AP·T and A·T base-pairs depends on an enzyme independent base-pair free energy difference $\Delta G = 1.1$ kcal/mol measured in vitro, and if the same molecular mechanism is operative as well in vivo, there are several definite predictions which can be tested. In what follows, we have chosen to approximate the L56 mutator and L141 antimutator polymerases as 'ideal' one- and two-step error-correcting enzymes respectively (Fig. 8.1). Exact equations for the fidelity of the various enzymes have been described (Galas and Branscomb, 1978; Clayton *et al.*, 1979) and can be applied (with effort) where more accurate quantitation is desired. Our purpose here is to sacrifice a small measure of mathematical fidelity to achieve greater physical clarity in relating *in vitro* and *in vivo* data.

The approximate relationship between fidelity of L56 and L141 polymerases, deoxynucleotide pool concentrations, and ΔG is given in the lower part of Fig. 8.1. Recalling that in the cell free measurements we have control over the dNTP pools, we can measure AP/A incorporation ratios and from these deduce a value for ΔG which should not depend on polymerase phenotype.

Taking the value of $\Delta G = 1.1$ kcal/mol obtained *in vitro* and using AP/A incorporation ratios in gene 43 mutator and antimutator allelic backgrounds measured *in vivo*, we can predict the dAPTP/dATP pool size ratios *in vivo* (Goodman, Hopkins and Gore, 1977). Following a standard mutagenesis protocol, the ratio of AP/A in packaged DNA isolated from mutator and antimutator phage is 1 AP/1200 A in L56 and 1 AP/6500 A in L141 (Goodman, Hopkins and Gore, 1977). Note that these values approximately parallel the ratios measured *in vitro* (Bessman *et al.*, 1974; Clayton *et al.*, 1979). With the AP/A incorporation ratio obtained *in vivo* and the ΔG value of 1.1 kcal/mol we can use the expressions for ideal one-step (L56) and ideal two-step (L141) polymerases given in Fig. 8.1 to compute dAPTP/dATP pool size ratios. The computed pool size ratios are 0.5% for mutator and 0.6% for antimutator. Preliminary pool size measurements place the dAPTP/dATP ratio at approximately 1–10% (Goodman, Hopkins and Gore, 1977). Measurements have recently been obtained (Hopkins and Goodman, 1985) which are in rough agreement with the above predictions.

8.3.3 AP·C HETERODUPLEX HETEROZYGOTES *in vivo*

A remarkable experimental versatility can be achieved with T4 by analysing the T4 mutator, wild type, antimutator system's response to 2-aminopurine. AP is known to drive both $A \cdot T \rightarrow G \cdot C$ and $G \cdot C \rightarrow A \cdot T$ transition mutation pathways (Fig. 8.8). $A \cdot T$ to $G \cdot C$ transitions arise in the following way. First, dAPTP competes with dATP for insertion at a site opposite a template T. The resulting formation of an AP·T base-pair is what we have been concerned with up to now. Once AP is stably incorporated opposite T, it serves as a template base during ensuing rounds of replication. The ambiguous base-pairing properties of AP permit H-bonding with T, and, at a frequency of about 5%, with C (Watanabe and Goodman, 1981). Thus AP·C base-pairs are formed at some of the AP template sites. During the following round of replication, the C will pair with a substrate G completing the $A \cdot T \rightarrow G \cdot C$ transition.

Because AP·C base-pairs are more severely mismatched than AP·T base-pairs, they are interesting to analyse and may prove more relevant in under-

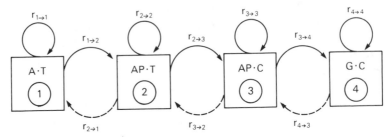

Figure 8.8 A closed loop model for generating base-pair populations for AP-induced A·T → G·C transitions. States 1 to 4 represent the population of A·T, AP·T, AP·C and G·C base-pairs, respectively. Interstate and intrastate transitions rates are designated by a set of rate constants $r_{i \to j}$, respectively. Broken lines reflect interstate and intrastate transition frequencies which in these studies do not contribute significantly to the state population towards which they move.

standing molecular mechanisms of DNA synthesis fidelity. Perhaps the AP·T pair is too similar to the A·T pair to reveal active selection mechanisms present in the polymerase. The AP·C base-pair should provide a more stringent test of the model since the ΔG value for AP·C versus AP·T (when AP serves as a template base) is expected to be significantly greater than 1.1 kcal/mol and more closely parallel to naturally occurring nucleotide mispairs, e.g. A·C (see Section 8.3.4). Another important point is that precisely the same analysis performed for the AP·T base-pair can also be made for the AP·C base-pair. The AP·C base-pair can be measured both *in vitro* (Watanabe and Goodman, 1981, 1982) and *in vivo* (Hopkins and Goodman, 1979) so that deoxynucleotide pool size and ΔG determinants of fidelity can be estimated independently.

In Fig. 8.9 an experiment is sketched in which we measured the population of AP mutagenized phage carrying an AP·C base-pair (r_{II}–r^+ heteroduplex heterozygote) at a defined r_{II} locus in T4 (Hopkins and Goodman, 1979). AP induces an A·T → G·C transition at that site. The basic idea is to restrict the large population of phage which carry A·T or AP·T base-pairs at the marker locus which, if allowed to propagate and then burst, would give rise to r_{II} phenotypic phage. Restriction of the A·T and AP·T base-pairs following standard AP mutagenesis is achieved by adsorption of the phage to CR(λ) bacteria. The small population of phage which are heteroduplex-hetero-zygotes (AP·C) or r^+ revertants (G·C) with respect to the locus can replicate their DNA in the CR(λ) host. A direct phenotypic analysis of single bursts from CR(λ) infection on a bacterial lawn of *E. coli* B is sufficient to distinguish between AP·C and G·C base-pairs. Phage carrying AP·C at the locus give rise to mixed phenotypic plaques of the r^+ and r_{II} types in a ratio of approximately one to one; phage carrying G·C at the marker locus give rise to r^+ plaques only. The genetically measured AP·C population can be

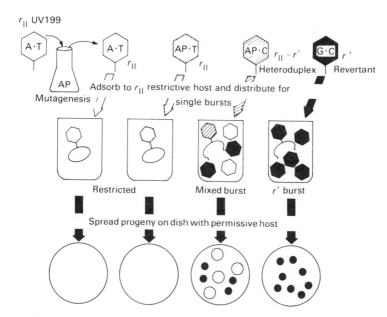

Figure 8.9 Experimental rationale for determining AP·C base-pairs. T4 phage carrying the r_{II} marker, UV199, are mutagenized with AP. The result is a population of mutagenized phage in which the marker exists in A·T, AP·T, AP·C or G·C states. A·T is mutant, G·C is revertant. We adsorb these mutagenized phage to the mutant-restrictive host. Phage carrying A·T or AP·T are restricted. Phage carrying AP·C or G·C survive. We isolate AP·C- or G·C-infected bacteria in separate tubes by dispensing samples so dilute that most do not contain a potential burst. Progeny from this one lytic cycle are plated on the permissive host. The AP·base-pair segregate to form 50% mutant and 50% wild-type phage. The G·C base-pair produces only wild-type phage.

accounted for if, in competition with dTTP, a substrate dCTP (HMdCTP for bacteriophage T4) is incorporated opposite a template AP site at a frequency of 2.4% per round of replication (Hopkins and Goodman, 1979). This surprisingly high observed frequency for forming an AP·C base-pair is consistent with an earlier prediction (Goodman, Hopkins and Gore, 1977) based on measuring AP·T base-pairs in DNA and resulting A·T → G·C reversion rates.

8.3.4 A·C AND AP·C BASE-MISPAIRS AND THE K_m DISCRIMINATION MODEL

In a series of recent experiments (Watanabe and Goodman, 1981, 1982) DNA copolymers containing A or AP have been used to measure the frequency of AP·C and A·C mispairs (Fig. 8.10). For example, templates were

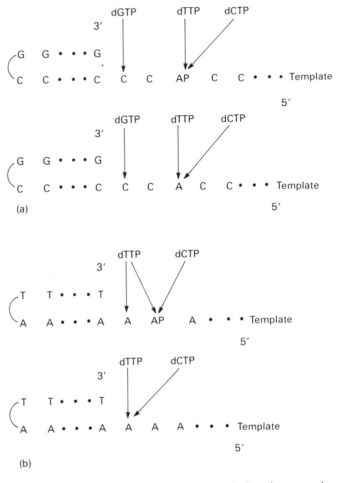

Figure 8.10 Templates and nucleotides present during incorporation assays measuring polymerization of competing nucleotides dCTP and dTTP. In part (a), templates composed primarily of dCMP with either dAPMP or dAMP interspersed are shown primed with covalently bonded oligo (dG). In these templates the presence of saturating levels of unlabelled dGTP is necessary to maximize rates of incorporation of the labelled nucleotides. In part (b) templates composed of dAMP alone, or primarily of dAMP with interspersed dAPMP are shown primed with covalently bonded oligo (dT). dTTP and dCTP are shown competing for polymerization opposite AP or A sites on all four templates.

constructed such that either AP or A, but not both, were interspersed among C residues and primed with a short oligo dG covalently bonded primer (Fig. 8.10(a)). We observed that dCMP is inserted about 5% (4.6 ± 2%, see Table 8.1) as frequently as dTMP at template AP sites, a value which is in excellent

Table 8.1 Incorporation ratios for A or AP templates. Double label deoxyribonucleotide incorporation experiments were carried out with mixed copolymers of adenine and cytosine or 2-aminopurine and cytosine as described in Watanabe and Goodman (1981). The template configurations are indicated in Fig. 8.10. Parallel tube assays contained 3 units ml^{-1} DNA polymerase α, 20 μM polymer and either 5 μM $[^3H]TTP$ (2.5×10^6 cpm nmol-1) plus 5 μMdCTP or 5 μM $[^3H]$ dCTP (1.9×10^7 cpm nmol^{-1}) plus 5 μM TTP. Assays with homopolymers of adenine or mixed copolymers of adenine plus 2-aminopurine were run in parallel tubes containing 3 units/ml polymerase, 0.1 mM template, and either 80 μM $[^3H]TTP$ (9.2×10^5 cpm nmol^{-1}) and 5 μM dCTP or 5 M $[^3H]dCTP$ (1.9×10^7 cpm nmol^{-1}) and 80 M TTP. Assays with equimolar concentrations of competing nucleotides were as described above but with 30 μM $[^3H]dCTP$ (1.9×10^7 cpm nmol^{-1}) and 30 μM TTP or 30 μM $[^3H]TTP$ (5.8×10^6 cpm nmol^{-1}) and 30 μM dCTP. Misincorporation frequencies for $p(dA) \cdot oligo(dT)$ or $p(dA) \cdot oligo(dT)$ are adjusted for any differences in concentrations of competing nucleotides by multiplying the ratio of dCMP incorporated per dTMP incorporated by the ratio $[dTTP]/[dCTP]$. Note that $\pm$ values represent 1 S.E.M.

Template	Incorporation of dCMP per incorporation of dTMP (%)	Template (%)	Incorporation of dCMP per incorporation of dTMP (%)
p(dC·dA)·oligo(dG)	0.13 ± 0.03	p(dC·dAP)·oligo(dG)	4.6 ± 2
p(dA)·oligo(dT)	<0.02*	p(dA·dAP)·oligo(dT) 40% AP	1.6 ± 0.9
		p(dA·dAP)·oligo(dT) 15% AP	0.6 ± 0.1

*0.02×10^{-2} represents an upper bound established from seven separate measurements under conditions described in Watanabe and Goodman (1981). Only two of the seven measurements yielded detectable $[^3H]dCMP$ incorporation; those measurements constituting the upper bound are listed here.

agreement with predicted misinsertion frequencies (Goodman, Hopkins and Gore, 1977) and with *in vivo* measurements of AP·C/AP·T misincorporation ratios at T4 r_{II} marker loci (Hopkins and Goodman, 1979). If AP does increase the probability of C misincorporation, then templates made up of both A and AP with covalently bonded T primer (Fig. 8.10(b)) should show an increasing I(C)/I(T) ratio with increasing proportions of AP. Results from templates with 0, 15% or 40% AP are shown in Table 8.1. Clearly I(C)/I(T) goes up with increasing AP; in fact, if one extrapolates from the mixed copolymers of A and AP, one would predict that I(C)/I(T) for a template made entirely of AP would be 4×10^{-2}, a value very close to that attained on the P(dC·dAP) templates with covalently bonded dG primer.

We also measured the competition between C and T for insertion opposite template A sites for the two different primer–template base contexts shown in Fig. 8.10. The C/T misinsertion ratios were about $0.13 \times 10^{-2} \pm 0.03 \times 10^{-2}$

for $p(dC \cdot dA) \cdot$ oligo dG and $<0.02 \times 10^{-2}$ for $p(dA) \cdot$ oligo (dT) template primers (Table 8.1). These ratios correspond to ΔG values of about 4 and >5 kcal/mol respectively. This difference in forming $A \cdot C$ base-mispairs may be caused by an increased base stacking energy for C next to G in preference to C next to T (Borer et al., 1974; Topal, DiGiuseppi and Sinha, 1980).

It is interesting that in contrast to $A \cdot C$ base-mispair formation, $AP \cdot C$ mispairing frequencies were the same for both G and T nearest neighbour primer configurations. Perhaps hydrogen bonding between $AP \cdot C$ base-pairs is sufficiently strong to mask any differences in AP–G, AP–T base stacking interactions. $AP \cdot C$ mispairs are formed at a frequency of about 5% whereas mispairings due to stacking may be occurring on the order of 0.1%.

Thus, using these polymers, we have addressed two basic questions: first, does the presence of AP in place of A enhance the misinsertion of C as predicted based on in vivo AP mutagenesis studies and, second, are the probabilities of inserting C versus T opposite AP explicable on the basis of the ratio of K_m values for each nucleotide?

As demonstrated above, the first question can clearly be answered in the affirmative: the ratio of dCMP/dTMP insertion is increased by factors ranging from 35–150 when AP replaces A as a template base.

Experiments have recently been performed (Watanabe and Goodman, 1982) to investigate the second question, and data from AP-containing templates indicate that the V_{max} values for the candidate nucleotides, dTTP and dCTP, are not significantly different (Fig. 8.11). The K_m values, however, do appear to differ significantly. We found that the ratio $K_m^T/K_m^C = 5\%$, which agrees with K_m discrimination model predictions.

We have just finished discussing the measurement of $AP \cdot C$ base mispairs in the 2-aminopurine-induced $A \cdot T \rightarrow G \cdot C$ mutational pathway where AP is situated on the template and dTTP and dCTP compete for insertion opposite AP. In a recent experiment (Mhaskar and Goodman, 1984), the reciprocal situation was examined, i.e. the AP-induced $G \cdot C \rightarrow A \cdot T$ mutational pathway, in which cytosine serves as a template base, and dAPTP and dGTP are present as substrates for DNA polymerase. Here, the mutagenic behaviour of AP is readily apparent in vitro since $AP \cdot C$ mispairs occur at a frequency two to three orders of magnitude greater than $A \cdot C$ mispairs (Mhaskar and Goodman, 1984). In fact, the frequency of forming $A \cdot C$ mispairs in competition with $G \cdot C$ base pairs is unexpectedly large, 3–6%, when dAPTP and dGTP compete at equimolar concentrations for insertion opposite C. The AP misinsertion frequency does not appear to depend significantly on which nearest neighbour base-stacking partner is on the 5′-primer side. However, proofreading of a misinserted AP nucleotide appears extremely sensitive to nearest-neighbour base-stacking partners; wild type T4 polymerase removes more than 85% of AP misinsertions occurring next to 5′-primer thymine residues, while removal of AP is about

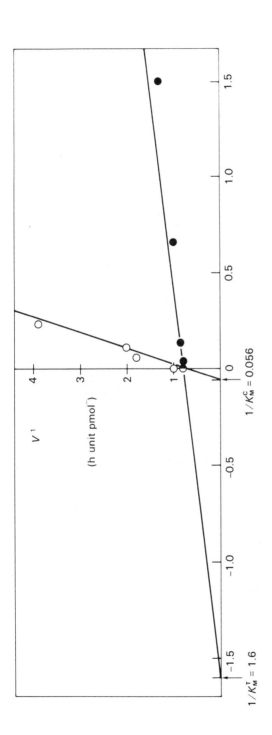

Figure 8.11 Double-reciprocal plot of V^{-1} versus $[\text{dNTP}]^{-1}$ for dTTP (●) and dCTP (○) incorporation on a poly (dC-dAP) · oligo dG hooked template–primer shown in Fig. 8.10(a) using DNA polymerase α purified from KB cells. V represents the initial velocity of incorporation of either dTTP of dCTP measured in the absence of the other at saturating concentration of dGTP as described in Watanabe and Goodman (1982). The data shown are for a single experiment. Based on duplicate experiments, the ratio is $K_m^C / K_m^T = 25 \pm 6$, which agrees with the observation that dTMP is inserted twenty-two-fold more frequently than dCMP when the two deoxynucleotides are allowed to compete with each other for insertion opposite AP, Watanabe and Goodman (1981).

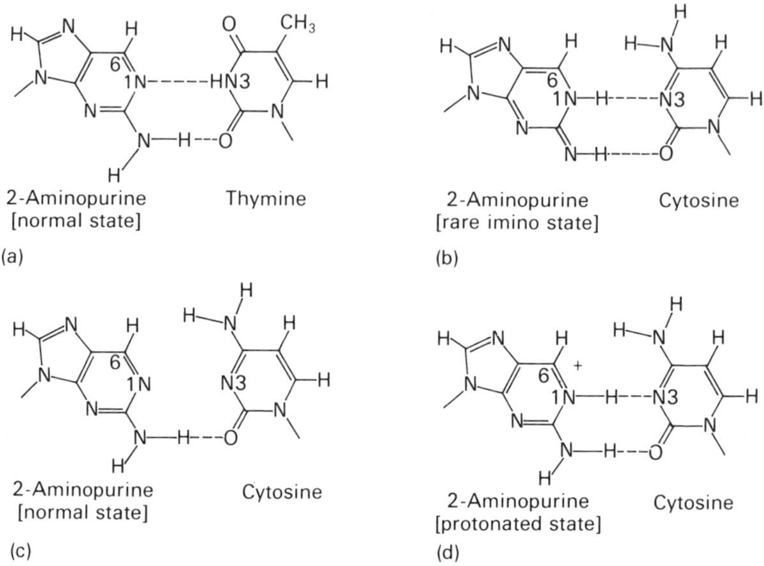

Figure 8.12 The base-pairing properties of 2-aminopurine. (a) Normal pairing with thymine by means of two hydrogen bonds; (b) pairing of disfavoured imino form with cytosine, a result of a tautomeric shift of a hydrogen to the N-1 position; (c) pairing of the normal amino form with cytosine by means of a single hydrogen bond; (d) pairing of 2-aminopurine acting as a proton acceptor with cytosine. Other possible base-pairing configurations not shown in the figure include the disfavoured imino and protonated forms of cytosine pairing with the amino form of 2-aminopurine.

five-fold less next to 5'-primer guanine residues (Mhaskar and Goodman, 1984).

The high frequencies of forming AP·C mispairs in the A·T → G·C and G·C → A·T pathways *in vitro* suggest that this mispair is relatively stable. One possible way in which mispair stabilization could occur is if an H-bond were formed between the 1–3 ring positions of AP and C (Fig. 8.12). A shared proton at the 1–3 positions would result if either AP or C underwent a tautomeric shift to a disfavoured imino form (Fig. 8.12(b) and (c)) or if one of the two bases were ionized (Fig. 8.12(d)). Both situations would result in the AP·C mispair being stabilized by two hydrogen bonds, at the 1–3 and 2–2 ring positions (Fig. 8.12(d)).

We reported observing the 1–3 H-bond based on an analysis of UV spectral shifts for melted and annealed synthetic DNA polymers containing AP·C base mispairs (Goodman and Ratliff, 1983). Janion and Shugar (1973) observed that, under acidic conditions, protonation of AP occurs at the 1-ring position (pK 3.4 for an AP monomer and 4.8 for AP in a DNA polymer). The key point to note is that a shift in the UV spectrum from 303 to 313 nm ccompanies protonation at the 1-position (Janion and Shugar, 1973). We ynthesized DNA polymers containing AP intermixed with guanine. UV pectra were obtained in the presence of poly(dC) at alkaline pH under onditions where the poly[d(AP, G)] was annealed to poly(dC) and under onditions where the two polymers were melted. A spectral shift accompanies ue melting–annealing reaction at pH 8.5 which is indistinguishable from the ift which occurs when AP becomes protonated at its 1-position in acid. The ata suggest that when AP pairs with C, a proton is, in fact, shared between ue 1- and 3-positions of the two bases (Fig. 8.12).

It is not possible to distinguish between ionized or disfavoured tautomer ases from the UV spectral data (Goodman and Ratliff, 1983). Since the uestion of rare tautomer involvement in mutagenesis has still not been :solved since the early days of molecular biology (Watson and Crick, 953a, b; Freese, 1959), it would seem opportune to attempt to determine the recise nature of the AP·C mispair given that the 1–3 H-bond acts to stabilize . We are currently investigating whether ionized or disfavoured tautomeric orms of AP or C are present in synthetic DNA polymers of defined sequence y means of ^{15}N-labelled NMR spectroscopy.

.4 Further predictive potential of the K_m model

ı the preceding sections we have presented and evaluated the K_m discrimin- tion model on an essentially qualitative level. Relatively few data yet exist hich support more quantitative tests. The model has the virtue, however, of ıaking a number of stringent quantitative predictions which may ultimately rove decisive. We want to indicate briefly here the nature and basis of these ıore quantitative implications of the model. Detailed discussions of the oints involved can be found in the reports by Galas and Branscomb (1978) ıd Clayton et al. (1979).

We have analysed a specific form of the K_m discrimination model whose inetic assumptions are embodied in the reaction scheme shown in Fig. 8.13. ll transitions between states are assumed to occur as competing, stochastic oisson processes where each possible transition is characterized by a Poisson ıte constant: the probability per unit of time that, if there were no competing rocess, the transition would occur. In state S, the dNTP site is empty and the rminal base is in its base-pairing position. A transition out of state S occurs ther by the association of a dNTP in the appropriate site (transition to state , rate constant a) or by the dissociation of the terminal base from its ıse-pairing position (transition to state E, rate constant δ_x). A transition om state E occurs either by the excision of the terminal base (rate constant e) r by reassociation of the terminal base to its pairing configuration (transition ı state S, rate constant α). A transition from state P occurs either by the ırmation of the phosphodiester bond between the nucleotide and the primer

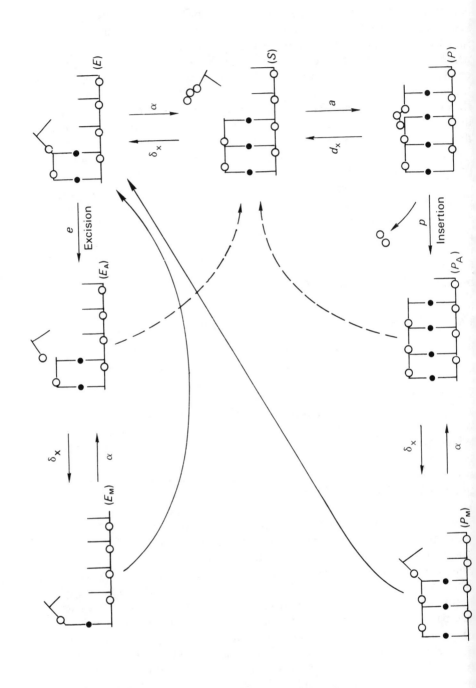

strand (rate constant p) or by the dissociation of the dNTP (transition to S, rate constant d_x). Only the two dissociation rates, d and δ, are assumed to depend on the correctness of the base. Following either excision or insertion, the primer terminus is assumed to reach melting equilibrium (again with rate constants α for annealing and δ_x for melting) in the transitions $E_A \leftrightarrow E_M$, and $P_A \leftrightarrow P_M$ before a reposition shift or base shift forward or backward is completed. This shift then allows the cycle to repeat. Thus at the beginning of each cycle the terminal base is assumed to be in an equilibrium distribution between states S and E prior fraying. We note the following points:

1) The distinction between right and wrong bases enters only in the off rate from the dNTP site, and in the melting-out rate for the primer terminus. Differences in these off rates reflect differences in the free energies of the reversible reactions involved and could be due to differences in hydrogen bonding energies, stacking energies, entropy changes, and enzyme binding energies.
2) Insertion and excision are in simultaneous competition for access to the primer terminus; the former requires an annealed terminus and the latter a melted-out one.
3) Following either the insertion (attachment) of a new base on the end of the primer or the excision of the present terminal base, the enzyme must reposition forward or backward one base before the process can begin again.
4) During enzyme reposition the primer terminus has time to reach equilibrium with respect to melting and annealing. This is the assumption we call 'prior fraying'.

A straightforward mass action analysis of this reaction diagram as drawn is not possible since the behaviour of the primer terminus depends on its content of correct and incorrect base pairs and thus on its history. Under the assumption that only the correctness of the last base on the primer need be considered and that only two alternatives exist, 'right' and 'wrong', we could expand the diagram of Fig. 8.13 to include separate loops for the two different forms of the primer and proceed with a mass action analysis. We have found it more illuminating and manageable to treat the process of template-directed DNA synthesis as a stochastic Turing machine which executes a one-dimensional random walk. This walk creates and/or destroys a binary tape (a mix of wrong and right bases) under the influence of its own history. The individual moves of excision or insertion are regarded as Poisson processes whose probability constants can be expressed in terms of the kinetic rate constants given in Fig. 8.12 and the correctness of the current terminal primer base. The problem then factors into two parts: (1) expressing the properties of the finished tape in terms of the individual step probabilities, and (2) expressing these latter probabilities in terms of the kinetic rate constants

(Galas and Branscomb, 1978; see also Bernardi *et al.*, 1979 and Chapter 10).

Such an analysis yields formulae which relate separately the frequencies of misincorporation, misinsertion and (correct and incorrect base) turnover to (1) differences in base pairing energies, (2) substrate concentrations for correct and incorrect bases, (3) base context differences, and (4) enzyme parameters. In addition, expressions can be obtained which give the effect of these factors on the amount, as distinct from the (error-rejecting) specificity, of exonuclease activity (Clayton *et al.*, 1979).

There follows an outline discussion of some of the issues that have been addressed thus far in attempting to check these more quantitative predictions.

(a) Equality of polymerase and exonuclease specificities
The polymerase and exonuclease activities of most enzymes appear to discriminate between right and wrong bases with the same intrinsic specificity. Moreover, this specificity is the same for enzymes which yield very different overall accuracies of replication (Galas and Branscomb, 1978; Clayton *et al.*, 1979). This implies that the same effective free energy differences can be attributed to both of these processes and construed as a property which is in some sense 'intrinsic' to a given set of competing base pairs. In part because the binding sites for dNTP and for the primer terminus might impose different steric constraints on base-pair formation, there is no general requirement that the effective right–wrong free energy difference be the same for both insertion and editing. However, it seems likely that an optimum steric geometry exists which we would expect to find embodied in both exonuclease and polymerase sites of all normal enzymes.

(b) Effect of substrate concentration on accuracy
The effect on the accuracy of incorporation at a given site of the concentration of dNTP substrates competing for that site and for the following site has been investigated theoretically. Figure 8.14 shows the calculated effect of next site substrate concentration for three variants of the T4 phage DNA polymerase: a mutator enzyme, wild type, and an antimutator enzyme. Increased substrate concentration decreases accuracy at the preceding site by suppressing exonuclease activity. At low concentrations the error frequency is inversely proportional to concentration. Under the same conditions the error frequency at a given site is also proportional to the ratio of the concentrations of wrong to right substrates competing for that site (Clayton *et al.*, 1979).

These predictions have been tested in an experiment by Fersht (1979) in a ϕX174 transfection system. Fersht used purified *E. coli* DNA polymerase III holoenzyme and controlled dNTP concentration to copy a ϕX DNA template mutant at a single site. The *in vitro* replicated DNA was observed genetically, revealing the dependence of Pol III fidelity on substrate concentration. The

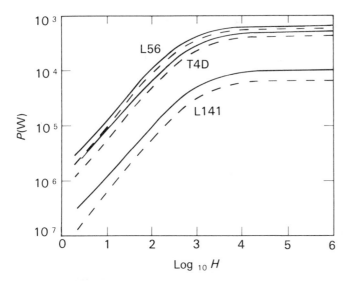

Figure 8.14 Misincorporation frequency versus substrate concentration for three enzymes. The incorporation rate is plotted as a function of a parameter, H, which is proportional to the total, affinity-weighted concentration of substrate competing for the next site: $H \propto \Sigma_i [\text{subst } (i)]/K_a(i)$. Curves are presented for the mutator (L56), wild type (T4D) and antimutator (L141) variants of the T4 polymerase. Values for enzyme-specific parameters were as determined in Clayton *et al.* (1979). In all curves a free energy difference between right and wrong substrates of 4.15 kcal/mol was assumed, corresponding to a value of 1000 for the parameter $s = \exp(\Delta G/kT)$. The solid curves are the results assuming no peelback, i.e. the one-step removal probability equals the ultimate removal probability. The dotted lines are with peelback included.

substrate competition at the i and $i+1$ sites in Fersht's experiment are shown in Fig. 8.15.

The wrong and right nucleotides at site i are dGTP and dATP respectively, while dGTP is the right nucleotide at site $i+1$. For this case the above considerations lead to the prediction that the error frequency is proportional to $[\text{dGTP}]^2/[\text{dATP}]$, which agrees very well with the experimental results obtained. We note that this same concentration dependence was predicted by Fersht (1979) from an analysis of a simple editing polymerase reaction scheme. While Fersht's model leads to a conclusion similar, in this case, to the one we have considered, it differs in some basic assumptions governing enzyme cycling, prior fraying, peelback (Goodman *et al.*, 1974; Galas and Branscomb, 1978), and exonuclease discrimination.

We emphasize, as did Fersht (1979), that the quadratic dependence of misincorporation of dGTP concentration occurs as a direct consequence of the presence of Pol III's 3'-exonuclease, even though it is much weaker than that of the T4 wild type enzyme. In contrast, a mammalian polymerase should

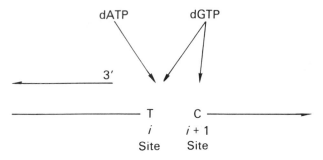

Figure 8.15   Competition between dATP and dGTP for incorporation opposite T when G is also the correct base for insertion at the next site opposite a C. This is the template configuration used by Fersht (1979) to measure the effect of deoxyribonucleotide concentration on *E. coli* Pol III accuracy in an *in vitro–in vivo* φX transfection system.

show an error rate which increases linearly with increasing dGTP concentration. Note, however, that the quadratic dependence of error frequency on the concentration of wrong nucleotide holds only when the wrong nucleotide at site i is also the *correct* nucleotide at site $i+1$. Thus, the Fersht (1979) experiment was particularly clever in revealing the importance of Pol III editing via the non-linear dependence on dGTP concentration.

(c) Preceding base context effects
The K_m model predicts preceding base context effects on the accuracy with which a given site is copied. As has been suggested by Bessman and Reha-Krantz (1977) a less stable base configuration in the bases preceding a given site should increase melting out and thus editing. In our model this corresponds to increasing the parameter δ_x (see discussion of base context effects in Section 8.5).

(d) Dependence on base-pairing free energy differences
In Fig. 8.16 we show the dependence of the misincorporation frequency on the free energy for the antimutator (L141) and mutator (L56) alleles. The parameter values characterizing the enzymes were as determined by Clayton *et al.* (1979). Curves for two values of the total (next site) substrate concentration are shown. The dotted, straight line shows the theoretical maximum accuracy limit for an ideal two-step editing. At very low substrate concentration, i.e. ten-fold below that necessary to maximize the velocity of the polymerizing process, a strong antimutator can, in the prior fraying model, deliver accuracies somewhat above the two-step limit. We also note that, given the parameter values used, there is no substrate concentration or free energy difference such that L141 and L56 differ by more than about twenty-fold in accuracy.

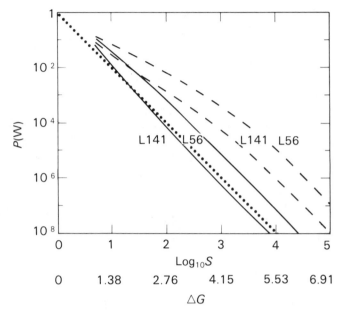

Figure 8.16 Misincorporation frequency versus discrimination free energy for the mutator (L141) and antimutator (L56) enzymes. Curves are plotted for two values to the dimensionless substrate concentration parameter H: H = 5 (——) and H = 1000 (– – – –). The straight, dotted line intersecting the ordinate at one and the abscissa at four represents the ideal, two-step editor. Otherwise, the comments in the caption to Fig. 8.15 apply.

(e) Quantitative prediction of mutation frequencies
While the above conclusion depends rather strongly on the values of the enzyme characterizing parameters obtained by Clayton *et al.* (1979), we have not found it possible to obtain a fit that appears acceptably consistent with the data presented in that paper while at the same time providing a difference in accuracies between L56 and L141 of more than about 100-fold. Thus, we have been able to show (Goodman, Hopkins and Gore, 1977), based on the net incorporation of AP into the DNA of T4 mutator, wild type, and antimutator bacteriophage and dAPTP/dATP pool size ratios measured *in vivo*, that the L141 antimutator replication system is significantly better than a one-step process while the L56 mutator system behaves closer to a one-step process (insertion without subsequent editing).

These conclusions are the same as those based on the *in vitro* measurements (Clayton *et al.*, 1979) for mutant and wild type polymerases for dAPTP and dATP completing at template T sites (see Fig. 8.4). These same conclusions, however, are *not* supported by the genetic data for forming AP·C base-mispairs (Hopkins and Goodman, 1979), see Section 8.3.3. In this case,

AP·C versus AP·T, we find that the L141 antimutator replication system provides an improvement over L56 mutator fidelity *in vivo* equivalent to a ΔG of 4–5 kcal mol^{-1}. It thus appears that the present model is inadequate by itself to explain the differences in replication accuracy between mutator and antimutator polymerases seen *in vivo*. What this has to tell us about the inadequacies of the present form of the K_m discrimination model can only be resolved by *in vitro* measurements in which the same AP·C versus AP·T competition is examined. *In vitro* comparisons using L141 and L56 polymerases to measure incorporation and removal of dTMP versus dCMP opposite template AP sites are currently in progress. These measurements should enable us to determine whether greater than 'ΔG' differences are evident for the two mutant polymerases *in vitro* as they are *in vivo* and whether the *in vitro–in vivo* relationships (Section 8.3) which appear to be satisfied for AP competition with A opposite template T are satisfied for the T versus C opposite template AP and A sites as well.

8.5 Concluding remarks

Models of the polymerization process, of the type we have here called illiterate, have the value that they place the burden of error discrimination in a physically plausible way on rather simple, independently measurable properties of the nucleotides. Such models readily yield quantitative predictions against which they can be tested. If, instead, polymerases prove to be literate devices, then the quantitative properties of the replication process will depend on relatively obscure internal mechanisms of the enzyme and will be much more difficult to elucidate.

We have focused our attention exclusively on a particular K_m discrimination polymerase model. No other has yet been developed in sufficient detail to allow explicit predictions to be made for the full range of available data. Many other models are possible, however. Several, differing from ours in various details, have been advanced. See for example Fersht (1979), Hopfield (1974, 1980), Bernardi *et al.* (1979). However, based on current data using the analogue 2-aminopurine to measure frequencies of forming the two mispaired intermediates, AP·T and AP·C, in the AP-induced A·T → G·C transition pathway, we find that the *in vitro* and *in vivo* experimental results are quantitatively consistent with K_m discrimination model predictions.

In the field of mutagenesis, the exciting prospect of being able to study the influence of base sequence on base-mispairing error frequencies is at hand. Using a modification of the Maxam-Gilbert sequencing technique, Pless and Bessman (1983) were able to measure the frequency of incorporating 2-aminopurine deoxyribonucleoside monophosphate in competition with dAMP opposite template thymine at 57 sites along a ϕX174 primer–template using purified DNA polymerases *in vitro*. The data indicate that the frequency of incorporating dAPMP opposite T can vary significantly even when nearest neighbour primer-strand nucleotides are the same. There does not appear to be any obvious relationship between ϕX174 DNA sequence and AP misincorporation frequency (Pless and Bessman, 1983).

We decided to analyse the site-dependent dAPMP incorporation data of Pless and Bessman by evaluating the effect of base composition (G·C or A·T content) on dAPMP incorporation (Petruska and Goodman, 1985). We constructed a simple model in which base composition was used to define the stability of a local region of DNA spanning both sides of the site of dAPMP incorporation. When we assume that the insertion frequency of dAPMP is influenced by the current primer terminus base, and that the editing efficiency is determined by the base composition on both sides of the site, we are able to predict dAPMP incorporation frequencies at each of the 57 sites along the ϕX174 DNA template strand. While far from perfect, this simple model gives a correlation coefficient of 0.83 with the Pless and Bessman data (1983) and a standard deviaton of ±2.8%. Based on the model, one can further deduce that the polymerase 'sees' four bases to the left and four bases to the right of the site of dAPMP incorporation in determining the fidelity at that site. The correlation coefficient is reduced significantly when greater or fewer than four bases are included in the analysis. The model is not necessarily limited to dAPMP data but may also be useful for the common nucleotides. With additional sequence data, it should soon be possible to make significant progress in obtaining deeper insight into the molecular mechanisms of mutagenesis.

What these observations imply for the K_m discrimination model is not yet clear. Our previous neglect of stacking energy (between the primer terminus and an incoming dNTP) was an approximation aimed at predicting average misincorporation frequencies. Clearly, however, site specific differences in stacking energies should produce corresponding differences in dNTP off rates and thus in insertion probabilities (Section 8.4). Provisionally, the observed effects do seem to be explainable in terms of known differences in stacking energies, though this remains to be investigated systematically. Also, as mentioned previously, the base-pairing stability of the primer terminus neighbourhood is expected to influence editing efficiency, by controlling melting-out rates, in the manner observed (though there was previously no indication as to how large this neighbourhood might be). It is less clear how the K_m model can rationalize the similar dependence of editing efficiency of the stability of the neighbourhood to the 3' side that these data also imply. One possibility is that a more stable 3' neighbourhood makes 'peelback' editing less likely (Galas and Branscomb, 1978; Goodman et al., 1974), though this should be quantitatively significant only under conditions of high overall turnover.

Finally, we bring to the reader's attention a series of recent biochemical studies on DNA polymerase III isolated from the highly mutagenic *mutD* (*dnaQ*) strains of *E. coli*. This mutagenic lesion was found to produce a defective $3' \rightarrow 5'$ exonuclease (Echols, Lu and Burgers, 1983; DiFrancesco *et al.*, 1984). It has recently been shown that the $3' \rightarrow 5'$ exonuclease activity is contained on a physically separate subunit, the ϵ subunit, of the Pol III core enzyme (Scheuermann and Echols, 1984). Hence there is now a good precedent indicating that polymerization and proofreading functions could exist in the cell as separate polypeptides. It is now perhaps more likely than ever that activities analogous to ϵ might be present in eukaryotic cells to help control the fidelity of replication.

References

Alberts, B., Morris, C. F., Mace, D., Sinha, N., Bittner, M. and Moran, L. (1975) Reconstruction of the T4 bacteriophage DNA replication apparatus from purified components. In *DNA Synthesis and Its Regulation*, ICN–UCLA Symposium on Molecular and Cellular Biology (eds M. Goulian, P. Hanawalt and C. F. Fox), Vol. 3, W. A. Benjamin, Inc., Menlo Park, pp. 241–269.

Bernardi, F., Saghi, M., Dorizzi, M. and Ninio, J. (1979) A new approach to DNA polymerase kinetics. *J. Mol. Biol.*, **129**, 93–112.

Bernstein, C., Bernstein, H., Mufti, S. and Strom, B. (1972) Stimulation of mutation in phage T4 by lesions in gene 32 and by thymidine imbalance. *Mutat. Res.*, **16**, 113–119.

Bessman, M. J. and Reha-Krantz, L. J. (1977) Studies on the biochemical basis of spontaneous mutation V. Effects of temperature on mutation frequency. *J. Mol. Biol.*, **116**, 115–123.

Bessman, M. J., Muzyczka, N., Goodman, M. F. and Schnaar, R. L. (1974) Studies on the biochemical basis of spontaneous mutation II. The incorporation of a base and its analogue into DNA by wild-type, mutator, and antimutator DNA polymerases. *J. Mol. Biol.*, **88**, 409–421.

Borer, P. N., Dengler, B., Tinoco Jr, I. and Uhlenbeck, O. C. (1974) Stability of ribonucleic acid double-stranded helices. *J. Mol. Biol.*, **86**, 843–853.

Brutlag, D. and Kornberg, A. (1972) Enzymatic synthesis of deoxyribonucleic acid, XXXVI. A proofreading function for the $3' \rightarrow 5'$ exonuclease activity in deoxyribonucleic acid polymerases. *J. Biol. Chem.*, **247**, 241–248.

Clayton, L. K., Goodman, M. F., Branscomb, E. W. and Galas, D. J. (1979) Error induction and correction by mutant and wild type T4 DNA polymerases. *J. Biol. Chem.*, **254**, 1902–1912.

Coulondre, C., Miller, J. H., Farabaugh, P. J. and Gilbert, W. (1979) Molecular basis of base substitution hotspots. In *Escherichia coli. Nature*, **274**, 775–780.

DiFrancesco, R., Bhatnagar, S. K., Brown, A. and Bessman, M. J. (1984) The interaction of DNA polymerase III and the product of the *Escherichia coli* mutator gene, *mutD. J. Biol. Chem.*, **259**, 5567–5573.

Drake, J. W. (1969) Comparative rates of spontaneous mutation. *Nature*, **221**, 1132.

Drake, J. W. and Allen, E. F. (1968) Antimutagenic DNA polymerases of bacteriophage T4. *Cold Spring Harbor Symp. Quant. Biol.*, **33**, 339–344.

Drake, J. W., Allen, E. F., Forsberg, S. A., Preparata, R. and Greening, E. O. (1969) Spontaneous mutation. Genetic control of mutation rates in bacteriophage T4. *Nature*, **221**, 1128–1131.

Echols, H., Lu, C. and Burgers, P. M. J. (1983) Mutator strains of *Escherichia coli*, *mutD* and *dnaQ*, with defective exonucleolytic editing by DNA polymerase III holoenzyme. *Proc. Natl Acad. Sci. USA*, **80**, 2189–2192.

Fersht, A. R. (1979) Fidelity of replication of phage φX174 DNA by DNA polymerase III holoenzyme; spontaneous mutation by misincorporation. *Proc. Natl Acad. Sci. USA*, **76**, 4946–4950.

Fisher, P. A., Wang, T. S.-F. and Korn, D. (1979) Enzymological characterization of DNA polymerase α. *J. Biol. Chem.*, **254**, 6128–6137.

Freese, E. (1959) The specific mutagenic effect of base analogues on phage T4. *J. Mol. Biol.*, **1**, 87–105.

Freese, E. B. and Freese, E. F. (1967) On the specificity of DNA polymerase. *Proc. Natl Acad. Sci. USA*, **57**, 650–657.

Galas, D. J. and Branscomb, E. W. (1978) Enzymatic determinants of DNA polymerase accuracy: theory of coliphage T4 polymerase mechanisms. *J. Mol. Biol.*, **124**, 653–687.

Glickman, B. W. and Radman, M. (1980) *Escherichia coli* mutator mutants deficient in methylation – instructed DNA mismatch correction, *Proc. Natl Acad. Sci. USA*, **77**, 1063–1067.

Goodman, M. F. and Ratliff, R. L. (1983) Evidence of 2-aminopurine·cytosine base mispairs involving two hydrogen bonds. *J. Biol. Chem.*, **258**, 12 842–12 846.

Goodman, M. F., Hopkins, R. and Gore, W. C. (1977) 2-Aminopurine-induced mutagenesis in T4 bacteriophage: a model relating mutation frequency to 2-aminopurine incorporation in DNA. *Proc. Natl Acad. Sci. USA*, **74**, 4806–4810.

Goodman, M. F., Watanabe, S. M. and Branscomb, E. W. (1982) Passive polymerase control of DNA replication fidelity: evidence against unfavored tautomer involvement in base transition mutations. In *Molecular and Cellular Mechanisms of Mutagenesis* (eds J. F. Lemontt and W. M. Generoso), Plenum Press, New York, pp. 213–229.

Goodman, M. F., Gore, W. C., Muzyczka, N. and Bessman, M. J. (1974) Studies on the biochemical basis of spontaneous mutation III. Rate model for DNA polymerase-effected nucleotide misincorporation. *J. Mol. Biol.*, **88**, 423–435.

Goodman, M. F., Keener, S., Guidotti, S. and Branscomb, E. W. (1983) On the enzymatic basis for mutagenesis by manganese. *J. Biol. Chem.*, **258**, 3469–3475.

Goodman, M. F., Hopkins, R. L., Watanabe, S. M., Clayton, L. K. and Guidotti, S. (1980) On the molecular basis of mutagenesis: enzymological and genetic studies with the bacteriophage T4 system. In *Mechanistic Studies of DNA Replication and Genetic Recombination*, ICN–UCLA Symposia on Molecular and Cellular Biology (eds B. Alberts and C. F. Fox), Vol. XIX, Academic Press, New York, pp. 685–705.

Hershfield, M. S. (1973) On the role of deoxyribonucleic acid polymerase in determining mutation rates. *J. Biol. Chem.*, **248**, 1417–1423.

Hibner, U. and Alberts, B. M. (1980) Fidelity of DNA replication catalyzed *in vitro* on a natural DNA template by the T4 bacteriophage multi-enzyme complex. *Nature*, **285**, 300–305.

Hopfield, J. J. (1974) Kinetic proofreading: a new mechanism for reducing errors in biosynthetic processes requiring high specificity. *Proc. Natl Acad. Sci. USA*, **71**, 4135–4139.

Hopfield, J. J. (1980) The energy relay: a proofreading scheme based on dynamic cooperativity and lacking in all characteristic symptoms of kinetic proofreading in DNA replication and protein synthesis. *Proc. Natl Acad. Sci. USA*, **77**, 5248–5252.

Hopkins, R. and Goodman, M. F. (1979) Asymmetry in forming 2-aminopurine. Hydroxymethylcytosine heteroduplexes; a model giving misincorporation frequencies and rounds of DNA replication from base-pair populations *in vivo. J. Mol. Biol.*, **135**, 1–22.

Hopkins, R. and Goodman, M. F. (1980) Deoxyribonucleotide pools, base pairing and sequence configuration affecting bromodeoxyuridine- and 2-aminopurine-induced mutagenesis. *Proc. Natl Acad. Sci. USA*, **77**, 1801–1805.

Hopkins, R. L. and Goodman, M. F. (1985) Ribonucleoside and deoxyribonucleoside triphosphate pools during 2-aminopurine mutagenesis in T4 mutator, wild type, and antimutator infected *Escherichia coli* cells. *J. Biol. Chem.*, **260**, 6618–6622.

Janion, C. and Shugar, D. (1973) Preparation and properties of poly 2-aminopurine ribotidylic acid. *Acta Biochimica Polonica*, **20**, 271–284.

Katritzky, A. R. and Waring, A. J. (1962) Tautomeric azines. I. Tautomerism of 1-methylarocil and 5-bromo-1-methyluracil. *J. Chem. Soc.*, Part 2, 1540–1544.

Koch, R. E. (1971) The influence of neighboring base pairs upon base-pair substitution mutation rates. *Proc. Natl Acad. Sci. USA*, **68**, 773–776.

Koch, A. L. and Miller, C. (1965) A mechanism for keeping mutations in check. *J. Theoret. Biol.*, **8**, 17–80.

Kornberg, A. (1974) *DNA Synthesis*, W. H. Freeman and Company, San Francisco, pp. 85.

Kornberg, A. (1980) *DNA Replication*, W. H. Freeman and Company, San Francisco, Chapters 4 and 5.

Liu, C. C., Burke, R. L., Hibner, U., Barry, J. and Alberts, B. M. (1978) Probing DNA replication mechanisms with the T4 bacteriophage *in vitro* system. *Cold Spring Harbor Symp. Quant. Biol.*, **43**, 469–487.

Loeb, L. A., Dube, D. K., Beckman, R. A., Koplitz, M. and Gopinathan, K. P. (1981) On the fidelity of DNA replication: nucleoside monophosphate generation during polymerization. *J. Biol. Chem.*, **256**, 3978–3987.

Mathews, C. K., North, T. H. and Reddy, G. P. V. (1979) Multienzyme complexes in DNA precursor biosynthesis. In *Advances In Enzyme Regulation* (ed. G. Weber), Vol. 17, Pergamon Press, Oxford, pp. 133–156.

Mhaskar, D. N. and Goodman, M. F. (1984) On the molecular basis of transition mutations: frequency of forming 2-aminopurine · cytosine base mispairs in the G·C → A·T mutational pathway by T4 DNA polymerase *in vitro. J. Biol. Chem.*, **259**, 11 713–11 717.

Mufti, S. (1979) Mutator effects of alleles of phage T4 genes 32, 41, 44, and 45 in the presence of an antimutator polymerase. *Virology*, **94**, 1–9.

Muzyczka, N., Poland, R. L. and Bessman, M. J. (1972) Studies on the biochemical basis of spontaneous mutation I. A comparison of the deoxyribonucleic acid polymerases of mutator, antimutator, and wild type strains of bacteriophage T4. *J. Biol. Chem.*, **247**, 7116–7122.

Nevers, P. and Spatz, H.-C. (1975) *Escherichia coli* mutants *uvr* D and *uvr* E deficient in gene conversion of λ-heteroduplexes. *Molec. Gen. Genet.*, **139**, 233–243.

Ninio, J. (1975) Kinetic amplification of enzyme discrimination. *Biochimie*, **57**, 587–595.

Nossal, N. G. (1979) DNA Replication with bacteriophage T4 proteins. *J. Biol. Chem.*, **254**, 6026–6031.

Nossal, N. G. and Hershfield, M. S. (1973) Exonuclease activity of wild type and mutant T4 DNA polymerases: hydrolysis during DNA synthesis *in vitro*. In *DNA Synthesis In Vitro* (eds R. D. Wells and R. B. Inman), University Park Press, Baltimore, pp. 47–62.

Petruska, J. and Goodman, M. F. (1985) Influence of neighboring bases on DNA polymerase insertion and proofreading fidelity. *J. Biol. Chem.*, **260**, 7533–7539.

Pless, R. C. and Bessman, M. J. (1983) Influence of local nucleotide sequence on substitution of 2-aminopurine for adenine during deoxyribonucleic acid synthesis *in vitro*. *Biochemistry*, **22**, 4905–4915.

Radman, M., Wagner, R. E., Jr, Glickman, B. W. and Meselson, M. (1980) DNA methylation, mismatch correction and genetic stability. In *Progress In Environmental Mutagenesis* (ed. M. Alacevic), Elsevier North Holland Bio Medical Press, Amsterdam, pp. 121–130.

Reddy, G. P. V. and Pardee, A. B. (1980) Multienzyme complex for metabolic channeling in mammalian DNA replication. *Proc. Natl Acad. Sci. USA*, **77**, 3312–3316.

Reha-Krantz, L. J. and Bessman, M. J. (1981) Studies on the biochemical basis of spontaneous mutation VI. Selection and characterization of a new bacteriophage T4 mutator DNA polymerase. *J. Mol. Biol.*, **145**, 677–695.

Ronen, A. (1979) 2-Aminopurine. *Mutat. Res.*, **75**, 1–47.

Ronen, A. and Rahat, A. (1976) Mutagen specificity and position effects on mutation in T4 *r*II nonsense sites. *Mutat. Res.*, **34**, 21–34.

Ronen, A., Rahat, A. and Halevy, C. (1976) Marker effects on reversion of T4 *r*II mutants. *Genetics*, **84**, 423–436.

Rydberg, B. (1977) Bromouracil mutagenesis in *Escherichia coli* evidence for involvement of mismatch repair. *Molec. Gen. Genet.*, **152**, 19–28.

Rydberg, B. (1978) Bromouracil mutagenesis and mismatch repair in mutator strains of *Escherichia coli*. *Mutat. Res.*, **52**, 11–24.

Schekman, R., Wickner, A. and Kornberg, A. (1974) Multienzyme systems of DNA replication. *Science*, **186**, 987–993.

Scheuermann, R. H. and Echols, H. (1984) A separate editing exonuclease for DNA replication: the ε-subunit of *Escherichia coli* DNA polymerase III holoenzyme. *Proc. Natl Acad. Sci. USA*, **81**, 7747–7751.

Sedwick, W. D., Wang, T. S.-F. and Korn, D. (1975) 'Cytoplasmic' deoxyribonucleic acid polymerase. *J. Biol. Chem.*, **250**, 7045–7056.

Speyer, J. F. (1965) Mutagenic DNA polymerase. *Biochem. Biophys. Res. Commun.*, **21**, 6–8.

Speyer, J. F., Karam, J. D. and Lenny, A. B. (1966) On the role of DNA polymerase in base selection. *Cold Spring Harbor Symp. Quant. Biol.*, **31**, 693–697.

Tomich, P. K., Chiu, C. S., Wovcha, M. G. and Greenberg, G. R. (1974) Evidence for a complex regulating the *in vivo* activities of early enzymes induced by bacteriophage

T4. *J. Biol. Chem.*, **249**, 7613–7622.

Topal, M. D. and Fresco, J. R. (1976) Complementary base pairing and the origin of substitution mutations. *Nature*, **263**, 285–289.

Topal, M. D., DiGiuseppi, R. and Sinha, N. K. (1980) Molecular basis for substitution mutations. *J. Biol. Chem.*, **255**, 11 717–11 724.

Wagner, R., Jr and Meselson, M. (1976) Repair tracts in mismatched DNA heteroduplexes. *Proc. Natl Acad. Sci. USA*, **73**, 4135–4139.

Wang, M.-L. J., Stellwagen, R. H. and Goodman, M. F. (1981) Evidence for the absence of DNA proofreading in He La cell nuclei. *J. Biol. Chem.*, **256**, 7097–7100.

Watanabe, S. M. and Goodman, M. F. (1978) Mutator and antimutator phenotypes of suppressed amber mutants in genes 32, 41, 44, 45, and 62 in bacteriophage T4. *J. Virol.*, **25**, 73–77.

Watanabe, S. M. and Goodman, M. F. (1981) On the molecular basis of transition mutations: the frequencies of forming and 2-aminopurine · cytosine adenine · cytosine base mispairs *in vitro*. *Proc. Natl Acad. Sci. USA*, **78**, 2864–2868.

Watanabe, S. M. and Goodman, M. F. (1982) Kinetic measurement of 2-aminopurine · cytosine and 2-aminopurine · thymine base pairs as a test of DNA polymerase fidelity mechanisms. *Proc. Natl Acad. Sci. USA*, **79**, 6429–6433.

Watson, J. D. and Crick, F. H. C. (1953a) Genetical implications of the structure of deoxyribonucleic acid. *Nature*, **171**, 964–967.

Watson, J. D. and Crick, F. H. C. (1953b) The structure of DNA. *Cold Spring Harbor Symp. Quant. Biol.*, **18**, 123–131.

Weissbach, A. (1977) Eucaryotic DNA polymerases. *Annu. Rev. Biochem.*, **46**, 25–47.

Wickner, S. and Hurwitz, J. (1976) Involvement of *Escherichia coli* dna Z gene product in DNA elongaton *in vitro*. *Proc. Natl Acad. Sci. USA*, **73**, 1053–1057.

Wolfenden, R. V. (1969) Tautomeric equilibria in inosine and adenosine. *J. Mol. Biol.*, **40**, 307–310.

9 Stability and change through DNA repair

S. G. SEDGWICK

9.1 Introduction

DNA is continually threatened by physical, chemical and biological factors which alter its structure, and so modify or block its biological activity. The role of DNA repair is to counter these changes. In conferring resistance to DNA damage, some repair systems simply remove DNA lesions, so regenerating the original DNA sequence. In other instances, lesions are not removed. Their effects are tolerated, or avoided, by recombination or changes in DNA replication, so that there is an intrinsic possibility of genetic rearrangement or sequence change as a result of repair. Thus, DNA repair mechanisms can be seen both as a means of maintaining DNA sequence and, on other occasions, as a source of variability or rearrangement.

This chapter describes why and how different repair systems operate. In so doing it shows how repair can conserve, or alter the character of the repaired DNA and the surviving cell, and perhaps of the whole organism. Examples are drawn from organisms as diverse as bacteriophage and humans, underlining the universal problem which is faced in maintaining DNA structure. First, however, it is worth examining the different factors influencing DNA structure. This illustrates the types of repair problems which arise, which in turn determine the type of repair response a cell adopts to survive.

9.2 Types of DNA damage and cellular responses

9.2.1 ORIGINS AND TYPES OF DNA DAMAGE

The sources of DNA damage are both intracellular and environmental. Inside the cell at physiological pH and temperature there are a number of spontaneous decay reactions affecting DNA. Apurinic (Lindahl and Nyberg, 1972) and apyrimidinic (AP) (Lindahl and Karlstrom, 1973) sites are formed by hydrolysis of the glycosylic bond linking bases to deoxyribose residues. The rate of purine loss would only produce 0.5 events per chromosome per 40 minute replication cycle in *E. coli*, but about 10 000 depurinations would occur in the larger human genome in its longer generation time (Lindahl, 1979). Pyrimidines are lost at approximately one twentieth the rate of purines. Hydrolytic deamination of cytosine also occurs forming uracil (Lindahl and Nyberg, 1974) which, if left *in situ*, will pair with adenine at the next replication cycle to cause a GC to AT change. A similar deamination of adenine producing hypoxanthine occurs at a fiftieth of this rate, but if left unrepaired would cause an AT to GC transition mutation (Karran and Lindahl, 1980).

DNA damage may also come from chance interactions with cellular metabolites. Oxygen metabolism is associated with rare genotoxic oxygen species having lethal and mutagenic effects on DNA (Fenn *et al.*, 1957; Bruyninckx *et al.*, 1978). Non-enzymic alkylation of DNA by S-adenosyl-methionine can produce 7-methylguanine, 3-methyladenine and traces of O^6-methylguanine (Rydberg and Lindahl, 1982). Of the methylated bases, 3-methyladenine has a particularly strong effect on cell viability, whilst O^6-methylguanine causes mutagenesis. A secondary consequence of base methylation is an increase in spontaneous hydrolytic cleavage of glycosylic bonds so increasing the chance of AP site formation (Lindahl, 1979, and references therein).

Normal DNA metabolism may also precipitate changes in DNA structure and sequence. The accuracy limitations of replicative complexes and changes in substrate pool size can lead to the incorporation and persistence of mismatched bases into DNA (see Chapter 8). Mismatched base-pairs can also be generated by homologous recombination where the single strand of one allelic form of a gene may be juxtaposed against a second allelic sequence. The need to unwind DNA and have, even transitory, tracts of single-stranded DNA further exposes the molecule to reactions which are precluded by the usual stacked double-strand structure. In one example rates of cytosine deamination are 200–300 times higher in single strand DNA (Lindahl, 1979). More localized features of the DNA sequence might also affect its suscept-ibility to certain damage and its repair. The idea that frameshift mutations occurred by 'slippage' of strands in monotonous sequences is almost twenty years old (Streisinger *et al.*, 1966). The occurrence of quasipalindromic sequences has also been proposed as a source of deletion, insertion and base change mutagenesis, depending upon the types of misalignment which occur and the way in which they are rectified (Todd and Glickman, 1982; Glickman and Ripley, 1984, and references therein). The efficiency of UV photo-product formation is also affected by the surrounding base composition (Haseltine *et al.*, 1980) and the condensation of eukaryotic DNA into domains can reduce its repairability (Leadon and Hanawalt, 1984).

A different and more variable threat comes from environmental chemicals and radiations. A description of the multitude of genotoxic agents, and the even greater variety of lesions which they make in DNA is outside the scope of this article. The lesions produced differ in severity from modifications, such as N[7]-methylation of guanine, which can be ignored, to fragmentation of chromosomes by double-strand breakage.

Much of the emphasis of studies on DNA-reactive agents has been with the effects of man-made chemicals and radiation sources. However, Ames (1983) has argued that these are secondary effects compared with the long-term exposure of DNA to naturally occurring genotoxic chemicals in food. Plants contain a large variety of toxic chemicals, possibly as a deterrent to consumption by bacteria, moulds and animals. A considerable number of these cause mutagenesis, indicating that they interact with DNA. Modification of food can also produce chemicals which damage DNA. The spoilage of grains, fruits and nuts by certain moulds can produce genotoxic agents. One such product, aflatoxin, is one of the most potent mutagens known. Rancidity of fats and oils is a dietary source of mutagenic oxidating agents, and in the human diet the amounts of these materials is further increased by cooking. Pyrolysis of proteins, amino acids and sugars by charring, roasting and grilling also produces potent mutagens.

Another environmentally imposed stress to DNA is naturally occurring radiation from sunlight, the cosmos, and the endogenous radioactivity of minerals and rocks. The absorption of these radiations can produce rearrangements of chemical bonding between the constituent atoms of DNA, either within one strand, or between strands to form cross-links, or with surrounding molecules to form adducts and hydration products. DNA strands may be fragmented either singly or doubly, producing different end groups with different repair requirements (Henner, Grunberg and Haseltine, 1983). With the more energetic radiations sufficient energy may be imparted to DNA to bring about a cluster of secondary ionizings causing saturations of double bonds, ring opening, and ring degradative reactions.

Obviously, this catalogue of threats to DNA structure would have a profound biological impact if there were no DNA repair systems to remove damage or nullify its effects.

Despite the somewhat bewildering diversity of DNA damage possible, two general repair problems can be recognized which can influence how a cell responds to the threat to its survival. One factor influencing the type of repair used is the degree of information loss caused by the lesion. A second factor which determines the amount of repair, at least in prokaryotes, is the frequency of the lesion. That is to say, DNA repair activities can be induced in response to the repair requirements of the cell.

9.2.2 STRATEGIES OF DNA REPAIR

There is a redundancy in genetic information inherent in DNA's duplex structure and in diploidy. Thus DNA damage can cause different extents of information change or loss, depending upon whether it affects one DNA strand, both strands and whether or not homologous duplexes are present. Three general strategies of repair can be recognized which cope with these three different degrees of genetic disablement.

Where damage is only in one strand, sufficient genetic information remains to enable a lesion removal strategy of repair to operate. In some cases the damage is enzymically reversed *in situ*. More often, specific damage recognition enzymes excise the damaged single strand of DNA. The duplex is reconstituted using the information encoded in the intact complementary strand as a template for repair DNA synthesis. In these ways, DNA repair restores DNA structure with little chance of sequence change or rearrangement.

Damage to both DNA strands necessitates a different strategy since in this case there is a total loss of information at the damaged site. The second strategy of repair by recombination enables the information of intact homologous duplexes to be used to restore the damaged duplex. Numerous mechanisms have been elaborated for this process. All involve information transfer by the physical exchange of DNA strands and/or by one duplex providing the templates for the repair replication of the other. This strategy of repair therefore produces heteroduplex DNA, and possibly strand exchange. Both of these products have the potential to change the character of the repaired DNA.

A third strategy operates where genetic information is absent in both strands and there is no means of interacting with a homologous duplex, should it even be present. In the face of a total loss of genetic information at a particular site there appear to be changes in either semi-conservative or repair replication which permit DNA synthesis past lesions. It is therefore not surprising that this tolerance strategy for dealing with damage appears to be a source of genetic change: in other words, mutagenesis.

The repair problems posed by a particular spectrum of DNA damage can further change before repair reaches completion by the interplay of repair systems with each other, and with semi-conservative DNA replication. For example, damage in one strand of DNA might encounter a replication fork before a lesion-removal enzyme. The lesion may pass through replication leaving a gap opposite it, so creating a double-strand loss of information in the daughter duplex. An extreme of this effect is demonstrated with excision-less mutants of bacteria, yeast and some mammalian cells. In the absence of lesion removal, more damage encounters replication forks, so modifying the repair problem and channelling it more into recombinational and replicative repair. This example serves as an important reminder that many lesions and repair intermediates can be the substrates for the action of several different enzymes. This stochastic element in DNA repair means that a single type of lesion might be processed through a branching array of steps. Although one branch might predominate, allowing us to categorize it as a repair 'pathway', there can be other rarer reactions making different repair products with possibly different biological outcomes.

9.2.3 REGULATION OF AMOUNTS OF DNA REPAIR

The first indications that repair capacity could change in response to the level of DNA damage came in 1953 with the experiments of Weigle. He showed that the survival and induced mutagenesis of UV-irradiated phage were increased by infecting *E. coli* which itself had been irradiated. Since then a common feature of many investigations into inducible repair has been to 'pre-induce' a cell with one dose of DNA damage. The idea is to test whether the cell is then more able to withstand a second challenge dose, or better able to reactivate damaged phage, viral or plasmid DNA. The advent of recombinant DNA techniques has since enabled direct tests of DNA damage-induced gene expression. To date the best characterizations of these inducible responses are in *E. coli* and these are described in depth in an excellent recent review (Walker, 1984). They will only be summarized briefly here along with evidence for similar systems in higher cell types. The general picture emerging is one where DNA repair may be only one part of a broader physiological response to DNA damage and its disruptive effect on DNA replication and cell division.

(a) The adaptive response

Alkylating agents add alkyl (methyl, ethyl, etc.) groups to numerous sites in DNA. They kill cells and cause mutations in the survivors. *E. coli* grown in low concentrations of these compounds is 'adapted' against the lethal and mutagenic effects of a subsequent challenge dose (Samson and Cairns, 1977). Adaption is now known to involve the synthesis of at least two repair enzymes (see Lindahl, 1982; Walker, 1984, for recent reviews). One removes 3-methyladenine, 3-methylguanine, O^2-methylcytosine and O^2-methylthymine from DNA. Amongst these lesions are those with a particularly strong lethal effect. The other repair protein is a methyltransferase which transfers alkyl groups from DNA to itself. O^6-methylguanine, O^4-methylthymine, and phosphotriesters are repaired in this way. It is the persistence of O^6-methylguanine which causes most mutagenesis by mispairing with T in DNA replication.

Mutants of the *ada* type have been isolated which are not adaptable and which do not make extra 3-methyladenine glycosylase and methyltransferase. At first they were thought to be defective in a third regulatory protein. However, cloning of this regulatory gene and immunological analysis of its product showed that it was in fact the methyltransferase (Teo *et al.*, 1984); the *ada* gene product is therefore bifunctional. It is a repair protein and also positively regulates its own synthesis and that of 3-methyladenine glycosylase.

Transcription is stimulated by alkylated Ada protein binding to a common sequence in the regulatory regions of *ada* and *alkA* genes (B. Sedgwick, I. Teo, M. Kilpatrick and T. Lindahl, personal communication). In addition there is induced synthesis of AlkB and AidB proteins whose functions are unknown (Volkert and Nguyen, 1984).

(b) The SOS response
The SOS response entails the damage induced expression of at least seventeen different genes (see Little and Mount, 1982; Walker, 1984, for recent reviews). It is induced by perturbations to the DNA's structure or replication cycle. Included in genes induced are ones with roles in excision, recombination and replicative repair. There are changes in cell division which can cause a temporary arrest in septation, alleviation of restriction, and changes in respiration. There are also *din* genes whose existence has only been detected by their *d*amaged *i*nduced expression and whose functions are unknown. Repair proteins of the adaptive response are not induced. Nevertheless alkylation damage can induce SOS functions, presumably by interfering with replication and causing bulky lesions with the addition of the larger alkyl groups (Boiteux, Huisman and Laval, 1984; Schendel *et al.*, 1983).

Despite their diverse physiological roles, the SOS induced genes share a common regulatory mechanism involving the *lexA*$^+$ and *recA*$^+$ gene products. In an undamaged cell, SOS gene expression is kept at low levels by LexA protein repressor binding to a consensus sequence, the SOS box, found in front of all SOS genes so far sequenced. The *lexA* gene itself is repressed in this way, so keeping LexA protein at a steady state level (Brent and Ptashne, 1980, 1981; Little, Mount and Yannisch-Perron, 1981). Another gene repressed is *recA*$^+$, which is expressed at low levels sufficient for homologous genetic recombination (Brent and Ptashne, 1981; Little, Mount and Yannisch-Perron, 1981). When DNA damage causes single-strandedness or produces DNA degradation products, the basal level of RecA protein becomes activated and brings about the specific proteolytic cleavage of the LexA protein (Little *et al.*, 1980). There is some question whether RecA protein is a protease itself or whether it is an allosteric effector of auto-digestion by LexA protein (Little, 1984). Whatever the mechanism, the resulting LexA cleavage fragments are unable to bind to the SOS box and gene expression ensues. More detailed accounts of *recA* and *lexA* regulation can be found elsewhere (Little, 1982, 1983; Brent, 1983a). Since the recA gene is itself repressed by LexA protein, there is also amplification of RecA protein synthesis. In fact, this is the most abundant protein of the SOS response and on some occasions may make up 5% of total protein after induction. RecA protein has a direct role in recombination repair. Thus both adaptive and SOS responses are both positively regulated by proteins which switch on gene expression and act directly in repair. A third activity of RecA protein in mutagenesis is also coming to light and this will be described in the section concerning replicative repair.

In addition to genes repressed by LexA, certain prophages are also induced as part of the SOS response. These encode their own repressor proteins which repress expression of most of the integrated phage genome. These repressors appear to represent a molecular example of convergent evolution. Although they differ in composition to LexA repressor, they undergo the same specific cleavage reaction at a single alanine–glycine bond which is surrounded by similar amino acids in both repressors. This convergence enables the prophage to 'keep an ear to the ground' so that its dormant state is aroused by conditions which threaten the future of its host (reviewed in Roberts and Devoret, 1983).

The intensity of the SOS response is governed by the type of damage and its duration before it is repaired. In the example of excision-less bacteria (see Section 9.2.2), the channelling of UV damage into recombinational repair creates more single-strandedness persisting for longer periods than when damage is simply excised. As a result, excisionless strains are approximately ten times more readily induced than wild type cells. When DNA damage levels are low, there is an intermediate amount of LexA protein inactivation. Since the binding of LexA has different affinities at different SOS genes, some SOS functions and not others can be induced (Brent, 1983a; Little, 1983). With the completion of repair, RecA protein is no longer activated, LexA protein levels rise, and the quiescent state of the regulatory network is restored. Thus the SOS response produces a complex, but transitory, change in cell character.

DNA damage can trigger expression of at least two other regulatory networks in *E. coli*. In one case, SOS inducing damage also induces part of the cells' response to heat shock (Krueger and Walker, 1984). In a second, peroxide damage induces repair capacity which is genetically distinguishable from either SOS or adaption responses (Demple and Halbrook, 1983).

(c) Inducible repair in eukaryotes
Results from physiological experiments with lower eukaryotes show that at certain stages of the cell cycle, certain recovery processes are improved by preinducing or by permitting *de novo* protein synthesis (Siede and Eckardt, 1984, and references therein; Fabre and Roman, 1977; Leaper *et al.*, 1980; Lee and Yarranton, 1983; Baker, 1983). Recently yeast has been directly tested for the presence of damage-inducible genes. In a clone bank of total yeast DNA, six random fragments have been identified which regulate damage induced expression of an adjacent *lac* structural gene (Ruby and Szostak, 1985). These clones were used to identify the damage induced message from their intact chromosomal sequences. Other experiments have

identified two different groups of sequences which show differential RNA synthesis after UV irradiation (McClanahan and McEntee, 1984; Rolfe, 1985b). The function of the genes producing these messages has not yet been determined. DNA damage has also been shown to induce expression from two cloned yeast genes, *RAD54* (Emery *et al.*, 1984) and the ligase gene *CDC9* (Peterson *et al.*, 1985; Barker *et al.*, 1984). UV induced proteins have also been detected in crude cell lysates (Rolfe, 1985a).

Mammalian cells can also be pretreated with DNA damaging agents so that the survival of damaged virus is increased. This is akin to Weigle reactivation of bacteriophage and similarly requires *de novo* protein synthesis (reviewed in Hall and Mount, 1981). Chinese hamster ovary (CHO) cells can be preconditioned so that their ability to repair breaks produced by replication on damaged templates is increased (D'Ambrossio, Aebersold and Setlow, 1978). Pretreatment of CHO cells with low doses of alkylating agents also diminishes the lethal effects of a large challenge dose of alkylation and reduces sister chromatid exchange induction (Samson and Schwartz, 1980). Although this preconditioning resembles features of the bacterial adaptive response no induced repair enzymes have yet been identified (see Section 9.3.2). UV-inducible proteins have been detected in crude extracts of human cells (Schorpe *et al.*, 1984). The same treatment also enhances SV40 transformation and expression (Hagedorn *et al.*, 1983), but as yet no damage-inducible genes have been identified.

9.3 Removal repair

The following section details different mechanisms for removal of damage to DNA. The examples range from processes reversing the changed covalent bonding of damaged bases to its original condition, to removing modifications from a base, to removing a damaged base, to removing a damaged nucleotide along with some undamaged neighbours. Each example uses proteins which recognize specific changes or distortions in DNA. In each case, sufficient information remains in the DNA for repair to be completed without interaction with other chromosomes. Apart from the systems of lesion reversal and lesion transfer to other proteins, lesion removal entails the release of damaged bases or nucleotides. The damaged strand is then restored by repair replication and ligation.

9.3.1 ENZYMATIC PHOTOREACTIVATION

Absorption of 220–300 nm ultraviolet light (UV) by double-stranded DNA can cause the linkage of adjacent pyrimidines by saturation of their 5,6 double bonds to give two pyrimidine rings in cis–syn configuration linked by a cyclobutane ring. A unique and specific mechanism for the repair of cis–syn pyrimidine dimers is performed by photoreactivation enzymes. The protein first binds to UV-irradiated DNA. A second irradiation with 300–500 nm light absorbed through a flavin group (Iwatsuki, Joe and Werbin, 1980; Sancar and Sancar, 1984) reverses the pyrimidine dimerization reaction so regenerating two unlinked pyrimidine bases (Sutherland, 1981, and references therein). This photoreversal is accompanied at the biological level with elimination of many of the lethal, mutagenic and recombinational effects of UV-irradiation. Photoreacting enzyme is found in prokaryotes, lower eukaryotes, insects, fish, amphibians, reptiles, birds and marsupial mammals (Sutherland, 1981, and references therein). Reports of its presence in placental mammals have not been confirmed by other workers (Sutherland, 1974; Mortelmans et al., 1977).

A second effect of photoreactivating enzyme is to increase survival of E. coli in the dark by increasing the efficiency of nucleotide excision repair by the UvrA, B and C proteins (Harm and Hillebrandt, 1962; Yamamoto, Satake and Shinagawa, 1984: Sancar, Smith and Sancar, 1984). Intriguingly, the reverse relationship was found in human Xeroderma pigmentosum (XP) cells where defective nucleotide excision was accompanied by reduced photo-reactivation activity (Sutherland, Rice and Wagner, 1975).

9.3.2 METHYLTRANSFERASE: REPAIR BY LESION TRANSFER

The mutagenic effect of alkylating agents in bacteria and mammalian cells is primarily linked with the formation of O^6-alkylguanine. This modified base is able to pair with thymine and so causes GC $\rightarrow$ AT transitions (Schendel and Robins, 1978; Newbold et al., 1980). The repair of O^6-alkyl guanine, and rarer O^4-alkyl thymine, occurs by removal of the offending alkyl group without any release of free bases or nucleotides, or apurinic site formation, or endonucleolytic scission. This hitherto unique repair strategy is performed by a protein called O^6-methylguanine-DNA methyltransferase. The mechanism of repair is best studied in E. coli but has an analogous counterpart in mammalian cells (Bogden, Eastman and Bresnick, 1981; Mehta et al., 1981; Waldstein, Cao and Setlow, 1982). E. coli methyltransferase, the product of the ada^+ gene, transfers methyl and ethyl groups from the O^6 position of guanine to itself. The alkyl group becomes attached to a specific cysteine residue in the protein which is not subsequently regenerated (Robins and Cairns, 1979; Lindahl, 1982). Methyltransferase activity is therefore stoichio-metric, rather than enzymic, requiring one protein molecule for the removal of one alkyl group from guanine (Robins and Cairns, 1979; Lindahl, Demple and Robins, 1982).

Recently Ada protein has been shown to remove phosphotriester alkyl groups from the DNA backbone. Although an alkyl-transfer mechanism is also employed for this lesion, it appears that a different cysteine is the acceptor (Margison, Cooper and Brennand, 1985; McCarthy and Lindahl, 1985). As outlined in a previous section, the Ada protein has a key role in switching on the adaptive response to alkylating agents. By stimulating its own synthesis it directly reduces the mutagenic effects of alkylation, and by switching on expression of other repair gene(s) it indirectly protects against the lethal effects of DNA alkylation. A distinctive feature of the alkylation mutagenesis is that it is clustered occurring at multiple sites in closely linked genes. One explanation is that methyltransferase does not remove lesions from the unwound DNA immediately preceding the replication fork. This idea is based on the thousand-fold reduced efficiency of methyltransferase activity on single-stranded DNA (Lindahl, Demple and Robins, 1982).

Several mammalian tumour cell lines exhibit the Mer⁻ phenotype which means they are sensitive to the killing and mutagenic effects of alkylating agents and act as poor hosts for similarly treated viruses. At the molecular level the Mer⁻ phenotype is linked to the absence of O^6-methylguanine DNA methyltransferase (Yarosh *et al.*, 1983; Harris, Karran and Lindahl, 1983). A different assay showed Mer⁻ cells to have this repair protein, but to be sluggish in generating more of it in response to the first bout of methyl acceptance (Waldstein, Cao and Setlow, 1982). However in other hands, this technique did not produce confirmatory results (Yarosh *et al.*, 1983), and adaptive synthesis of the protein could not be seen (Foote and Mitra, 1984).

9.3.3 BASE EXCISION REPAIR

Enzymes recognizing certain damaged bases and cleaving the sugar–base bond are called glycosylases. Their activity generates apurinic and apyrimidine (AP) sites which in turn are subject to recognition and elimination by endonucleolytic scission. The simplest glycosylases leave AP site recognition and removal to a second enzyme but others have their own AP endonuclease activity. Thus, different types of damage recognized by an array of glycosylases are channelled into a common repair mechanism.

Altered bases removed by glycosylases include uracil (Lindahl *et al.*, 1977), hypoxanthine (Karran and Lindahl, 1978, 1980), 3-methyladenine, 7-methylguanine, O^2-methylpyrimidines (McCarthy, Karran and Lindahl, 1984; Riazzudin and Lindahl, 1978; Karran, Hjelmgren and Lindahl, 1982; Evensen and Seeberg, 1982) and pyrimidine dimers (Haseltine *et al.*, 1980; Radany and Friedberg, 1980; Demple and Linn, 1980; Nakabeppu and Sekiguchi, 1981). Oxidative base damage producing ring opened 7-methyladenine and 7-methylguanine (Chetsanga and Lindahl, 1979; Breimer, 1984), thymine glycol (Demple and Linn, 1980; Breimer and Lindahl, 1984), urea, 5-hydroxy-5-methylhydantoin and methyltartronylurea (Breimer and Lindahl, 1980, 1984) are also removed by glycosylase action. Similarly small and typically highly specific enzymes have been found in mammalian cells (Lindahl, 1982).

(a) Uracil glycosylase
The biological role of uracil glycosylase is to guard against the mutagenic effects of uracil produced by hydrolytic deamination of cytosine (Lindahl and

Nyberg, 1974) or misincorporation of dUTP (Tye *et al.*, 1978). This can be seen in the high spontaneous mutagenesis of *ung* mutants, lacking uracil glycosylase, where C to T base changes are particularly abundant. They are presumably absent in wild type cells because deaminated cytosines are removed before they have the chance to pair with adenine (Duncan and Miller, 1980). In wild type cells the remaining hot spots are at 5-methyl-cytosines which, upon deamination, would be converted to thymine and therefore would be unrecognizable (Coulondre *et al.*, 1978). It has been proposed that the deamination of C to form U has provided the selective force against using U as a normal base for DNA. The inclusion of U in DNA would have eliminated the ability to discriminate between deamination products of cytosine and a correctly occurring uracil (Lindahl *et al.*, 1977).

(b) 3-Methyladenine glycosylases
In *E. coli* there are two 3-methyladenine glycosylases (reviewed in Lindahl, 1982). One, present constitutively, removes 3-methyladenine, *tag* mutants lack this enzyme. The second, the product of the *alkA*$^+$ gene, is synthesized in response to alkylation damage in the adaptive response. It is less specific, removing 3-alkylpurines and O^2-alkylpyrimidines. Mutational inactivation of either *tag* or *alkA* genes renders *E. coli* particularly sensitive to killing by alkylating agents. These glycosylases therefore deal with lesions having lethal, rather than mutagenic, effects.

(c) Pyrimidine dimer – DNA glycosylases
Two enzymes with similar activities for pyrimidine dimer photoproducts have been isolated from *Micrococcus luteus* and T4 infected *E. coli*. Both enzymes have glycosylase activity towards the 5′ glycosylic bond of a pyrimidine dimer, followed by endonuclease action at the phosphodiester bond 3′ to the (AP) site (Haseltine *et al.*, 1980; Radany and Friedberg, 1980; Demple and Linn, 1980; Nakabeppu and Sekiguchi, 1981). AP endonuclease activity is weak compared with glycosylic cleavage and can be replaced in *E. coli* by other enzymes such as exonuclease III (see below). This two-step endonucleotic scission opens the DNA to excision of a short oligonucleotide with a 5′ pyrimidine dimer attached to it.

(d) Endonuclease III
E. coli endonuclease III has a similar mechanism of action to *M. luteus* and T4 pyrimidine dimer glycosylases, except that its activity is directed towards hydration, oxidation and ring opened derivatives of thymine (Demple and Linn, 1980; Breimer and Lindahl, 1984; Katcher and Wallace, 1983). The wide range of damage recognized is rather untypical of a glycosylase. It may recognize a structural feature common to all oxidized pyrimidine derivatives,

possibly the absence of a 5,6 double bond, or the non-planar structure of base derivatives lacking this bond (Breimer and Lindahl, 1984; Katcher and Wallace, 1983). A similar calf thymus enzyme with AP activity and glycosylase activity towards thymine glycol and urea has been found.

9.3.4 APURINIC AND APYRIMIDIC (AP) ENDONUCLEASE

AP sites can be generated physicochemically and as an intermediate step in repair after glycosylase activity. The base leaving reaction can also be accelerated by methylation and other chemical and radiation modifications (Lawley and Brookes, 1963; Drinkwater, Miller and Miller, 1980; Sage and Haseltine, 1984).

Apurinic endonucleases cleave either the 3′ phosphodiester bond following an AP site (type I) or the 5′ bond preceding it (type II). The major AP endonuclease activity in *E. coli* is type II provided by exonuclease III (Rogers and Weiss, 1980) (sometimes called endonuclease II, see Lindahl, 1979). This complex enzyme has in addition, RNase H, 3′–5′ exonuclease, and phosphatase activities. Type II cleavage provides the correct 3′-OH to prime repair replication which displaces the base-less deoxyribose residue at the 5′ end of a short oligonucleotide. This reaction is not possible by type I cleavage alone because the protruding 3′-deoxyribose group is unable to prime DNA replication. This may be removed by other 3′–5′ exonucleases in the cell. Alternatively, type I and II enzymes may act together to sever connection of the deoxyribose from the phosphodiester chain at either side (Lindahl, 1979; Mosbaugh and Linn, 1980).

The absence of exonuclease III in *xth* mutants causes sensitivity to agents like H_2O_2 which produce oxidative damage and enzymic AP site formation, and to agents like MMS where methylation stimulates physicochemical base loss (Demple, Halbrook and Linn, 1983, and references therein).

The mutagenic effect of AP sites is only realized in *E. coli* when the cells are independently induced for the mutagenic repair activity provided by the *umuCD* operon (see later). Details of other *E. coli* AP endonucleases are reviewed elsewhere (Lindahl, 1979, 1982).

In humans and other mammals a single type II AP endonuclease is usually found which sometimes also recognizes thymine glycols (Nes, 1980; Brent, 1983b) and UV and X-ray damage (possibly 6-hydroxy 5,6-dihydrothymine) (Bacchetti and Benne, 1975). However, an additional type I enzyme has been found in normal human fibroblasts, but is reported absent from group D cells of the UV sensitive syndrome, Xeroderma pigmentosum (Mosbaugh and Linn, 1980). Two other human syndromes have been tested for AP endonuclease activity and found proficient. These were ionizing radiation sensitive Ataxia telangiectasia and Bloom's syndrome which has elevated sister chromatid exchange.

9.3.5 NUCLEOTIDE EXCISION

A more enzymically complex system of pyrimidine dimer removal is found in *E. coli*. Damage recognition requires at least three proteins, the products of the *uvrA*, *B*, and *C* genes (Seeberg, Nissen-Meyer and Strike, 1976; Sancar and Rupp, 1983. Until recently models for excision repair envisaged a single endonucleolytic break 5' to a pyrimidine dimer (for example Hanawalt *et al.*, 1979). However, *in vitro* cleavage analysis of a defined UV-irradiated substrate showed that the UvrA B C protein complex makes two incisions spanning a pyrimidine dimer. One cut is at the eighth phosphodiester bond 5' to the pyrimidine dimer, the other at the fourth or fifth bond at the 3' side (Sancar and Rupp, 1983). Unlike the T4 and *M. luteus* pyrimidine dimer recognition enzymes, the UvrABC complex can recognize a variety of adducts and cross-links. This was initially indicated by the sensitivity of *uvrA*, *B* or *C* mutants to UV, mitomycin C, psoralens and 4-nitroquinoline oxide, and later confirmed in *in vitro* assays. The 6–4 photoproduct, which has a particularly strong effect on a UV-induced mutagenesis (see later sections) is also removed by this system (Sancar and Rupp, 1983; Haseltine, 1983). The separation of double endonucleolytic cuts by approximately one superhelical turn may indicate that the repair complex acts at one face of the damaged duplex, which need not include the specific site of the lesion (Sancar, Smith and Sancar, 1984). Thus the complex may recognize a more general type of helix distortion rather than the particular features of a pyrimidine dimer or any other lesion.

Pyrimidine dimer removal has been found in yeast (Unrau, Wheatcroft and Cox, 1971), *Ustilago maydis* (Unrau, 1975), *Tetrahymena* (Brunk and Hanawalt, 1967), *Drosophila melanogaster* (Boyd *et al.*, 1983, and references therein), mammals to varying extents, and humans (Regan, Trosko and Carrier, 1968). In contrast, excision repair capacity is low in reptiles and fish (Woodhead, Setlow and Grist, 1980; Regan *et al.*, 1982), in marsupials (Buhl, Setlow and Regan, 1974) and rodents (see Section 9.8.3).

To date, the specific mechanism employed in eukaryotes has not been worked out, but there are hints of a multiprotein complex analogous to the *E. coli* system. For instance, several of the mutants of yeast, *Drosophila*, Chinese hamster ovary (CHO) cells, and humans lack lesion incision activity and fall into multiple complementation groups. This indicates a multiprotein incision step unlike the single glycosylase protein used by T4 and *M. luteus*. In the yeast *Saccharomyces cerevisiae* at least five loci, *RAD 1,2,3,4*, and *10* are required to make the initial incision step (Reynolds and Friedberg, 1981; Reynolds, Love and Friedberg, 1981; Wilcox and Prakash, 1981). In *Drosophila*, *mei-9^{D2}* and *mus(2) 201* mutants fail to make the strand breaks associated with incision near lesions (Harris and Boyd, 1980; Boyd *et al.*, 1982). The human mutants are individuals with photosensitive syndrome Xeroderma pigmentosum (XP) in which eight complementation groups have been identified (Robbins *et al.*, 1983). Within these XP groups the extent of repair deficiency varies. However, the general reduction of lesion removal (Paterson, Lohman and Slutyer, 1973) and concomitant repair replication (Cleaver, 1968; Cleaver and Bootsma, 1975) is linked with a defect in the initial incision of damaged DNA (Setlow *et al.*, 1969; Fornace, Kohn and Karn, 1976; Erixon and Ahnstrom, 1979). A defect in the early stage of excision repair is also indicated by repair replication resuming when T4 UV endonuclease is introduced into XP cells, nuclei, or lysates (Tanaka, Sekiguchi and Okada, 1975; Smith and Hanawalt, 1978; Ciarrocchi and Linn, 1978).

Two groups of experiments point to the possibility that the XP defect is not simply an endonuclease deficiency. Extracts of XP A, C and G could excise lesions from UV irradiated DNA but not from their own chromatin. Certain combinations of XP extracts and different XP chromatin also permitted excision (Mortelmans *et al.*, 1976; Fujiwara and Kano, 1983). Whether these reactions represent actual *in vivo* events has to be resolved, but they do hint at a pre-incisional function which makes DNA 'eligible' for repair. Substantially equivalent results have been obtained in a study of five different complementation groups of UV sensitive Chinese hamster ovary cells (CHO) where again the defect in lesion removal was at, or before, the incision stage (Thompson and Carrano, 1983, and references therein).

The yeast, *Drosophila*, human, and CHO excision repair systems further resemble the *E. coli* UvrA B C system in the broad range of sensitivity produced when the systems are defective (Boyd *et al.*, 1983; Setlow, 1978; Thompson and Carrano, 1983, and references therein). This again contrasts with a specific type of UV sensitivity which would be expected of a deficiency in a T4 type enzyme.

Following incision, the offending segment of DNA is eliminated and replaced by repair replication. In *E. coli* the displacement of this segment might be performed by DNA polymerase I which performs most of the repair replication and has 5'–3' exonuclease activity. An ancillary activity might be provided by helicase II, the product of the SOS inducible *uvrD* gene, in unwinding the damaged region (Hickson *et al.*, 1983; Kumura and Sekiguchi, 1984). Mutations at the *uvrD* locus reduce the rate of pyrimidine dimer release from DNA and the rejoining of repaired DNA to parental. They do not block incision (Van Sluis, Mattern and Patterson, 1974; Sinzinis, Smirnov and Saenko, 1973). There is also *in vitro* evidence that UvrD protein aids the incision of DNA by UvrABC complex (Kumura *et al.*, 1985). Other exonucleases, such as exo V and VIII, might also displace the damaged DNA strand in lieu of DNA polymerase I action (Hanawalt *et al.*, 1979, and references therein). However, considerable evidence points to the efficiency of a concerted displacement repair synthesis reaction by DNA polymerase I compared with a two-step reaction performed by two different enzymes.

The size of repair replication patches is heterogeneous. Approximately 99% are about twenty bases long showing that exonucleolytic degradation flanking the double endonucleolytic cuts is quite restrained. These patches require DNA polymerase I action. The remaining patches may be up to 1500 bases long. Although they can be best seen in *polA* mutants where their formation uses DNA polymerase II or III they may also use DNA polymerase I in wild type cells (Cooper, 1982, and references therein). Physical assays of long patch repair replication show that it requires induction of the SOS regulatory network and post-irradiation growth conditions (Hanawalt *et al.*, 1979, and references therein). The bulk of repair replication performed by DNA polymerase I did not show these requirements. However, more sensitive assays of *uvrA*, *B C* and *D* gene expression have shown that they too are under control of the SOS regulatory system (Walker, 1984, and references therein).

The subsequent repair replication stage of excision repair in mammalian cells takes two forms (Regan and Setlow, 1974a). In one, patches three to four bases long are made in response to X-rays and alkylating agents, and possibly apurinic sites. The duration of this repair is short and usually complete within two hours. A second type of repair replication occurs in response to UV, *N*-acetoxy acetylaminofluorine and similar agents producing 'bulky' lesions. Patches of repair replication are between 14 and 120 bases long, depending upon which agent is used and which experimental protocol is used.

The final stage of repair employs DNA ligase to link the repair patch to the pre-existing DNA (Youngs and Smith, 1977).

9.3.6 MISMATCH REPAIR

Mismatched base-pairs are thought to be generated in both recombination and replication. In recombination they may arise after single-strand exchange produces a heteroduplex of strands from two different alleles which have mismatched bases in regions of allelic differences of sequence. Replicative mismatches are thought to occur by the persistence of erroneously incorporated bases which escape elimination by DNA polymerase proofreading activity (see Chapter 8).

Early evidence for mismatch repair came from genetic analyses where the processes of gene conversion, high negative interference and map expansion were attributed to mismatch repair of recombinational heteroduplex (Holliday, 1964; White and Fox, 1974; Fincham and Holliday, 1970). Mismatch repair has also been implicated for the production of pure mutant clones in UV irradiated *U. maydis* (Holliday, 1962) and yeasts (Nasim and Auerbach, 1967; Haefner, 1968; Eckardt and Haynes, 1977; James and Kilbey, 1977). Since UV photoproducts usually occur in one DNA strand, the idea was that mutations initially were made in one strand, then transferred to both by mismatch correction before chromosomal replication.

More direct characterization of the genetic and molecular nature of mismatch repair has come from heteroduplex transfection experiments. Heteroduplexes have been reconstituted from genetically distinguishable strands. Typically the progeny of a single transfection shows a predominance of one parental type, the other being eliminated by mismatch correction. This process has been detected in *B. subtilis* (Spatz and Trautner, 1970), and with SV40 (Lai and Nathans, 1975; Wilson, 1977), polyoma (Miller, Cooke and Fried, 1976) and plasmids (Folger *et al.*, 1985) introduced into mammalian cells. Correction of heteroduplex occurs in *E. coli* and is reduced by *mutS*, *mutL*, *mutH* and *uvrE* mutations (Nevers and Spatz, 1975; Wildenberg and Meselson, 1976; Rydberg, 1978; Bauer, Krammer and Knippers, 1981; Pukkila *et al.*, 1983). The high spontaneous mutation frequencies of these mutants show the role of mismatch correction in eliminating replicative errors in chromosomal DNA.

The way *E. coli* distinguishes the 'good' base from the 'bad' in a replicative mismatch is through the transient undermethylation of new DNA. This only becomes methylated in the wake of replication by the *dam* modification system acting at the 6 position of adenine in GATC sequences (see Marinus, 1984, for a current review). Heteroduplex made from *dam* and *dam*-3 phage is biased in mismatch correction in favour of the methylated strand and cannot work at all on fully methylated DNA (Radman *et al.*, 1980; Pukkila *et al.*, 1983).

In yeast the co-conversion of linked markers several hundred nucleotides apart first indicated that mismatch repair extended over a large region (Fogel and Mortimer, 1969). In transfection experiments with *E. coli* co-conversion was seen with markers up to 2000 or 3000 bases apart (Wildenberg and Meselson, 1976; Wagner and Meselson, 1976). Similarly in SV40, co-conversion occurred over 600 bases (Miller, Cooke and Fried, 1976). Since co-conversion occurred on the same strand, a model has arisen for an excisional mechanism of mismatch repair where the patches of repair replication extend over several hundred base pairs.

An *in vitro* mismatch repair system mimics the *in vivo* requirements for *mut H, L* and *S* functions and is biased to making repair replication in the undermethylated strand of a heteroduplex (Lu, Clark and Modrich, 1983). Cross-complementation seen by mixing extracts promises to open the way to biochemical purification and characterization of these gene products. In *Ustilago maydis* DNase I recognizes mismatched bases in DNA and this enzyme may be involved in gene conversion during mitotic allelic recombination (Ahmad, Holloman and Holliday, 1975; Pukkila, 1978).

9.4 Recombinational repair

Some DNA damage is not amenable to the simple removal strategies of repair. Usually this is when there is loss of genetic information in both DNA strands; good examples are double-strand breaks and inter-strand cross-links. Two-stranded information loss also occurs in *E. coli* when a single-strand lesion meets a replication fork before a removal enzyme. With UV, the pyrimidine dimer photoproducts dislocate the processivity of chromosomal replication. A gap in the newly synthesized DNA is formed opposite the pyrimidine dimer, and extends some thousands of bases (reviewed in Hall and Mount, 1981). Thus the repair problem is compounded from a loss of coding in one strand to loss of coding in both. A major bacterial mechanism for overcoming these types of repair problems uses recombinational exchange (the possibility that a small, physically undetectable fraction is repaired by *de novo* replication will be discussed later). The general strategy is to donate an intact DNA strand into the damaged duplex from a homologous sister or daughter chromosome. The donating duplex is simply restored by repair replication.

9.4.1 POST-REPLICATION REPAIR OF UV DAMAGE

Our understanding of the physical mechanism of recombinational repair of UV post-replication gaps is partly derived from studies of the related process of homologous genetic recombination, which shares the requirement for the recombinogenic RecA protein. After single-strand transfer into the daughter strand gap, continued co-operative binding of RecA protein brings all four strands into synapsis creating the classic Holliday recombination intermediate. The junction may migrate by reciprocal strand displacement until isomerization and endonucleolytic cleavage of the crossover terminates the liaison between the two duplexes (West, Countryman and Howard-Flanders, 1983; Howard-Flanders, West and Stasiak, 1984). The separation of the four-stranded junction occurs in two alternative pair-wise permutations of strands so that there is a 50% chance that flanking sequences to the left and right of the repaired site are recombined (Ganesan, 1974).

In UV-irradiated mammalian cells there is a post-replicative process for making full-size new DNA, but this appears replicative rather than recombinational (see next section). Nevertheless, multiplicity reactivation and enhanced recombination do occur in UV-irradiated SV40 (Dubbs, Rachmeler and Kit, 1974; Upcroft, Carter and Kidson, 1980; Gentil, Margot and Sarasin, 1983), Herpes simplex (Selsky *et al.*, 1979; Hall, Featherston and Almy, 1980; DasGupta and Summers, 1980), and adenovirus (Young and Fisher, 1980). Irradiation of hosts did not stimulate recombination which remained normal in XP, XP variant, Fanconi's anaemia and Ataxia telangiectasia cells. Only in one case of Bloom's syndrome, and not others, was multiplicity reactivation reduced (Selsky *et al.*, 1979). Bloom's syndrome is renowned for its harlequin chromosomes showing a high level of sister chromatid exchange. However, the possibility that this is caused by abnormalities in recombination which also affect multiplicity reactivation remains conjectural.

9.4.2 CROSS-LINK REPAIR

DNA cross-links can be repaired by a pre-replicative recombination process, provided that an excision event releases the damaged base in one of the DNA strands. This produces an adduct gap structure which is equivalent to the pyrimidine dimer gap presented to post-replication repair. Coding is lacking in both strands, but can be restored by recombinational exchange (Lin, Bardwell and Howard-Flanders, 1977; Sinden and Cole, 1978). In humans, Fanconi's anaemia cells are sensitive to cross-linking agents and are defective in making the first half-incision at one end of the crosslink (Fujiwara and Tatsumi, 1977). Whether the later steps of repair in human cells also resemble the bacterial cut and recombine system is not however known.

9.4.3 DOUBLE-STRAND BREAK REPAIR

Double-strand break repair was first identified in the extremely resistant organism, *Micrococcus radiodurans* (now known as *Deinococcus radiodurans*) (Kitayama and Matsuyama, 1968). Later it was detected in *E. coli*, provided they were $recA^+$ and had multiple chromosomal copies (Krisch, Krasin and Sauri, 1976; Krasin and Hutchinson, 1977). Double strand break repair requires *de novo* protein synthesis and intervention of the SOS response (Krasin and Hutchinson, 1981). The extra radiation resistance conferred by the SOS response has been linked with increased RecA protein synthesis (Krasin and Hutchinson, 1981; Pollard, Fluke and Kazanis, 1981) although SOS induction of *recN* and other genes may also be required (Picksley, Ahfield and Lloyd, 1984). SOS induction was first linked to the induction of an inducible inhibitor of DNA degradation which in turn permitted more repair (Pollard and Randall, 1973). Similar arguments have been made for SOS induction increasing post-replicative recombination repair of UV damage (Volkert, George and Witkin, 1976). However, it is equally plausible that it is efficient repair itself which precludes degradation (Krasin and Hutchinson, 1981). This seems even more likely now that purified RecA protein has been found to have *in vitro* properties consistent with its roles in double-strand break repair (West and Howard-Flanders, 1984) and post-replication repair (Howard-Flanders, West and Stasiak, 1984; Radding, 1982). A further indication of a direct SOS-induced recombinational role of RecA protein in these repair mechanisms is that increased RecA protein levels will even increase recombination in undamaged molecules (Lloyd, 1978).

Double-strand break repair has also been detected in yeast and *Ustilago maydis*, and is again linked with proficiency in homologous recombination (Ho, 1975; Resnick, 1978; Szostak, Orr-Weaver and Rothstein, 1983, and references therein). As with *E. coli*, the presence of homologous duplexes in yeast enhances survival and increases double-strand break repair (Brunborg, Resnick and Williamson, 1980), but not in the sensitive *rad52* mutant which lacks induced recombination. Similarly, double-strand

leavage of the yeast segment of plasmids stimulates integrative
ecombination and again RAD52 is required (Szostak, Orr-Weaver and
Rothstein, 1983).

Double-strand break repair has also been found in mammalian cells (Corry
nd Cole, 1973), and again there are suggestions of a recombinational
mechanism. Heteroduplex formation, presumably by recombination, has
been physically detected after treatment with ionizing radiation, mitomycin-
C (a cross-linking agent) and UV (Resnick and Moore, 1979; Fonck, Barthel
nd Bryant, 1984; Moore and Holliday, 1976; Rommelaere and Miller-
Faures, 1975). Also the frequency of interplasmid recombination was
ncreased by making double strand gaps with restriction endonucleases
Kucherlapati *et al.*, 1984).

Mammalian mutants deficient in double-strand break repair have been
solated in CHO cells (Jeggo and Kemp, 1983; Kemp, Sedgwick and Jeggo,
984). These *xrs* mutants are also sensitive to peroxide, possibly because of
coincident endonucleolytic cutting at base damage on opposite strands
Kemp, 1985).

The absence of double-strand break repair and cellular sensitivity is
ccompanied with a five-fold increase in chromatid gaps and breaks in CHO
rs mutants (Kemp, 1985) and increased chromosome loss in yeast *rad52*
mutants (Mortimer, Contopoulou and Schild, 1981).

In all of the recombinational repair mechanisms outlined above there are
wo potential sources of genetic change. One is the initial formation of
heteroduplex which provides a substrate for mismatch repair and so can cause
gene conversion. The different ways in which the reciprocal cross-over
unction is undone then determine whether heteroduplex is accompanied or
not by reciprocal exchange of flanking sequences. Readers interested in a
description and discussion of some current recombination models are
directed to recent reviews (Szostak, Orr-Weaver and Rothstein, 1983; Orr-
Weaver and Szostak, 1985)

.5 Replicative repair and induced mutagenesis

The third strategy of repair appears to act by increasing the tolerance of repair
nd semi-conservative replication to lesions in template DNA. It is the least
well characterized and least accessible to physical assays. As the process func-
ions around changes in DNA polymerase activities it can alter base sequence.
Thus, mutations form a large part of our information on this subject, since
hey are the 'tracks' left by the passing of this rather elusive recovery process.

5.1 DIRECT AND INDIRECT MUTAGENESIS

Direct mutagenesis occurs when an agent chemically transmutes the coding
roperties of one base into another. An example already encountered is O^6-
methylation of guanine which permits pairing with thymine.

In many more instances, DNA damage is not specific or subtle enough to convert directly the pairing properties of one base to another. Nevertheless, it can be mutagenic. This mutagenesis arises during the cell's response to the damages. Such indirect mutagenesis was first identified by the non-mutability of certain *lexA* and *recA* mutants (Witkin, 1967; Miura and Tomizawa, 1968; Bridges, Law and Munson, 1968), which we now know are unable to induce the SOS response and its numerous component genes. Later, mutations at the *umuCD* operon were found which specifically blocked the mutagenic component of the SOS response without affecting other parts of it (Kato and Shinoura, 1977; Steinborn, 1978). This identified *umuCD* as the possible structural gene for induced mutagenesis. The moderate sensitivity of *umuCD* mutants to many types of DNA damage reinforced earlier seminal ideas that certain repair processes were intrinsically mutagenic (Witkin, 1967). A similar situation exists in the yeast *S. cerevisiae* where the *rev3* mutation prevents UV-induced mutagenesis and causes mild UV sensitivity (Lemontt, 1971a), again pointing to the existence of a mutagenic repair system.

9.5.2 MECHANISMS OF MUTAGENIC REPAIR: MICROORGANISMS

Several forms of evidence indicate that mutagenic repair is distinct from excision and recombinational repair systems so far described. These appear different because they are proficient in non-mutable *E. coli umuC* mutants (Kato, 1977), in naturally unmutable *Proteus mirabilis* (Hofemeister and Eitner, 1981), and in yeast *rev3* mutants (Lemontt, 1971b, c; Prakash, 1981). Regulatory mutants of the SOS network can also be obtained which permit recombinational repair without UV-induced mutagenesis. Conversely, recombination activity can be blocked in *E. coli* and *rad52* yeast without preventing induced mutagenesis (McKee and Lawrence, 1979; reviewed in Sedgwick and Goodwin, 1985).

9.5.3 MUTAGENESIS IS TARGETED AT LESIONS

The evidence that indirect mutagenesis occurs at, or near, lesions comes from genetic and nucleotide sequencing data showing that each mutagen causes a distinctive distribution and type of mutation. The differences in spectra between two mutagens can therefore be attributed to different mutational sites (Miller, 1982, 1983). These results argue against the induction of a general mutator activity. They also show that SOS-induced mutagenesis can bring about a wider variety of changes than the direct mispairing type of mutation produced by alkylation agents. Mutagenesis by transition, transversion, deletion, duplication and frameshifting have all been observed. The elucidation of these changes, briefly reviewed below, can be used to envisage both the mechanisms and lesions which provoked them. (The literature on this topic is substantial and not always unequivocal. Papers by Le Clerc *et al.*

(1984), Miller (1983) and Kunkel (1984) provide good introductory discussions to the field.)

9.5.4 AP SITE MUTAGENESIS

One of the simpler mutational spectra is produced in depurinated ΦX174 (Schaaper and Loeb, 1981; Schaaper, Glickman and Loeb, 1982; Schaaper, Kunkel and Loeb, 1983) and M13 (Kunkel, 1984) infecting SOS-induced cells. The predominance of G to T and A to T transversions suggests replicative insertion of adenine opposite putative depurination sites. Indeed DNA polymerases do have a greater affinity for purines (England et al., 1969; Strauss et al., 1982). On damaged template DNA polymerase I (Schaaper, Kunkel and Loeb, 1983; Kunkel, Schaaper and Loeb, 1983; Boiteux and Laval, 1982) or Klenow fragment (Strauss et al., 1982) preferentially inserted A at apurinic and apyrimidinic sites. Polymerization by the true replication enzyme, Pol III, also caused mutations in depurinated DNA that required similar changes in guanine residues (Kunkel et al., 1983).

Apurinic sites may prove particularly important to SOS mutagenesis since they may be a common premutational lesion produced by a wide variety of chemical mutagens including benzo (a) pyrene, N-acetoxy acetylaminofluorine, aflatoxin B1, B-propriolactone and epoxycyclopenta (cd) pyrene (Schaaper and Glickman, 1982; Eisenstadt et al., 1982; Foster, Eisenstadt and Miller, 1983). There are two reasons for this idea. First, these agents all produce mutagenic spectra with a predominance of G to T and A to T transversions like genuine depurination. Second, the mutagenic potency of such agents has been qualitatively related to their ability to cause AP sites (Drinkwater, Miller and Miller, 1980; and discussed in Schaaper and Glickman, 1982). Remember that modification at N^7 and N^3 positions of purines labilizes glycosylic bonds, and that AP sites are produced by a wide variety of glycosylases. Thus the finding that AP sites are mutagenic in SV40 raises the possibility that this lesion is a potent source of mutations both in prokaryotic and eukaryotic cells (Gentil, Margot and Sarasin, 1984).

Individual effects with some agents have been interpreted by adducts rotating so that they pair with adenine, but it is difficult to see this as a general mechanism (Eisenstadt et al., 1982). Another interpretation is that preferential purine insertion occurs on any non-instructional template (Strauss et al., 1982). The important feature of all models is, nevertheless, a changed DNA polymerase activity.

9.5.5 UV MUTAGENESIS

Polymerase bias may prove to be one factor in determining mutational spectra, but is insufficient to explain many of the changes detected. With UV the mutational spectrum is more complex and includes transitions, transversions, double mutants (both tandem and non-tandem), duplications, deletions and frameshifts. Transitions are the most frequent contributing to about two-thirds of the total (LeClerc *et al.*, 1984; Shinura *et al.*, 1983; Coulondre and Miller, 1977; Wood, Skopek and Hutchinson, 1984). In *E. coli* there are approximately equal numbers of T to C and C to T changes (LeClerc *et al.*, 1984; Wood *et al.*, 1984), and about twice as many T to C as C to T changes in yeast (Lawrence and Christensen, 1979). Three-quarters or more of transition mutations occurred at sites with adjacent pyrimidines where it was usually the 3′ base of the pair which was mutated. Where a dipyrimidine was created by changing an ACA sequence to TCA there was a fifteen-fold increase in mutagenesis (Miller, 1983).

Two examples where UV mutagenesis did not occur at dipyrimidines in single-stranded DNA (Brandenburger *et al.*, 1981; Schaaper and Glickman, 1982) may be atypical for technical reasons (discussed in LeClerc *et al.*, 1984). At first sight these results might naïvely have been expected because the photoreversibility of mutation has long implicated cyclobutane pyrimidine dimers as premutagenic lesions. However, a comparison of mutation frequencies in adjacent pyrimidines in the *lacI* gene shows a marked preference for TC and CC (Brash and Haseltine, 1982). Although this system does not detect CT changes, these are underrepresented in the λ cI gene (Wood, Skopek and Hutchinson, 1984). This discrepancy is not due to preferential excision of TT and CT pyrimidine dimers (Todd and Glickman, 1982) but rather appears to be caused by a second dipyrimidine photoproduct. In this lesion pyrimidines are linked by a single 6–4 bond. The 6–4 photoproduct is most frequent at TC sites which proved most mutated in the *E. coli lacI* gene (Brash and Haseltine, 1982). In the cI gene approximately half of the transitions occurred at TT sites. Given the initial low estimates for 6–4 photoproducts at TT, this might indicate that pyrimidine dimers are the mutational lesions. However, the absence of mutations produced under special conditions making only pyrimidine dimers and an upwards revision of 6–4 photoproduct yield at TTs (Franklin, Low and Haseltine, 1982), further strengthens the case for mutagenesis at 6–4 photoproducts.

The 6–4 photoproduct is removed by the *uvrABC* excision repair system in accord with the enhanced mutability of excisionless strains (Sancar and Rupp, 1983; Haseltine, 1983). In contrast, 6–4 photoproduct is not removed by enzymic photoreactivation (Haseltine, 1983). The apparent paradox that mutagenesis is photoreactivated is explained by proposing that the removal of pyrimidine dimers would curtail SOS-induced expression of mutagenic repair activity from the *umuCD* operon (Haseltine, 1983). Similarly, UV-irradiated and photoreactivated *E. coli* retains some lesion which enhances mutagenesis after a further irradiation producing only pyrimidine dimers (Mennigman, 1972).

The results with UV thus far described fall into the general class described for other mutagens. Mutagenesis might occur by preferential purine insertion at non-coding lesions, be they 6–4 photoproducts or pyrimidine dimers. Indeed, *in vitro*, DNA polymerase I could be made to incorporate bases opposite dipyrimidine photoproducts and did so with a preference for purine insertion (Strauss *et al.*, 1982; Moore *et al.*, 1981). However, this simple unified mechanism is unable to account for UV-induced transversions, deletions, double mutations and duplications, neither can it account for many of the minor classes of mutation produced by other agents. To some extent this reflects our incomplete knowledge of minor premutagenic lesions. A comparison of the distribution and types of minor mutations produced by several agents suggests targeted random base insertion rather than purine insertion (Foster, Eisenstadt and Cairns, 1982). A second possibility is that UV damage can also cause 'semi-targeted' mutagenesis where changes occur around, but not exactly at a lesion, producing the double, but not necessarily tandem mutations in UV irradiated DNA (Schaaper and Glickman, 1982). These, and the superimposition of a number of other factors, may combine to diversify the mutagenic spectra of many agents. For UV it is known that dipyrimidine photoproduct yields are influenced by flanking bases and that AT richness 5' to a site correlates with its mutability (Brash and Haseltine, 1982; Wood, Skopek and Hutchinson, 1984). The possibility of forming hairpin loops in quasipalindromes may produce base substitutions, deletions, insertions and frameshifts depending upon the type of misalignment and the way it is corrected (Todd and Glickman, 1982; Glickman and Ripley, 1984).

9.5.6 DNA POLYMERASES AND INDUCED MUTAGENESIS

In general, the analyses of mutational changes point to the generation of mutations by changed DNA polymerase activity. A more direct indication of DNA polymerase involvement is that processing of potentially mutagenic lesions into heritable changes in genotype required DNA polymerase III (Bridges, Mottershead and Sedgwick, 1976). An altered DNA polymerase I has been isolated from SOS-induced *E. coli* and has reduced fidelity in replication on synthetic templates (Lackey, Krauss and Linn, 1982). However, it is difficult to reconcile the possible involvement of this altered DNA polymerase with the hypermutability of *E. coli polA* mutants with only 1 or 2% of active enzyme (Kondo *et al.*, 1970; Witkin and George, 1973).

DNA polymerase behaviour on damaged templates might also give an insight into how activities might be changed to cope with damaged template. *E. coli* DNA polymerases I and III stopped chain elongation at, or just before, lesions in DNA templates damaged by several mutagens (Moore *et al.*, 1981, 1982; Strauss *et al.*, 1982; Kunkel, Schaaper and Loeb, 1983; Schaaper, Kunkel and Loeb, 1983). Rather than simply stopping, the polymerase 'idles' in a cyclic reaction, inserting a base opposite the lesion and then removing it with its 3'–5' exonuclease (Villani, Boiteux and Radman, 1978). Thus, one proposal was that mutations are produced by reducing 3'–5' proof reading, so increasing the chances of a base remaining opposite a lesion. Indeed, 3'–5' exonuclease is reduced *in vitro* by RecA protein and deoxynucleoside monophosphates that *in vivo* would be expected to accumulate in DNA damaged cells (Kunkel *et al.*, 1983; Fersht and Knill-Jones, 1983). In other *in vitro* experiments bases were stably positioned opposite lesions by substituting Mn for Mg and increasing deoxyribonucleoside triphosphate concentrations (Strauss *et al.*, 1982; Moore *et al.*, 1981, 1982). In this case Mn could exert its effect with Klenow fragment which lacks 5'–3' exonuclease.

9.5.7 UNTARGETED MUTAGENESIS

The observation of so-called untargeted mutagenesis might also be taken as an indication that SOS induction changes replicative DNA synthesis. However, care must be taken with this topic as untargeted mutagenesis now seems to describe two quite different processes. In one case, mutagenesis is enhanced when undamaged phage infects UV-irradiated *E. coli* (Ichikawa-Ryo and Kondo, 1975). In the second, bacterial mutagenesis is increased by genetic manipulations which induce the SOS response without an *external* source of DNA damage (Witkin, 1976). Untargeted phage mutagenesis is typified by the production of mixed bursts meaning that the mutations arise at any time throughout the repeated cycles of DNA replication preceding lysis. Further indication that these mutations are replication errors is that their yield is increased by inactivation of the mismatch repair system (Caillet-Fauquet, Maenhaul-Michel and Radman, 1984). Bacterial 'untargeted' mutagenesis requires *umuCD*, *lexA* and *recA* genes (Ciesla, 1982; Witkin and Kogoma, 1984), but in the phage system untargeted mutagenesis can occur in the absence of *umuCD* and *lexA*. Even in *recA* mutants some phage mutagenesis occurs and this can be increased to wild type levels with an additional *recB* mutation (Ichikawa-Ryo and Kondo, 1975; Defais, 1983; Wood and Hutchinson, 1984).

The mutation types produced in phage and bacterial systems are also different. With λ, three-quarters of the changes were frameshifts (Wood and Hutchinson, 1984). In *E. coli tif* mutants the mutagenic changes were not frameshifts, nor were they a random spread of changes expected of a general mutator activity. Rather, they displayed the now familiar 'footprint' of GC to TA transversions typical of many of the chemicals thought to make apurinic sites (Miller and Low, 1984). Thus the so-called untargeted mutagenesis in bacteria appears to be, in fact, targeted at cryptic lesions. These lesions may be unable to trigger SOS induction but may be susceptible to its activity induced by other means.

9.5.8 SITES OF MUTAGENESIS

A number of *in vivo* situations arise where DNA polymerases might be presented with damaged templates. Obviously semi-conservative replication of damaged DNA is one, and replication of depurinated ϕX174 with the *E. coli* replicative enzyme does cause mutations (Kunkel *et al.*, 1983). Furthermore, mutagenic activity provided by the *muc* operon, a plasmid analogue of *umuCD* (see Walker, 1984, for a review), changes the replicative mutagenesis of the base analogue mutagen 2,6 diaminopurine (Mattern *et al.*, 1985). However, in UV-irradiated *E. coli*, the timing of mutation fixation by loss of photoreversibility points more to altered DNA polymerase acting in repair replication either before or after semi-conservative replication (Nishioka and Doudney, 1969, 1970; Doubleday, Bridges and Green, 1975).

In post-replication repair, gaps are generated opposite lesions. Relatively crude physical assays indicate that most are filled by recombinational repair and are unaffected by *umuCD* mutations (Kato, 1977). However, *umuCD* involvement in repair of a small number cannot be excluded. In excision repair, damage in one strand might complicate the repair replication of a nearby lesion in the opposite strand (Bresler, 1975) although the linear kinetics for the production of such mutants are not consistent with such a two-hit model.

In summary, the mutagenic DNA polymerase activity may occur both in semi-conservative and repair replication depending upon the type of lesion. Since the dénouement of untargeted mutagenesis in *E. coli*, there appears to be no evidence for a general mutator activity although there may be cases of mutations near, but not at, DNA lesions.

9.5.9 ROLES OF *recA* AND *umuCD* IN MUTAGENESIS

Given the possible involvement of altered DNA polymerase activity, what are the roles of *recA*, and *umuC* and *D* gene products? Little is yet known about the UmuC and D proteins, apart from their molecular weights, inducibility, and nucleotide-sequence derived amino acid composition (Walker, 1984; Kitagawa *et al.*, 1985). The suspected alteration of DNA polymerase could theoretically be caused by *umuCD* encoding a new polymerase, by modifying the properties of a pre-existing polymerase, or by regulating some other factor which in turn influences polymerase activity. Resolution of these possibilities awaits biochemical characterization of UmuC and D proteins.

However, new *in vivo* experiments on *umuCD* and *recA* involvement in mutagenesis have suggested a different possibility. Bridges and Woodgate (1984, 1985) have shown that the requirement for *umuCD* in UV-induced mutagenesis of *E. coli* can be partially waived by photoreactivating *umuC* or *D* mutants some time after they have been UV-irradiated. They propose a two-step model of induced mutagenesis. First, an altered polymerase inserts bases opposite DNA lesions, and then DNA elongation continues on from the misincorporated base. From the induction of mutations by delayed photo-reactivation they propose that UmuCD proteins participate in the second continuation step. In other words, the combination of a base opposite a lesion presents a problem for priming ongoing replication, and is dealt with by *umuC* and *D* activity. Similar experiments with strains having RecA proteins with different degrees of activation indicate a direct role for activated RecA protein in mutagenesis (Bridges and Woodgate, 1985; Witkin and Kogoma, 1984), possibly at the first misincorporation step. It is notable that earlier work had shown that *in vitro* replication was unable to continue on from a base inserted opposite a lesion (Strauss *et al.*, 1982).

This two-step model therefore presents the first unified explanation for *umuCD* and *recA* activity in targeted mutagenesis and promises to stimulate vigorous future research.

9.5.10 REPLICATIVE REPAIR IN EUKARYOTES

A recovery system with superficial resemblance to the bacterial post-replication repair system is found in both lower and higher eukaryotes. UV-irradiated cells produce newly synthesized DNA of lower than normal molecular weight which increases to full size during subsequent incubation. In brief, there is no single explanation for the way in which the changed pattern of DNA synthesis occurs or how it recovers, although events are clearly not like the bacterial repair system. The complex interpretations of current work have been discussed in depth elsewhere (Hall and Mount, 1981). In summary, the contentious points are as follows. In certain eukaryotes, the size of newly synthesized DNA corresponds to that expected if an interruption in DNA synthesis occurred at every pyrimidine dimer. However, in other cell lines this correspondence is not found. The DNA is smaller, but not small enough to support the idea that each pyrimidine dimer causes discontinuity. Furthermore, some experiments indicate that gapped daughter strands are produced by the resumption of DNA synthesis downstream from a lesion. But other experiments with different cells indicate gaps being formed, but not opposite pyrimidine dimers. Yet another group of experiments favours a model with no gaps formed at all, rather, replication stalls at certain lesions.

Another major difference seen in some cells is that DNA synthesized sometime after irradiation can be large, even though the template continues to contain pyrimidine dimers (Meyn and Humphrey, 1971; Lehmann and Kirk-Bell, 1972; Buhl, Setlow and Regan, 1973).

The putative mechanisms for elongation of DNA to full size are equally contentious, but are at least united in showing that the *de novo* DNA synthesis is involved with little, if any, reciprocal recombination exchange (Lehmann, 1972; Buhl, Setlow and Regan, 1972; Lehmann and Kirk-Bell, 1978). Changed DNA synthesis might either fill gaps, if they exist, or extend continuous semi-conservative replication past a lesion.

Whatever the mechanics involved in post-replication repair, there are examples of mutations in *Drosophila* (Boyd and Shaw, 1982) and humans (Lehmann *et al.*, 1975) where its alteration causes increased sensitivity to DNA damage. The human example is the variant category of Xeroderma pigmentosum (XPv) whose cells are proficient in excision repair. After UV-irradiation new DNA is even smaller than in normal cells, more time is required for it to reach its maximal size, and this may still not be as large as normal.

Xeroderma variant cells and classical XP cells yield greater numbers of mutants induced per unit dose than in normal cells. In a different analysis of these data XPv cells show a greater number of mutants for any given level of survival than normal or XP cells (Maher *et al.*, 1976). Classic XP cells resembled normal cells in showing similar amounts of mutagenesis at *similar levels* of survival (Maher and McCormick, 1976).

Therefore, the classic XP defect has a simple dose-modifying effect on cell survival without affecting the mutagenic potential of any lesions persisting in DNA. XPv cells differ by increasing the mutagenic potential of damage persisting after excision, and this difference has been linked to their altered post-replication repair behaviour.

9.5.11 MUTATIONAL CHANGES IN EUKARYOTES

An earlier section showed how analysing mutational spectra could be used to piece together possible molecular mechanisms of induced mutagenesis. In higher eukaryotes this approach is also being used, although it is technically more difficult and complicated by changes in phenotype caused by gene amplification, genome rearrangement or deletion, and heritable changes in gene expression. Nevertheless, there is considerable evidence for gene mutation and some of the systems employed to study it are reviewed. In an early example, hamster cells became resistant to 8-azaguanine after MNNG or EMS treatment (Baudet, 1973). Such cells are deficient in hypoxanthine guanine phosphoribosyl transferase (HPRT) activity. Some isolates contained HPRT cross-reactive material, even though they were enzymically inactive. Radio-immune assays also recognized mutated forms of HPRT from hamster, human and mouse cells after chemical treatment. Composition analyses revealed amino acid changes. Base substitution mutagenesis was also indicated by changes in electrophoretic mobility, altered substrate affinities of residual HPRT, and by thermolability (Fenwick *et al.*, 1977; Milman *et al.*, 1976; Epstein *et al.*, 1977; Sharp, Capecchi and Capecchi, 1973). Nonsense mutations also appear after methyl nitrosoguanidine treatment and can be suppressed by microinjection of suppressor tRNA (Capecchi *et al.*, 1977). Ouabain resistance is also thought to involve base substitution mutagenesis to produce an active Na/K adenosine-5-triphosphatase with reduced affinity for ouabain.

A system promising to identify specific mutational spectra in a CHO genomic APRT sequence is being developed (Nalbantoglu, Gonclaves and Meuth, 1983; Gonclaves, Drobetsky and Meuth, 1984). Using a cloned APRT gene, hybridization tests with digested genomic DNA can detect

APRT mutants with restriction site polymorphism. The mutated genes can themselves be cloned with the intention of defining the exact nature of the mutational changes.

A similar approach employs an integrated bacterial *gpt* gene as the target for mutagenesis rather than the large HPRT locus which it can replace and which is difficult to analyse physically (Thacker *et al.*, 1983; Tindall *et al.*, 1984). Although mutations could be scored, their frequencies were sometimes higher than expected for normal chromosomal genes. Furthermore, frequencies of mutation were also elevated in HPRT DNA transfected into a hprt cell line. The conclusion drawn from these results is that different chromosomal sites may have differing susceptibilities to mutagenesis, linked in some way to the susceptibility for integrative transfection. In contrast to mutagenesis of 'true' genomic sequences, the mutagenesis of transplanted genes was predominantly by deletion. However, more recent experiments have shown that many integrated *gpt* genes become spontaneously inactive through some epigenetic effect rather than mutation (Ashman and Davidson, 1984).

In mammalian cells 'pre-induction' with DNA damage increases the survival and mutagenesis of damaged viruses (Dasgupta and Summers, 1978; Lytle, Goddard and Lin, 1980; Sarasin and Benoit, 1980). In UV-irradiated SV40, reversion of two temperature-sensitive mutations occurred by second-site base substitution mutagenesis at a site with adjacent pyrimidines, that is, a site for dipyrimidine photoproduct formation (Sarasin *et al.*, 1981). Thus the type of mutation arising resembles that anticipated by models of transdimer replication on damaged template DNA. Indeed, earlier physical experiments with SV40 DNA showed that pre-inducing hosts prior to infection increased the template capacity of the damaged DNA without changing lesion content or causing recombinational rearrangement of the progeny virus (Sarasin and Hanawalt, 1980).

Early attempts using mammalian shuttle vector plasmids as similar substrates for induced mutagenesis were fraught with problems. Even with undamaged plasmid, mutation frequencies of 1% or more were found by passaging *lac* or *galk* vectors through mammalian cells (Calos, Lebkowski and Botcham, 1983; Razzaque *et al.*, 1984, and references therein). In addition to deletions, duplications and insertions, there was an approximately equal incidence of point mutation including some suppressible by bacterial tRNA suppressors.

Following transfection there was massive destruction and rearrangement of the plasmid DNA en route to the nucleus. Such rapacious degradation might form the basis of the high frequencies of rearrangement if partially degraded plasmid were able to penetrate the nucleus.

More recently a human cell line has been discovered which replicates

SV40-based vectors without such extensive mutagenesis. LacI mutations in the SV40 shuttle plasmid are analysed by retransformation of *E. coli*. With UV the mutation spectrum was remarkably similar to that obtained with *lacI* mutated in *E. coli*. There was a predominance of GC to AT transitions occurring more than 90% of the time at dipyrimidines. Tandem base changes occurred and two of the three most mutated sites were also hotspots in *E. coli* (Letkowski *et al.*, 1986).

9.6 DNA damage and epigenetic change

Thus far the generation of phenotypic change by DNA damage has been discussed within the framework of alterations in nucleotide sequence produced by recombination and mutagenesis. However, there is an increasing number of examples were changed phenotype stems from a heritable, epigenetic change in the level of gene expression.

In a genetically manipulated strain of *E. coli*, SOS-induced expression can become 'locked on' to give a heritable epigenetic change (Toman, Dambly and Radman, 1980). In λ lysogenic induction, rerepression is prevented by *cro* protein binding to sites once occupied by repressor, so committing the phage irreversibly to lytic growth. Thus when bacterial genes are put in place of the phage ones, they become irreversibly induced by SOS induction and remain expressed long after the damage has gone.

In eukaryotic cells there is considerable evidence for heritable patterns of DNA methylation influencing gene expression (reviewed in Doerfler, 1983; Riggs and Jones, 1983; Holliday, 1986). To understand how DNA damage may affect methylation, current ideas on the inheritance of methylation patterns must be described a little more fully. Experiments with transfecting DNA have shown that methylation patterns are maintained for many divisions (Compere and Palmiter, 1981; Harland, 1982). Since methylation is a post-synthetic modification of DNA, it follows that it is performed by methyltransferases operating on hemi-methylated DNA. Indeed, *in vitro* methyltransferase activity in mammalian cell extracts showed a preference for hemimethylated DNA as substrate (Jones and Taylor, 1980; Gruenbaum, Cedar and Razin, 1982). Thus, once a site in a DNA strand had become demethylated and had passed through DNA replication, it would become unmethylated in both strands, and would remain unmethylated because of the methyltransferase's preference for hemimethylated substrate. How might DNA damage cause demethylation of either one, or both DNA strands? One proposal, that alkylation agents reduce methylation by directly inactivating methyltransferase, has gained some experimental support (Wilson and Jones,

1983). Hemimethylated DNA also had reduced substrate activity *in vitro* for mouse spleen methyltransferase after depurination, single strand breakage and alkylating, arylating and cross-linking treatments. Double-strand breaks and UV-irradiation had little effect. Nevertheless UV has been shown to cause expression of mouse metallothionein I genes with concomitant demethylation (Lieberman, Beach and Palmiter, 1983).

A third possibility is that demethylation occurs during DNA repair. Excision of a DNA lesion and repair replication would produce a short patch of unmethylated DNA. If replicated, this patch would be fixed as a fully demethylated site in both strands of the daughter DNA duplex (Holliday, 1979). Support for this model comes from measurements of methylation rates in repair replication following UV, methylnitrosourea, and N-acetoxy amino-fluorine treatments in human diploid fibroblasts (Kastan, Gowans and Lieberman, 1982). In confluent cells remethylation was prolonged over several days. Under these conditions hemimethylated sites would be expected to accumulate so increasing the probability that an increased number might be fixed into fully demethylated sites by replication. However, it remains to be determined whether the gross measures of DNA methylation employed in these experiments do in fact reflect specific changes in the heritable pattern of methylation.

9.7 Evolution of indirect mutagenesis

It will be remembered that indirect mutagenesis arises in a cell's response to damage. Since this response, at least in bacteria and yeast, is known to confer resistance to damage it has an obvious and immediate selective value. What then is the function, if any, of the intrinsic mutagenesis of the process? In general, mutation rates can be viewed as being either optimal, or the lowest physiologically possible (Maynard-Smith, 1978). An optimum rate would be determined by increased fitness from the generation of variation being balanced by loss of fitness through excessive deleterious mutagenesis. In this view, the *E. coli umuCD* sequence would gain its selective advantage by providing an optimum level of induced mutagenesis. An embellishment to this model explains the inducibility of mutagenic repair as an inducible source of variation which would provide a burst of genetic variation in times of stress (Echols, 1981). Such an argument for the selection of *umuCD*-like sequences through mutational benefits has been supported by the greater relative effect that *umuC* and *D* mutations have on induced mutagenesis compared with cellular survival. However, examination of the mechanisms by which induced mutagenesis is proposed to confer a selective advantage to the *umu CD* sequence reveals several problems.

In a mechanism where induced mutagenesis confers an immediate selective advantage, mutations would be produced which give greater protection against the lethal effects of the mutagenic environment. Indeed there are examples where regimes of repeated irradiation, with intervening growth periods, have generated resistant lines in several bacteria (Erdman, Thatcher and McQueen, 1961; Wright and Hill, 1968; Davies and Sinskey, 1973; Bresler, Verbenko and Kalinin, 1980). However, these experiments do not identify whether the resistant mutants were produced by mutagenic repair of radiation damage, or by selection of spontaneous mutants. Even if such selection were to operate, the fortuitous production of a mutation with the correct change to confer an immediate increase in resistance would be rare. Far fewer cells would benefit from this outcome of mutagenic repair than cells whose survival was increased by the direct effects of the process.

Another view of optimum mutation rates envisages a more general diversification of genotype which at some future time might confer a selective advantage. Thus, the selective advantage of *umuCD* might be deferred until one of the mutations it had produced in the past fortuitously increased fitness in a future changed environment. For maximal effect, such a mechanism requires an asexual organism where recombination and segregation are unable to separate the mutator gene from its potentially advantageous mutation elsewhere in the genome. Even in bacteria, which can be considered as clonal on a generation-to-generation basis, there are countless plasmids, transposons and bacteriophages with differing sexual proclivities. The evolutionary impact of these elements on bacterial diversification may be considerable. A second, more fundamental problem is best summarized by a quotation from Gould (1983) in his discussion of the evolutionary importance of a gene duplication which 'may make future evolutionary change possible, but selection cannot preserve it unless it confers an immediate significance. Future utility is an important consideration in evolution but it cannot be the explanation for current preservation. Future utilities can only be fortuitous effects of other direct reasons for immediate favour.'

In an optimum mutation rate model the actual level of mutagenesis would be related to the instability of the environment. Clearly in an unchanging environment there would be less opportunity for genetic change to be advantageous, more chance for it to be deleterious, and hence a selection for lower optimum mutation frequency. Whether *E. coli* and related species occupy a stable or unstable environment is debatable. Nevertheless, finding quite different levels of induced mutability of enterobacteria from the same environment, stable or not, does question the idea of there being optimal levels of this mutability (Sedgwick and Goodwin, 1985).

An alternative view is that the selective advantage of mutagenic repair is conferred by its survival enhancement, and that mutagenesis is simply a byproduct of this process. This is the minimal-rate-possible view of mutagenesis. The selective advantage would be a determined balance of survival enhancement by translesion replication against loss of fitness by gene inactivation. One model for the evolution of such a mechanism is based on the possibility that *umuCD* is, or was, part of a transposable element (Sedgwick and Goodwin, 1985, and references therein). For such a gene there would be more advantage to be gained by providing a strategy of repair not already employed by the cells into which it transposes. As *E. coli* has efficient lesion-excision and recombinational-avoidance systems, the only remaining strategy would be to provide some means of directly tolerating lesions. This carries a finite probability of mutagenesis, but may be the cost of a selective advantage provided by survival enhancement. The contribution that mutagenic repair makes to survival has often been discounted because *umuC* or *D* or yeast *rev3* mutants are not as sensitive to damage as many other repair-deficient cells. To some extent this difference may be more apparent than real. In the usual logarithmic presentation of surviving fraction there is a considerable difference between, say, 90% killing of a *umuC* mutant compared with 99.999% killing of some other more sensitive mutant. However, on an arithmetic basis this difference is much less apparent and shows that appreciable *numbers* of cells rely on mutagenic repair as well as other repair systems. Thus the enhancement of survival by mutagenic repair can affect a large *number* of cells and would confer an immediate selective advantage.

9.8 DNA repair effects in multicellular organisms

The previous sections have shown how DNA repair systems can influence the survival and properties of single cells. This final section briefly recounts how some of these cellular changes can bring about changes in the character of a multicellular organism.

9.8.1 SURVIVAL

Perhaps the best known example of the link between DNA repair and the well-being of a whole-organism is provided by the properties of Xeroderma pigmentosum individuals (reviewed in Cleaver and Bootsma, 1975; Robbins, 1983). These people generally display sun sensitivity and necrosis of exposed tissue in parallel to the UV-sensitivity of their cultured cells, and their excision repair defect. The sun sensitivity of Cockayne's syndrome (CS) patients is also manifested at the cellular level (Wade and Chu, 1979). Unlike XP, these cells are proficient in excision repair. They have other abnormalities in post-irradiation DNA and RNA synthesis and in reactivation of damaged viral DNA (reviewed in Robbins, 1983). A third human syndrome, Ataxia telangiectasia (AT) displays ionizing sensitivity at both the cellular level and in person. It was, in fact, the lethal effects of attempted radiotherapy of tumours in AT individuals which introduced AT to the field of DNA repair (Taylor *et al.*, 1975).

The influence of defects in cellular repair processes on the survival of whole organisms can also be seen in the fruit fly, *Drosophila* (reviewed in Boyd *et al.*, 1983). In humans, particular emphasis has been put on the importance of DNA repair in protecting non-regenerating cells, such as nerves. This idea stems from the presence of neurological abnormalities and degenerations in AT, CS and some XP individuals (Robbins, 1983; Kidson *et al.*, 1983). XP neurological problems arise in the more UV-sensitive complementation groups A, B, D and G. In general, the appearance and onset of neurological decline approximate to the severity of the repair defect, although there are some inconsistencies depending upon which technique is used to quantitate DNA repair. Such observations have prompted more extensive surveys of other neurological, retinal, and skeletal degenerative diseases to assess their DNA repair ability. Thus far, ionizing radiation sensitivity has been reported in at least ten other neuronal degenerative diseases, in Usher retinal dystrophy and several muscular dystrophies. It must be emphasized that sensitivity was not always found and, when detected, was not as great as with AT. It was even less than AT heterozygotes, which do not display neurological disorders. Thus repair defects may precipitate neurological disorders in some instances; however, in others the problem may be in other metabolic steps which have only a secondary effect on cell survival.

9.8.2 DNA DAMAGE, REPAIR AND CARCINOGENESIS

Carcinogenesis is considered to be a multistep process in which some events may be influenced by DNA damage and repair. There are several lines of evidence that damage-induced mutagenesis is one of these steps. Firstly, a correlation between mutagenic and carcinogenic potency has been found with so many agents that a causal relationship seems likely (McCann *et al.*, 1975). Secondly, organisms whose cells have increased mutation frequencies have an increased incidence of tumours. This effect is seen in small rodents whose cells have poor excision repair and higher mutation frequencies than other mammals. From thirty to thirty-three months, over 30% of mice have one, if not more, tumours (Zurcher *et al.*, 1982), a level not reached in humans until late in life (Dix and Cohen, 1980).

In the classic and variant forms of Xeroderma pigmentosum, induced cellular mutation frequencies are increased in parallel with these individuals' increased susceptibility to melanoma. Furthermore, the melanomas arise mainly on exposed, sun-sensitive parts of the body, although there may also be some increase in internal tumour frequency (Kraemer, Lee and Scotto, 1984). In normal individuals, the incidence of melanomas is also related to the degree of exposure determined partly by latitude and solar incidence (Lancaster, 1956; Lee and Merril, 1971; Magnus, 1973). Melanomas therefore appear, like mutations, to be caused by an environmental stress to DNA which if left unremoved is particularly likely to cause tumours. In fish, UV-induced tumours can be produced, but are not seen if the fish cells are photoreactivated after UV-irradiation. This shows that it is the pyrimidine dimer photoproduct which causes tumours (Hart, Setlow and Woodhead, 1977).

More direct evidence for a mutational involvement in cellular transformation has come with the discovery of cellular oncogenes. In a number of cases single base-pair substitutions arising spontaneously are sufficient for transformation to occur (Krontiris *et al.*, 1985, and references therein). In one, possibly two examples, where the inducing agent might be identified, the mutational change is one which might have been predicted from our earlier knowledge of mutagenic mechanisms. In one case, rat mammary tumours were induced with methylnitrosourea, a potent alkylation mutagen. A tumour was identified with a single G to A transition, just the type of change expected from known, direct mutagenic effects of O^6-methylguanine mispairing with thymine (Sukumar *et al.*, 1983). Indeed, variations in levels of O^6- methylguanine methyltransferase in different tissues have been related to tissue specificity for tumourigenesis (Pegg, 1984).

A mutationally altered *c-Ha-ras* oncogene has also been identified in a melanoma from a normal Japanese patient (Sekiya *et al.*, 1984). Bearing in mind that melanomas can be induced by sunlight, one is left to ponder on the significance of the mutated base being the 3' member of a potential dipyrimidine photoproduct site.

A second mechanism of oncogene activation may be to stimulate gene expression. Genome rearrangements, either moving the proto-oncogene into a more heavily transcribed region, or bringing more active transcriptional regulators near the gene, have been detected (summarized in Bishop, 1983). Some of these events may be produced by DNA damage. In this respect it is worth emphasizing the recombinogenic potential of many types of DNA damage. Thus, as a general rule, where DNA repair by lesion removal is absent, an increase in tumourigenicity would be expected, just as there is an increase in mutagenesis and recombination at the cellular level. In the ionizing radiation sensitive syndrome, Ataxia telangiectasia, there is a tendency to develop certain types of tumour, particularly of the lymphatic system. The cells of these individuals do not display a marked increase in mutability in line with their increased radio-sensitivity. However, the increased incidence of DNA rearrangements evidenced by sister chromatid exchange, may play a part in changing patterns of gene expression. An important feature of AT is that the increased tumour incidence is manifested on a smaller scale in AT heterozygotes. Given the greater frequency of these individuals, AT heterozygosis may contribute to a significant proportion of tumours in the population at large (Swift *et al.*, 1976).

Another of the possible stages in the multistep process of transformation could be seen when the human bladder carcinoma oncogene *c-Ha-ras* was reintroduced into rodent cells to see whether it caused malignant transformation. It did not, except when the cells were immortalized by introducing other oncogenes (Land, Parada and Weinberg, 1983; Ruley, 1983), or by treatment of the cells with known carcinogens (Newbold and Overell, 1983). The chemicals used included benzo(a)pyrene and the alkylating agents, dimethylsulphate and methylnitrosourea. Whatever the mechanism of immortalization, the efficiency of these agents would be expected to be reduced by the repair proteins known to exist for removing DNA damage.

9.8.3 DNA DAMAGE, REPAIR AND AGEING

It might be expected that long-lived organisms might either accumulate more DNA damage than short-lived ones, or might possess more proficient DNA repair systems (see Chapter 12). These ideas have been tested by examining cells from individuals of different age, from different 'aged' tissue cultures, from individuals with premature ageing diseases or defects in DNA repair, and from cells of species with different lifespans.

The evidence for cumulative DNA damage comes from ageing cells or terminally differentiated tissue and indicates an increased frequency of DNA strand breakage as judged by DNA polymerase priming activity, alkaline sucrose sedimentation and electrophoretic mobility. Where alkaline conditions were employed there is also a possibility that alkali-labile sites might have accumulated and contributed to the strand breakage detected (reviewed in Hart et al., 1979). Changes in histone properties with age have also been invoked as a possible factor in changing accessibility of chromatin to DNA repair enzymes (Hart et al., 1979).

Efforts to find a relationship between cellular age and repair proficiency are not consistent. In examining the repair of single-strand breaks no differences were found in late passage human lung cells (Clarkson and Painter, 1974), but in other human tissues young cells showed more rapid, strong joining than older cells (Epstein, Williams and Little, 1974). However, removal of X-ray damaged bases did decline rapidly in late passage fibroblasts (Mattern and Cerutti, 1975), but again there is conflict between reports of decreased repair replication in late passage cells (Hart and Setlow, 1976) and proficient repair replication up to the final stages of cell growth (Clarkson and Painter, 1974).

After UV radiation, ageing cultures of human fibroblasts retained their ability to perform repair replication in some experiments (Painter et al., 1973; Smith and Hanawalt, 1976), but not others (Hart and Setlow, 1976). Other examples of varying repair proficiencies are discussed in more detail elsewhere (Hart et al., 1979). In the experiments of Hart and Setlow (1976), proficiency in UV and N-acetoaminofluorine repair replication declined with a comparable decrease in overall replicative DNA synthesis. These experiments, therefore, exemplify the difficulty in assigning any causal relationship between repair and ageing since many aspects of macromolecular synthesis may be altered with ageing (see Chapter 12). Similar reservations apply to reports of decreased semi-conservative and repair replication in differentiating skeletal muscle (Stockdale, 1971) and decreased polymerase fidelity in old spleen (Barton and Young, 1975) and ageing fibroblasts (Linn, Kairis and Holliday, 1976).

A number of human syndromes present symptoms of 'premature ageing' and cells from these individuals have been tested for repair proficiency. Hutchinson-Gilford Progeria cells were found to be deficient in single-strand break repair following X-irradiation (Epstein, Williams and Little, 1974, 1979), but confirmatory experiments using Progeria cells and Werners cells showed no deficiency (Regan and Setlow, 1974b). Similarly, conflicts arise in examining lifespans of syndromes exhibiting sensitivity to DNA damage. Reduced *in vitro* lifespans were found for cultured cells from Fanconi's anaemia, Bloom's syndrome and Ataxia telangiectasia (Thompson and Holliday, 1983). However, cells from Xeroderma individuals with a defined deficiency in repair have the same longevity in culture as normal human fibroblasts (Goldstein, 1971). In normal human cells, no difference was found in UV repair-replication in cells from human babies and adults (Liu, Parsons and Hanawalt, 1982).

A third approach has been to compare repair capacities of animals with different lifespans. Fibroblasts were tested for UV repair activity using an autoradiographic assay for replication, presumed to be the restorative part of nucleotide excision repair. The amount of repair replication in five genera of placental mammals was related to the log of the donor animal's expected lifespan (Hart and Setlow, 1974). Later experiments using more refined assays of nucleotide excision repair extended this correlation to include primates and rodents with different life expectancies (Paffenholz, 1978; Sacher and Hart, 1978; Hall, Bergman and Walford, 1981; Hall et al., 1984). A stricter correlation of lifespan and repair replication was found in lens epithelium, which has the advantage of having no complication of normal replication (Treton and Courtois, 1982). However, in other studies this correlation is less apparent. In one study of twenty-one mammals, repair replication was measured by the size of repair replication patches and their number, the product of the two being the total amount of repair replication. The size of repaired regions varied from species to species, but fell within the range observed with different human isolates. In contrast, the number of repaired sites and the total amount of repair replication did show some relationship to lifespan (Francis, Lee and Regan, 1981). However, the variation in these data contrasts markedly with earlier reports. In yet another examination no correlation between unscheduled DNA synthesis and lifespan could be found (Kato et al., 1980). However, these experiments suffer from using a mixture of primary and established cell cultures, from the relatively crude autoradiographic assay of repair replication, and possibly from the inclusion of hibernating species whose lifespans are not experimentally known (Hall *et al.*, 1984). Francis, Lee and Regan (1981) do point out, however, that 'the cells derived from the animal sources common to all these structures showed about the same relationship between DNA repair and lifespan; the disagreement therefore is partially dependent upon the choice of animals'.

In these types of study it is usually the rodent cells which display the lowest repair activity. Thus it is intriguing to learn that prenatal tissue displays high levels of repair (Peleg, Raz and BenIshai, 1976) which only declines to the typical low level after serial passage *in vitro*. Nothing is known of what developmental factors bring about this change or what developmental processes favour such a change. Even though rodent cells lose excision repair activity, they are not unduly UV-sensitive compared with species more proficient in excision repair. Presumably the rodent cells place greater reliance upon post-replication repair, but the proficiency of this process has not yet been related to lifespan or ageing.

This chapter has outlined the molecular events of DNA repair and indicated several ways in which the well-being of a multicellular organism may be affected. As with any process of detection it is always attractive to make such chains of events. But in piecing together cohesive theories from incomplete evidence there is always the risk of making links which are more apparent than real. This problem has been clearly put in a recent best-selling novel, appropriately a labyrinthine detective story set in mediaeval Italy (*The Name of the Rose* by Umberto Eco). The leading investigator explains to his young understudy that 'The order that our mind imagines is like a net, or like a ladder, built to attain something. But afterward you must throw the ladder away, because you discover that even if it was useful, it was meaningless.'

References

Ahmad, A., Holloman, W. K. and Holliday, R. (1975) Nuclease that preferentially inactivates DNA containing mismatched bases. *Nature*, **258**, 54–56.

Ames, B. N. (1983) Dietary carcinogens and anticarcinogens. *Science*, **221**, 1256–1264.

Ashman, C. R. and Davidson, R. L. (1984) High spontaneous mutation frequency in shuttle vector sequences recovered from mammalian cellular DNA. *Mol. Cell. Biol.*, **4**, 2266–2272.

Bacchetti, S. and Benne, R. (1975) Purification and characterization of an endonuclease from calf thymus acting on irradiated DNA. *Biochim. Biophys. Acta*, **390**, 285–297.

Baker, T. I. (1983) Inducible nucleotide excision repair in *Neurospora*. *Mol. Gen. Genet.*, **190**, 295–299.

Barker, D. G., White, J., Johnson, A. L. and Johnston, L. H. (1984) Structure and sequence of DNA ligase genes from *S. cerevisiae* (*CDC9*) and *S. pombe* (*CDC17*), and their induction by UV-irradiation. In *Proceedings of XII International Conference on Yeast Genetics and Molecular Biology*, Edinburgh.

Barton, R. W. and Yang, W.-K. (1975) Low molecular weight DNA polymerase: decreased activity in spleens of old Balb/c mice. *Mech. Ageing Dev.*, **4**, 123–136.

Baudet, A. L., Roufa, D. J. and Caskey, C. T. (1973) Mutations affecting the structure of hypoxanthine:guanine phosphoribosyltransferase in cultured Chinese hamster cells. *Proc. Natl Acad. Sci. USA*, **70**, 320–324.

Bauer, J., Krammer, G. and Knippers, R. (1981) Asymmetric repair of bacteriophage T7 heteroduplex DNA. *Mol. Gen. Genet.*, **181**, 541–547.

Bishop, J. M. (1983) Cellular oncogenes and retroviruses. *Ann. Rev. Biochem.*, **52**, 301–354.

Bogden, J. M., Eastman, A. and Bresnick, E. (1981) A system in mouse liver for the repair of O^6-methyguanine lesions in methylated DNA. *Nucl. Acids Res.*, **9**, 3089–3103.

Boiteux, S. and Laval, J. (1982) Coding properties of poly (deoxycytidylic acid) templates containing uracil or apyrimidimic sites: *In vitro* modulation of mutagenesis by deoxyribonucleic acid repair enzymes. *Biochemistry*, **21**, 6746–6751.

Boiteux, S., Huisman, O. and Laval, J. (1984) 3-methyladenine residues in DNA induce the SOS function *sfiA* in *Escherichia coli*. *EMBO J.*, **3**, 2569–2573.

Boyd, J. B. and Shaw, K. E. S. (1982) Postreplication repair defects in mutants of *Drosophila melanogaster*. *Mol. Gen. Genet.*, **186**, 289–294.

Boyd, J. B., Harris, P. V., Presley, J. M. and Narachi, M. (1983) *Drosophila melanogaster*: A model eukaryote for the study of DNA repair. In *Cellular Responses to DNA Damage* (eds E. C. Friedberg and B. A. Bridges), Alan R. Liss Inc., New York, pp. 107–123.

Boyd, J. B., Snyder, R. D., Harris, P. V., Presley, J. M., Boyd, S. F. and Smith, P. D. (1982) Identification of a second locus in *Drosophila melanogaster* required for excision repair. *Genetics*, **100**, 239–257.

Brandenburger, A., Godson, G. N., Radman, M., Glickman, B. W., van Sluis, C. A. and Doubleday, O. P. (1981) Radiation-induced base substitution mutagenesis in single-stranded DNA phage M13. *Nature*, **294**, 180–182.

Brash, D. E. and Haseltine, W. A. (1982) UV-induced mutation hotspots occur at DNA damage hotspots. *Nature*, **298**, 189–192.

Breimer, L. H. and Lindahl, T. (1980) A DNA glycosylase from *Escherichia coli* that releases free urea from polydeoxyribonucleotide containing fragments of base residues. *Nucl. Acids Res.*, **8**, 6199–6211.

Breimer, L. H. and Lindahl, T. (1984) DNA glycosylase activities for thymine residues damaged by ring saturation, fragmentation, or ring contraction are functions of Endonuclease III in *Escherichia coli*. *J. Biol. Chem.*, **259**, 5543–5548.

Breimer, L. H. (1984) Enzymatic excision from γ-irradiated polydeoxyribonucleotides of adenine residues whose imidazole rings have been ruptured. *Nucleic Acids Res.*, **12**, 6359–6367.

Brent, R. (1983a) Regulation of the *E. coli* SOS response by the *lexA* gene product. In *Cellular Responses to DNA Damage* (eds E. C. Friedberg and B. A. Bridges), Alan R. Liss Inc., New York, pp. 361–368.

Brent, R. B. and Ptashne, M. (1980) The *lexA* gene product represses its own promotor. *Proc. Natl Acad. Sci. USA*, **77**, 1932–1936.

Brent, R. and Ptashne, M. (1981) Mechanism of action of the *lexA* gene product. *Proc. Natl Acad. Sci. USA*, **78**, 4204–4208.

Brent, T. P. (1983b) Properties of a human lymphoblast AP-endonuclease associated with activity for DNA damaged by ultraviolet light, X-rays, or osmium tetroxide. *Biochemistry*, **22**, 4507–4512.

Bresler, S. E. (1975) Theory of misrepair mutagenesis. *Mutation Res.*, **29**, 467–472.

Bresler, S. E., Verbenko, V. N. and Kalinin, V. L. (1980) Mutants of *Escherichia coli* K-12 with enhanced resistance to ionizing radiation. I. Isolation and study of cross-resistance to various agents. *Soviet Genetics*, **16**, 1094–1101.

Bridges, B. A. and Woodgate, R. (1984) Mutagenic repair in *Escherichia coli X*. The *umuC* gene product may be required for replication past pyrimidine dimers but not for the coding error in UV-mutagenesis. *Mol. Gen. Genet.*, **196**, 364–366.

Bridges, B. A. and Woodgate, R. (1985) Mutagenic repair in *Escherichia coli*. *RecA* and *umuC,D* gene products act at different steps in UV mutagenesis. *Proc. Natl Acad. Sci. USA*, **82**, 4193–4197.

Bridges, B. A., Law, J. and Munson, R. J. (1968) Mutagenesis in *Escherichia coli* II. Evidence for a common pathway for mutagenesis by ultraviolet light, ionizing radiation and thymine deprivation. *Mol. Gen. Genet.*, **103**, 266–273.

Bridges, B. A., Mottershead, R. P. and Sedgwick, S. G. (1976) Mutagenic DNA repair in *Escherichia coli* III. Requirement for a function of DNA polymerase III in ultraviolet-light mutagenesis. *Mol. Gen. Genet.*, **144**, 53–58.

Brunborg, G., Resnick, M. A. and Williamson, D. H. (1980) Cell-cycle – specific repair of DNA double-strand breaks in *Saccharomyces cerevisiae*. *Radiation Res.*, **82**, 547–558.

Brunk, C. F. and Hanawalt, P. C. (1967) Repair of damaged DNA in a eucaryotic cell: *Tetrahymena pyriformis*. *Science*, **158**, 663–664.

Bruyninckx, W. J., Mason, H. S. and Morse, S. A. (1978) Are physiological oxygen concentrations mutagenic? *Nature*, **274**, 606–607.

Buhl, S. N., Setlow, R. B. and Regan, J. D. (1972) Steps in DNA chain elongation and joining after ultraviolet irradiation of human cells. *Int. J. Radiat. Biol.*, **22**, 417–424.

Buhl, S. N., Setlow, R. B. and Regan, J. D. (1973) Recovery of the ability to synthesize DNA in segments of normal size at long times after ultraviolet irradiation of human cells. *Biophys. J.*, **13**, 1265–1275.

Buhl, S. N., Setlow, R. B. and Regan, J. D. (1974) DNA repair in *Potorous tridactylus*. *Biophys. J.*, **14**, 791–803.

Caillet-Fauquet, P., Maenhaul-Michel, G. and Radman, M. (1984) SOS mutator effect in *E. coli* mutants deficient in mismatch correction. *EMBO J.*, **3**, 707–712.

Calos, M. P., Lebkowski, J. S. and Botcham, M. R. (1983) High mutation frequency in DNA transfected into mammalian cells. *Proc. Natl Acad. Sci. USA*, **80**, 3015–3019.

Capecchi, M. R., von der Haar, R. A., Capecchi, N. E. and Sveda, M. M. (1977) The isolation of a suppressable nonsense mutant in mammalian cells. *Cell*, **12**, 371–381.

Chetsanga, C. J. and Lindahl, T. (1979) Release of 7-methylguanine residues whose imidazole rings have been opened from damaged DNA by a DNA glycosylase from *Escherichia coli*. *Nucl. Acids Res.*, **6**, 3673–3687.

Ciarrocchi, G. and Linn, S. (1978) A cell-free assay measuring repair DNA synthesis in human fibroblasts. *Proc. Natl Acad. Sci. USA*, **75**, 1887–1891.

Ciesla, Z. (1982) Plasmid pKM101-mediated mutagenesis in *Escherichia coli* is inducible. *Mol. Gen. Genet.*, **186**, 298–300.

Clarkson, J. R. and Painter, R. B. (1974) Repair of X-ray damage in ageing WI-38 cells. *Mutation Res.*, **23**, 107–112.

Cleaver, J. E. (1968) Defective repair replication of DNA in Xeroderma Pigmentosum. *Nature*, **218**, 652–656.

Cleaver, J. E. and Bootsma, P. (1975) Xeroderma pigmentosum: Biochemical and genetic characteristics. *Ann. Rev. Genet.*, **9**, 19–38.

Compere, S. J. and Palmiter, R. D. (1981) DNA methylation controls the inducibility of the mouse metallothionein-1 gene in lymphoid cells. *Cell*, **25**, 233–240.

Cooper, P. K. (1982) Characterization of long patch excision repair of DNA in ultraviolet-irradiated *Escherichia coli*: An inducible function under Rec-Lex control. *Mol. Gen. Genet.*, **185**, 189–197.

Corry, P. M. and Cole, A. (1973) Double strand rejoining in mammalian DNA. *Nature New Biology*, **245**, 100–101.

Coulondre, C. and Miller, J. H. (1977) Genetic studies of the *lac* repressor. IV. Mutagenic specificity in the *lac* I gene of *Escherichia coli*. *J. Mol. Biol.*, **117**, 577–606.

Coulondre, C., Miller, J. H., Farabaugh, P. and Gilbert, W. (1978) Molecular basis of base substitution hotspots in *Escherichia coli*. *Nature*, **274**, 775–780.

D'Ambrossio, S. M., Aebersold, P. M. and Setlow, R. B. (1978) Enhancement of postreplication repair in ultraviolet-light-irradiated Chinese hamster cells by irradiation in G_2 or S-phase. *Biophys. J.*, **23**, 71–78.

Dasgupta, U. B. and Summers, W. C. (1978) Ultraviolet reactivation of *Herpes simplex* virus is mutagenic and inducible in mammalian cells. *Proc. Natl Acad. Sci. USA*, **75**, 2378–2381.

Dasgupta, U. B. and Summers, W. C. (1980) Genetic recombination of *Herpes simplex* virus, the role of the host cell and UV-irradiation of the virus. *Mol. Gen. Genet.*, **178**, 617–623.

Davies, R. and Sinskey, A. J. (1973) Radiation-resistant mutants of *Salmonella typhimurium* LT2: Development and characterization. *J. Bacteriol.*, **113**, 133–144.

Defais, M. (1983) Role of the *E. coli umuC* gene product in the repair of single-stranded DNA phage. *Mol. Gen. Genet.*, **192**, 509–511.

Demple, B. and Halbrook, J. (1983) Inducible repair of oxidative DNA damage in *Escherichia coli*. *Nature*, **304**, 466–468.

Demple, B. and Linn, S. (1980) DNA N-glycosylases and UV repair. *Nature*, **287**, 203–208.

Demple, B., Halbrook, J. and Linn, S. (1983) *Escherichia coli xth* mutants are hypersensitive to hydrogen peroxide. *J. Bacteriol.*, **153**, 1079–1082.

Dix, D. and Cohen, P. (1980) On the role of ageing in cancer incidence. *J. Theor. Biol.*, **83**, 163–173.

Doerfler, W. (1983) DNA methylation and gene activity. *Ann. Rev. Biochem.*, **52**, 93–124.

Doubleday, O. P., Bridges, B. A. and Green, M. H. L. (1975) Mutagenic DNA repair in *Escherichia coli* II. Factors affecting loss of photoreversibility of UV induced mutations. *Mol. Gen. Genet.*, **140**, 221–230.

Drinkwater, N. R., Miller, E. C. and Miller, J. A. (1980) Estimation of apurinic/apyrimidinic sites and phosphotriesters in deoxyribonucleic acid treated with electrophilic carcinogens and mutagens. *Biochemistry*, **19**, 5087–5092.

Dubbs, D. R., Rachmeler, M. and Kit., S. (1974) Recombination between temperature sensitive mutants of Simian Virus 40. *Virology*, **57**, 161–174.

Duncan, B. K. and Miller, J. H. (1980) Mutagenic deamination of cytosine residues in DNA. *Nature*, **287**, 560–561.

Echols, H. (1981) SOS functions, cancer and inducible evolution. *Cell*, **25**, 1–2.

Eckardt, F. and Haynes, R. H. (1977) Induction of pure and sectored mutant clones in excision proficient and deficient strains of yeast. *Mutation Res.*, **43**, 327–338.

Eisenstadt, E., Warren, A. J., Porter, J., Atkins, P. and Miller, J. H. (1982) Carcinogenic epoxides of benzo (a) pyrene and cyclopenta (cd) pyrene induce base substitutions via specific transversions. *Proc. Natl Acad. Sci. USA*, **79**, 1945–1949.

Emery, H. S., Schild, D., Kellogg, D. E. and Mortimer, R. K. (1984) Studies of the structure and regulation of the yeast *RAD54* and *RAD52* genes. In *Proceedings of the XII International Conference on Yeast Genetics and Molecular Biology*, Edinburgh.

England, P., Huberman, J., Jovin, T. M. and Kornberg, A. (1969) Enzymic synthesis of deoxyribonucleic acid – binding of triphosphates to deoxyribonucleic acid polymerase. *J. Biol. Chem.*, **244**, 3038–3044.

Epstein, J., Williams, J. R. and Little, J. B. (1973) Deficient DNA repair in human progeroid cells. *Proc. Natl Acad. Sci. USA*, **70**, 977–982.

Epstein, J., Williams, J. R. and Little, J. B. (1974) Rate of DNA repair in progeric and normal human fibroblasts. *Biochem. Biophys. Res. Commun.*, **59**, 850–857.

Epstein, J., Leyra, A., Kelley, W. N. and Littlefield, J. W. (1977) Mutagen-induced diploid human lymphoblast variants containing altered hypoxanthine-guanine phosphoribosyl transferase. *Somatic Cell Genet.*, **3**, 135–148.

Erdman, I. E., Thatcher, F. S. and McQueen, K. F. (1961) Studies on the irradiation of microorganisms in relation to food preservation II. Irradiation resistant mutants. *Can. J. Microbiol.*, **7**, 207–215.

Erixon, K. and Ahnstrom, G. (1979) Single strand breaks in DNA during repair of UV induced damage in normal human and xeroderma pigmentosum cells as determined by alkaline DNA unwinding and hydroxylapatite chromatography. Effects of hydroxyurea, 5-fluorodeoxyuridine and 1-β-D-arabinofuranosylcytosine on the kinetics of repair. *Mutation Res.*, **59**, 257–271.

Evensen, G., and Seeberg, E. (1982) Adaptation to alkylation resistance involves the induction of DNA glycosylase. *Nature*, **296**, 773–775.

Fabre, F. and Roman, F. (1977) Genetic evidence for inducibility of recombination competence in yeast. *Proc. Natl Acad. Sci. USA*, **74**, 1667–1671.

Fenn, W. O., Gerschman, R., Gilbert, D. L., Terwilliger, D. E. and Cothran, F. V. (1957) Mutagenic effects of high oxygen tensions on *Escherichia coli*. *Proc. Natl Acad. Sci. USA*, **43**, 1027–1032.

Fenwick, R. G., Sawyer, T. H., Kruh, G. D., Astrin, K. H. and Caskey, C. T. (1977) Forward and reverse mutations affecting the kinetics and apparent molecular weight of mammalian HGPRT. *Cell*, **12**, 383–391.

Fersht, A. R. and Knill-Jones, J. W. (1983) Contribution of 3′→5′ exonuclease activity of DNA polymerase III holoenzyme from *Escherichia coli* to specificity. *J. Mol. Biol.*, **165**, 669–682.

Fincham, J. R. S. and Holliday, R. (1970) An explanation of fine structure map expansion in terms of excision repair. *Mol. Gen. Genet.*, **109**, 309–322.

Fogel, S. and Mortimer, R. K. (1969) Informational transfer in meiotic gene conversion. *Proc. Natl Acad. Sci. USA*, **62**, 96–103.

Folger, K. M., Thomas, K. and Capecchi, M. R. (1985) Efficient correction of mismatched bases in plasmid heteroduplexes injected into cultured mammalian cell nucleii. *Mol. Cell. Biol.*, **5**, 70–74.

Fonck, K., Barthel, R. and Bryant, P. E. (1984) Kinetics of recombinational hybrid formation in X-irradiated mammalian cells a possible first step in the repair of DNA double strand breaks. *Mutation Res.*, **132**, 113–118.

Foote, R. S. and Mitra, S. (1984) Lack of induction of O^6-methyguanine–DNA methyltransferase in mammalian cells treated with N-methyl-N'-nitro-N-nitro-soguanidine. *Carcinogenesis*, **5**, 277–281.

Fornace, A. J., Kohn, K. W. and Karn, H. E. (1976) DNA single strand breaks during repair of UV damage in human fibroblasts and abnormalities of repair in xeroderma pigmentosum. *Proc. Natl Acad. Sci. USA*, **73**, 39–43.

Foster, P. L., Eisenstadt, E. and Cairns, J. (1982) Random components of mutagenesis. *Nature*, **299**, 305–367.

Foster, P. L., Eisenstadt, E. and Miller, J. H. (1983) Base substitution mutations induced by metabolically activated aflatoxins, B. *Proc. Natl Acad. Sci. USA*, **80**, 2695–2698.

Francis, A. A., Lee, W. H. and Regan, J. D. (1981) The relationship of DNA excision repair of ultraviolet-induced lesions to the maximum life span of mammals. *Mech. Ageing Development*, **16**, 181–189.

Franklin, W. A., Low, K. M. and Haseltine, W. A. (1982) Alkaline lability of novel fluorescent photoproducts produced in ultraviolet light irradiated DNA. *J. Biol. Chem.*, **257**, 13 535–13 543.

Fujiwara, Y. and Kano, Y. (1983) Characteristics of thymine dimer excision from xeroderma pigmentosum chromatin. In *Cellular Responses to DNA Damage* (eds E. C. Friedberg and B. A. Bridges), Alan R. Liss, New York, pp. 215–224.

Fujiwara, Y. and Tatsumi, M. (1977) Cross-link repair in human cells and its possible defect in Fanconi's anemia cells. *J. Mol. Biol.*, **113**, 635–649.

Ganesan, A. K. (1974) Persistence of pyrimidine dimers during post-replication repair in ultraviolet light-irradiated *Escherichia coli*. K12. *J. Mol. Biol.*, **87**, 103–119.

Gentil, A., Margot, A. and Sarasin, A. (1983) Effect of UV-irradiation on genetic recombination of simian virus 40 mutants. In *Cellular Responses to DNA Damage* (eds E. C. Friedberg and B. A. Bridges), Alan R. Liss Inc., New York, pp. 385–396.

Gentil, A., Margot, A. and Sarasin, A. (1984) Apurinic sites cause mutations in simian virus 40. *Mutation Res.*, **129**, 141–147.

Glickman, B. W. and Ripley, L. S. (1984) Structural intermediates of deletion mutagenesis: A role for palindromic DNA. *Proc. Natl Acad. Sci. USA*, **81**, 512–516.

Goldstein, S. (1971) The role of DNA repair in ageing of cultured fibroblasts from xeroderma pigmentosum and normals. *Proc. Soc. Exp. Biol. Med.*, **137**, 730–734.

Gonclaves, O., Drobetsky, E. and Meuth, M. (1984) Structural alterations of the *aprt* locus induced by deoxyribonucleoside triphosphate pool imbalances in Chinese hamster ovary cells. *Mol. Cell. Biol.*, **4**, 1792–1799.

Gould, S. J. (1983) In *Hens Teeth and Horses Toes*, W. W. Norton and Co., USA.

Gruenbaum, Y., Cedar, H. and Razin, A. (1982) Substrate and sequence specificity of a eukaryotic DNA methylase. *Nature*, **295**, 620–622.

Haefner, K. (1968) Concerning the mechanism of ultraviolet mutagenesis. A micromanipulatory pedigree analysis in *Schizosaccharomyces pombe*. *Genetics*, **57**, 169–178.

Hagedorn, R., Thielman, H. W., Fischer, H. and Schroeder, C. H. (1983) SV40 induced transformation and T-antigen production is enhanced in normal and repair-deficient human fibroblasts after pretreatment of cells with UV light. *J. Cancer Res. Clin. Oncol.*, **106**, 93–96.

Hall, J. D. and Mount, D. W. (1981) Mechanisms of DNA replication and mutagenesis in ultraviolet-irradiated bacteria and mammalian cells. *Prog. Nucl. Acids Res. Mol. Biol.*, **25**, 53–126.

Hall, J. D., Featherston, J. D. and Almy, R. E. (1980) Evidence for repair of ultraviolet light-damaged Herpes virus in human fibroblasts by a recombination mechanism. *Virology*, **105**, 490–500.

Hall, K. Y., Bergman, K. and Walford, R. L. (1981) DNA repair, H-2, and ageing in NZB and CBA mice. *Tissue Antigens*, **17**, 104–110.

Hall, K. Y., Hart, R. W., Benirschke, A. K. and Walford, R. L. (1984) Correlation between ultraviolet-induced DNA repair in primate lymphocytes and fibroblasts and species maximum achievable lifespan. *Mech. Ageing Dev.*, **24**, 163–173.

Hanawalt, P. C., Cooper, P. K., Ganesan, A. K. and Smith, C. A. (1979) DNA repair in bacteria and mammalian cells. *Ann. Rev. Biochem.*, **48**, 783–836.

Harland, R. M. (1982) Inheritance of DNA methylation in microinjected eggs of *Xenopus laevis*. *Proc. Natl Acad. Sci. USA*, **79**, 2323–2327.

Harm, W. and Hillebrandt, B. (1962) A non-photoreactivable mutant of *E. coli B*. *Photochem. Photobiol.*, **1**, 271–272.

Harris, A. L., Karran, P. and Lindahl, T. (1983) O^6-methylguanine-DNA methyl-transferase of human lymphoid cells: Structural and kinetic properties and absence in repair-deficient cells. *Cancer Res.*, **43**, 3247–3252.

Harris, P. V. and Boyd, J. B. (1980) Excision repair in *Drosophila*. Analysis of strand breaks appearing in DNA of *mei-9* mutants following mutagen treatment. *Biochim. Biophys. Acta*, **610**, 116–129.

Hart, R. W., D'Ambrossio, S. M., Ng, K. J. and Modak, S. (1979) Longevity, stability and DNA repair. *Mech. Ageing Dev.*, **9**, 203–223.

Hart, R. W. and Setlow, R. B. (1974) Correlation between deoxyribonucleic acid excision-repair and lifespan in a number of mammalian species. *Proc. Natl Acad. Sci. USA*, **71**, 2169–2173.

Hart, R. W. and Setlow, R. B. (1976) DNA repair in late passage human cells. *Mech. Ageing Dev.*, **5**, 67–77.

Hart, R. W., Setlow, R. B. and Woodhead, A. D. (1977) Evidence that pyrimidine dimers in DNA can give rise to tumours. *Proc. Natl Acad. Sci. USA*, **74**, 5574–5578.

Haseltine, W. A. (1983) Site specificity of ultraviolet light induced mutagenesis. In *Cellular Responses to DNA Damage* (eds E. C. Freidberg and B. A. Bridges), Alan R. Liss Inc., New York, pp. 3–22.

Haseltine, W. A., Gordon, L. K., Lindan, C. P., Grafstrom, R. H., Shaper, M. L. (1980) Cleavage of pyrimidine dimers in specific DNA sequences by a pyrimidine dimer DNA-glycosylase of *M. luteus*. *Nature*, **285**, 634–641.

Henner, W. D., Grunberg, S. M. and Haseltine, W. A. (1983) Enzyme action at 3' termini of ionizing radiation-induced DNA strand breaks. *J. Biol. Chem.*, **258**, 15 198–15 205.

Hickson, I. D., Arthur, H. M., Bramhill, D. and Emmerson, P. T. (1983) The *E. coli uvrD* gene product is DNA helicase II. *Mol. Gen. Genet.*, **190**, 265–270.

Ho, K. S. Y. (1975) Induction of DNA double-strand breaks by X-rays in a radiosensitive strain of the yeast *Saccharomyces cerevisiae*. *Mutation Res.*, 30 327–334.

Hofemeister, J. and Eitner, G. (1981) Repair and plasmid R46 mediated mutation requires inducible functions in *Proteus mirabilis. Mol. Gen. Genet.*, 183, 369–375.

Holliday, R. (1962) Mutation and replication in *Ustilago maydis. Genet. Res., Camb.* 3, 472–486.

Holliday, R. A. (1964) The mechanism of gene conversion in fungi. *Genet. Res. Camb.*, 5, 282–304.

Holliday, R. (1979) A new theory of carcinogenesis. *Brit. J. Cancer.*, 40, 513–522.

Holliday, R. (in press, 1986) The inheritance of epigenetic defects. *Science*.

Howard-Flanders, P., West, S. C. and Stasiak, A. (1984) Role of RecA protein spiral filaments in genetic recombination. *Nature*, 309, 215–220.

Ichikawa-Ryo, H. and Kondo, S. (1975) Indirect mutagenesis in phage lambda by ultraviolet preirradiation of host bacteria. *J. Mol. Biol.*, 97, 77–92.

Iwatsuki, N., Joe, C. O. and Werbin, H. (1980) Evidence that deoxyribonucleic acid photolyase from baker's yeast is a flavoprotein. *Biochemistry*, 19, 1172–1176.

James, A. P. and Kilbey, B. J. (1977) The timing of UV mutagenesis in yeast: A pedigree analysis of induced recessive mutation. *Genetics*, 87, 237–248.

Jeggo, P. A. and Kemp, L. M. (1983) X-ray-sensitive mutants of Chinese hamster ovary cell line. Isolation and cross-sensitivity to other DNA-damaging agents *Mutation Res.*, 112, 313–327.

Jones, P. A. and Taylor, S. M. (1980) Cellular differentiation, cytidine analogs and DNA methylation. *Cell*, 20, 85–93.

Karran, P., Hjelmgren, T. and Lindahl, T. (1982) Induction of a DNA glycosylase for N methylated purines is part of the adaptive response to alkylating agents. *Nature* 296, 770–773.

Karran, P. and Lindahl, T. (1978) Enzymatic excision of free hypoxanthine from polydeoxynucleotides and DNA containing deoxyinosine monophosphate residues *J. Biol. Chem.*, 253, 5877–5879.

Karran, P. and Lindahl, T. (1980) Hypoxanthine in deoxyribonucleic acid: Generation by heat-induced hydrolysis of adenine residues and release in free form by a deoxyribonucleic acid glycosylase from calf thymus. *Biochemistry*, 19, 6005–6011.

Kastan, M. B., Gowans, B. J. and Lieberman, M. W. (1982) Methylation of deoxycytidine incorporated by excision-repair synthesis of DNA. *Cell*, 30, 509–516.

Katcher, H. L. and Wallace, S. S. (1983) Characterization of the *Escherichia coli* X-ray endonuclease, Endonuclease III. *Biochemistry*, 22, 4071–4081.

Kato, H., Harada, M., Tsuchiya, K. and Moriwaki, K. (1980) Absence of correlation between DNA repair in ultraviolet irradiated mammalian cells and lifespan of the donor species. *Japan. J. Genetics*, 55, 99–108.

Kato, T. (1977) Effects of chloramphenicol and caffeine on postreplication repair in *uvrA umuC⁻* and *uvrA⁻recF⁻* strains of *Escherichia coli. Mol. Gen. Genet.*, 156 115–120.

Kato, T. and Shinoura, Y. (1977) Isolation and characterization of mutants of *Escherichia coli* deficient in induction of mutations by ultraviolet light. *Mol. Gen. Genet.*, 156, 121–131.

Kemp, L. M. (1985) PhD thesis, Council for National Academic Awards, London.

Kemp, L. M., Sedgwick, S. G. and Jeggo, P. A. (1984) X-ray sensitive mutants of Chinese hamster ovary cells defective in double-strand break rejoining. *Mutation Res.*, **132**, 189–196.

Kidson, C., Chen, P., Imray, F. P. and Gipps, E. (1983) Nervous system disease associated with dominant cellular radiosensitivity. In *Cellular Responses to DNA Damage* (eds E. C. Friedberg and B. A. Bridges), Alan R. Liss Inc., New York, pp. 721–729.

Kitagawa, Y., Akaboshi, E., Shinagawa, H., Horii, T., Ogawa, H. and Kato, T. (1985) Structural analysis of the *umu* operon required for inducible mutagenesis in *Escherichia coli. Proc. Natl Acad. Sci. USA*, **82**, 4336–4340.

Kitayama, S. and Matsuyama, A. (1968) Possibility of the repair of double-strand scissions in *Micrococcus radiodurans* DNA caused by gamma rays. *Biochem. Biophys. Res. Commun.*, **33**, 418–422.

Kondo, S., Ichikawa, H., Iwo, K. and Kato, T. (1970) Base-change mutagenesis and prophage induction in strains of *Escherichia coli* with different repair capacities. *Genetics*, **66**, 187–217.

Kraemer, K. H., Lee, M. M. and Scotto, J. (1984) DNA repair protects against cutaneous and internal neoplasia, evidence from xeroderma pigmentosum. *Carcinogenesis*, **5**, 511–514.

Krasin, F. and Hutchinson, F. (1977) Repair of DNA double-strand breaks in *Escherichia coli*, which requires *recA* function and the presence of a duplicate genome. *J. Mol. Biol.*, **116**, 81–98.

Krasin, F. and Hutchinson, F. (1981) Repair of DNA double-strand breaks in *Escherichia coli* cells requires synthesis of proteins that can be induced by UV light. *Proc. Natl Acad. Sci. USA*, **78**, 3450–3453.

Krisch, R. E., Krasin, F. and Sauri, C. J. (1976) DNA breakage, repair, and lethality after ^{125}I decay in rec^+ and *recA* strains of *Escherichia coli. Int. J. Radiat. Biol.*, **29**, 37–50.

Krontiris, T. G., DiMartino, N. A., Colb, M. and Parkinson, D. R. (1985) Unique allelic restriction fragments of the human Ha-*ras* locus in leukocyte and tumour DNAs of cancer patients. *Nature*, **313**, 369–374.

Krueger, J. H. and Walker, G. C. (1984) *groEL* and *dnaK* genes of *Escherichia coli* are induced by UV irradiation and nalidixic acid in a $htpR^+$ – dependent fashion. *Proc. Natl Acad. Sci. USA*, **81**, 1499–1503.

Kucherlapati, R. S., Eves, E. M., Song, K.-Y., Morse, B. S. and Smithies, O. (1984) Homologous recombination between plasmids in mammalian cells can be enhanced by treatment of input DNA. *Proc. Natl Acad. Sci. USA*, **81**, 3153–3157.

Kumura, K. and Sekiguchi, M. (1984) Identification at the *uvrD* gene product of *Escherichia coli* as DNA Helicase II and its induction by DNA damaging agents. *J. Biol. Chem.*, **259**, 1560–1565.

Kumura, K., Sekiguchi, M., Steinum, A.-L. and Seeberg, E. (1985) Stimulation of the UvrABC enzyme-catalysed repair reactions by the UvrD protein (DNA helicase II). *Nucl. Acids Res.*, **13**, 1483–1492.

Kunkel, T. (1984) Mutational specificity of depurination. *Proc. Natl Acad. Sci. USA*, **81**, 1494–1498.

Kunkel, T., Schaaper, R. M. and Loeb, L. A. (1983) Depurination induced infidelity of deoxyribonucleic acid synthesis with purified deoxyribonucleic acid replication proteins *in vitro*. *Biochemistry*, **22**, 2378–2384.

Lackey, D., Krauss, S. W. and Linn, S. (1982) Isolation of an altered form of DNA polymerase I from *Escherichia coli* cells induced for *recA/lexA* functions. *Proc. Natl Acad. Sci. USA*, **79**, 330–334.

Lai, C.-J. and Nathans, D. (1975) A map of temperature-sensitive mutants of simian virus 40. *Virology*, **66**, 70–81.

Lancaster, H. O. (1956) Some geographical aspects of the mortality from melanoma in Europeans. *Med. J. Aust.*, **1**, 1082–1087.

Land, H., Panada, L. F. and Weinberg, R. A. (1983) Tumorigenic conversion of primary embryo fibroblasts requires at least two co-operating oncogenes. *Nature*, **304**, 596–602.

Lawley, P. D. and Brookes, P. (1963) Further studies on the alkylation of nucleic acids and their constituent nucleotides. *Biochem. J.*, **89**, 127–138.

Lawrence, C. W. and Christensen, R. B. (1979) Absence of relationship between UV-induced reversion frequency and nucleotide sequence at the *cyc1* locus of yeast. *Mol. Gen. Genet.*, **177**, 31–38.

Leadon, S. and Hanawalt, P. C. (1984) Ultraviolet irradiation of monkey cells enhances the repair of DNA adducts in alpha DNA. *Carcinogenesis*, **5**, 1505–1510.

Leaper, S., Resnick, M. A. and Holliday, R. (1980) Repair of double-strand breaks and lethal damage in DNA of *Ustilago maydis*. *Genet. Res., Camb.*, **35**, 291–307.

Lebkowski, J. S., Clancy, S., Miller, J. H. and Calos, M. P. (1986) The *lacI* shuttle: rapid analysis of the mutagenic specificity of ultraviolet light in human cells. *Proc. Natl Acad. Sci. USA* (in press).

LeClerc, J. E., Istock, N. L., Saran, B. R. and Allen, R. (1984) Sequence analysis of ultraviolet-induced mutations in M13 *lacZ* hybrid phage DNA. *J. Mol. Biol.*, **180**, 217–237.

Lee, J. A. and Merril, J. M. (1971) Sunlight and the aetiology of malignant melanoma: a synthesis. *Med. J. Aust.*, **2**, 846–851.

Lee, M. G. and Yarranton, G. T. (1983) Inducible DNA repair in *Ustilago maydis*. *Mol. Gen. Genet.*, **185**, 245–250.

Lehmann, A. R. (1972) Postreplication repair of DNA in ultraviolet-irradiated mammalian cells. *J. Mol. Biol.*, **66**, 319–337.

Lehmann, A. R., Kirk-Bell, S., Arlett, C. F., Paterson, M. C., Lohman, P. H. M., de Weerd-Kastelein, E. A. and Bootsma, D. (1975) Xeroderma Pigmentosum cells with normal levels of excision repair have a defect in DNA synthesis after UV irradiation. *Proc. Natl Acad. Sci. USA*, **72**, 219–223.

Lehmann, A. R. and Kirk-Bell, S. (1972) Post replication repair of DNA in ultraviolet-irradiated mammalian cells. No gaps on DNA synthesized late after ultraviolet irradiation. *Eur. J. Biochem.*, **31**, 438–445.

Lehmann, A. R. and Kirk-Bell, S. (1978) Pyrimidine dimer sites associated with the daughter DNA strands in UV-irradiated human fibroblasts. *Photochem. Photobiol.*, **27**, 297–307.

Lemontt, J. F. (1971a) Mutants of yeast defective in mutation induced by ultra-violet light. *Genetics*, **68**, 21–33.

Lemontt, J. F. (1971b) Pathways of ultraviolet mutability in *Saccharomyces cerevisiae* 1. Some properties of double mutants involving *uvs9* and *rev*. *Mutation Res.*, **13**, 311–317.

Lemontt, J. F. (1971c) Pathways of ultraviolet mutability in *Saccharomyces cerevisiae*
11. The effect of *rev* genes on recombination. *Mutation Res.*, **13**, 319–326.

Lieberman, M. W., Beach, L. R. and Palmiter, R. D. (1983) Ultraviolet radiation-
induced metallothionein-1 gene activation is associated with extensive DNA
demethylation. *Cell*, **35**, 207–214.

Lin, P. F., Bradwell, E. and Howard-Flanders, P. (1977) Initiation of genetic
exchanges in phage-prophage crosses. *Proc. Natl Acad. Sci. USA*, **74**, 291–295.

Lindahl, T. (1979) DNA glycosylases, endonucleases for apurinic/apyrimidinic sites
and base excision repair. *Prog. Nucl. Acid Res. Mol. Biol.*, **22**, 135–192.

Lindahl, T. (1982) DNA repair enzymes. *Ann. Rev. Biochem.*, **51**, 61–87.

Lindahl, T., Demple, B. and Robins, P. (1982) Suicide inactivation of the *E. coli*
O^6-methylguanine-DNA methyltransferase. *EMBO J.*, **1**, 1359–1363.

Lindahl, T. and Karlstrom, O. (1973) Heat-induced depyrimidination of deoxyribonucleic acid in neutral solution. *Biochemistry*, **12**, 5151–5154.

Lindahl, T. and Nyberg, B. (1972) Rate of depurination of native deoxyribonucleic
acid. *Biochemistry*, **11**, 3610–3617.

Lindahl, T. and Nyberg, B. (1974) Heat-induced deamination of cytosine residues in
deoxyribonucleic acid. *Biochemistry*, **13**, 3405–3410.

Lindahl, T., Sedgwick, B., Demple, B. and Karran, P. (1983) Enzymology and
regulation of the adaptive response to alkylating agents. In *Cellular Responses to
DNA Damage* (eds E. C. Friedberg and B. A. Bridges), Alan R. Liss Inc., New
York, pp. 241–253.

Lindahl, T., Ljungquist, S., Siegert, W., Nyberg, B. and Sperens, B. (1977) DNA
N-glycosidases. Properties of uracil-DNA glycosidase from *Escherichia coli J. Biol.
Chem.*, **252**, 3286–3294.

Linn, S., Kairis, M. and Holliday, R. (1976) Decreased fidelity of DNA polymerase
activity isolated from ageing human fibroblasts. *Proc. Natl Acad. Sci. USA*, **73**,
2818–2822.

Little, J. W. (1982) The SOS regulatory system: control of its state by the level of *recA*
protease. *J. Mol. Biol.*, **167**, 791–808.

Little, J. W. (1983) Variations in the *in vivo* stability of LexA repressor during the SOS
regulatory cycle. In *Cellular Responses to DNA Damage* (eds E. C. Friedberg and B.
A. Bridges), Alan R. Liss Inc., New York, pp. 369–378.

Little, J. W. (1984) Autodigestion of LexA and phage repressors. *Proc. Natl Acad. Sci.
USA*, **81**, 1375–1379.

Little, J. W. and Mount, D. W. (1982) The SOS regulatory system of *Escherichia coli.
Cell*, **29**, 11–22.

Little, J. W., Edmiston, S. H., Pacelli, L. Z. and Mount, D. W. (1980) Cleavage of the
Escherichia coli lexA protease by the *recA* protease. *Proc. Natl Acad. Sci. USA*, **77**,
3225–3229.

Little, J. W., Mount, D. W. and Yannisch-Perron, C. R. (1981) Purified *lexA* protein
is a repressor of the *recA* and *lexA* genes. *Proc. Natl Acad. Sci. USA*, **78**, 4199–4203.

Liu, S.-C. C., Parsons, C. S. and Hanawalt, P. C. (1982) DNA repair response in
human epidermal keratinocytes from donors of different age. *J. Invest. Dermatol.*,
79, 330–335.

Lloyd, R. G. (1978) Hyper-recombination in *Escherichia coli* K-12 mutants
constitutive for protein X synthesis. *J. Bacteriol.*, **134**, 929–935.

Lu, A.-L., Clark, S. and Modrich, P. (1983) Methyl-directed repair of DNA base-pair mismatches *in vitro*. *Proc. Natl Acad. Sci. USA*, **80**, 4639–4643.

Lytle, C. D., Goddard, J. G. and Lin, C. H. (1980) Repair and mutagenesis of herpes simplex virus in UV-irradiated monkey cells. *Mutation Res.*, **70**, 139–149.

McCann, J., Choi, E., Yamasaki, E. and Ames, B. N. (1975) Detection of carcinogens as mutagens in the *Salmonella*/microsome test: Assay of 300 chemicals. *Proc. Natl Acad. Sci. USA*, **72**, 5135–5139.

McCarthy, T. V. and Lindahl, T. (1985) Methyl phosphotriesters in alkylated DNA are repaired by the Ada regulatory protein of *E. coli. Nucl. Acids Res.*, **13**, 2683–2698.

McCarthy, T. V., Karran, P. and Lindahl, T. (1984) Inducible repair of O^6-alkylated DNA pyrimidines in *Escherichia coli. EMBO J.*, **3**, 545–550.

McClanahan, T. and McEntee, K. (1984) Specific transcripts are elevated in *Saccharomyces cerevisiae* in response to DNA damage. *Mol. Cell. Biol.*, **4**, 2356–2363.

McKee, R. H. and Lawrence, C. W. (1979) Genetic analysis of gamma ray mutagenesis in yeast. I. Reversion in radiation sensitive genes. *Genetics*, **93**, 361–373.

Magnus, K. (1973) Incidence of malignant melanoma of the skin in Norway, 1955–1970. Variations in time and space and solar radiation. *Cancer*, **32**, 1275–1286.

Maher, V. M. and McCormick, J. J. (1976) Effect of DNA repair on the cytotoxicity and mutagenicity of UV irradiation and of chemical carcinogens in normal and xeroderma pigmentosum cells. In *Biology of Radiation Carcinogenesis* (eds J. M. Yuhas, R. W. Tennant and J. B. Regan), Raven Press, New York, pp. 129–145.

Maher, V. M., Ouellette, L. M., Curren, R. D. and McCormick, J. J. (1976) Frequency of ultraviolet light-induced mutations is higher in xeroderma pigmentosum variant cells than in normal cells. *Nature*, **261**, 593–595.

Margison, G. P., Cooper, D. P. and Brennand, J. (1985) Cloning of the *E. coli* O^6-methylguanine and methylphosphotriester methyltransferase gene using a functional DNA repair assay. *Nucl. Acids Res.*, **13**, 1939–1952.

Marinus, M. G. (1984) *DNA methylation* (eds A. Razin, H. Ceder, and A. D. Riggs), Springer Series in Molecular Biology, Springer Verlag, New York.

Mattern, I. E., Olthoff-Smit, F. P., Jacobs-Meijsing, B. L. M., Enger-Valk, B. E., Pouwels, P. H. and Lohman, P. H. M. (1985) A system to determine basepair substitutions at the molecular level, based on restriction enzyme analysis; influence of *muc* genes of pKM101 on the specificity of mutation induction in *E. coli. Mutation Res.*, **148**, 35–45.

Mattern, M. R. and Cerutti, P. A. (1975) Age-dependent excision repair of damaged thymine from γ-irradiated DNA by isolated nucleic from human fibroblasts. *Nature*, **254**, 450–452.

Maynard-Smith, J. (1978) *The Evolution of Sex*, Cambridge University Press, Cambridge.

Mehta, J. R., Ludlum, D. B., Renard, A. and Verly, W. G. (1981) Repair of O^6-ethylguanine in DNA by a chromatin fraction from rat liver: Transfer of the ethyl group to an acceptor protein. *Proc. Natl Acad. Sci. USA*, **78**, 6766–6770.

Mennigman, H.-D. (1972) Pyrimidine dimers as pre-mutational lesions in *Escherichia coli* WP2 Hcr⁻. *Mol. Gen. Genet.*, **117**, 167–186.

Meyn, R. E. and Humphrey, R. M. (1971) Deoxyribonucleic acid synthesis in ultraviolet light irradiated Chinese hamster cells. *Biophys. J.*, **11**, 295–301.

Miller, J. H. (1982) Mutational specificity in bacteria. *Ann. Rev. Genet.*, **17**, 215–238.

Miller, J. H. (1983) Carcinogens induce targeted mutations in *Escherichia coli*. *Cell*, **31**, 5–7.

Miller, J. H. and Low, K. B. (1984) Specificity of mutagenesis resulting from the induction of the SOS system in the absence of mutagenic treatment. *Cell*, **37**, 675–682.

Miller, L. K., Cooke, B. E. and Fried, M. (1976) Fate of mismatched base-pair regions in polyoma heteroduplex DNA during infection of mouse cells. *Proc. Natl Acad. Sci. USA*, **73**, 3073–3077.

Milman, G., Krauss, S. W. and Olsen, A. S. (1977) Tryptic peptide analysis of normal and mutant forms of hypoxanthine phosphoribosyltransferase from HeLa cells. *Proc. Natl Acad. Sci. USA*, **74**, 926–930.

Milman, G., Lee, E., Changas, G. S., McLaughlin, J. R. and George, M. G. (1976) Analysis of HeLa cell hypoxanthine phosphoribosyltransferase mutants and revertants by two dimensional polyacrylamide gel electrophoresis: Evidence for silent gene inactivation. *Proc. Natl Acad. Sci. USA*, **73**, 4589–4593.

Miura, A. and Tomizawa, J.-I. (1968) Studies on radiation sensitive mutants of *E. coli* III. Participation of the Rec system in induction of mutation by ultraviolet irradiation. *Mol. Gen. Genet.*, **103**, 1–10.

Moore, P. D. and Holliday, R. (1976) Evidence for the formation of hybrid DNA during mitotic recombination in Chinese hamster cells. *Cell*, **8**, 573–579.

Moore, P., Bose, K., Rabkin, S. and Strauss, B. (1981) Sites of termination of *in vitro* DNA synthesis on ultraviolet and N-acetylaminofluorene treated φX174 templates by prokaryotic and eukaryotic DNA polymerases. *Proc. Natl Acad. Sci. USA*, **78**, 110–114.

Moore, P. D., Rabkin, S. D., Osborn, A. L., King, C. M. and Strauss, B. S. (1982) Effect of acetylated and deacetylated 2-aminofluorene adducts on *in vitro* DNA synthesis. *Proc. Natl Acad. Sci. USA*, **79**, 7166–7170.

Mortelmans, K., Cleaver, J. E., Friedberg, E. C., Paterson, M. C., Smith, B. P. and Thomas, G. H. (1977) Photoreactivation of thymine dimers in UV-irradiated human cells: unique dependence on culture conditions. *Mutation Res.*, **44**, 433–446.

Mortelmans, K., Friedberg, E. C., Slor, H., Thomas, G. and Cleaver, J. E. (1976) Defective thymine dimer excision by cell-free extracts of xeroderma pigmentosum cells. *Proc. Natl Acad. Sci. USA*, **73**, 2757–2761.

Mortimer, R. K., Contopoulou, R. and Schild, D. (1981) Mitotic chromosome loss in a radiation sensitive strain of the yeast *Saccharomyces cerevisiae*. *Proc. Natl Acad. Sci. USA*, **78**, 5778–5782.

Mosbaugh, D. W. and Linn, S. (1980) Further characterization of human fibroblast apurinic/apyrimidinic DNA endonucleases. *J. Biol. Chem.*, **255**, 11 743–11 752.

Nakabeppu, Y. and Sekiguchi, M. (1981) Physical association of pyrimidine dimer glycosylase and apurinic/apyrimidinic DNA endonuclease essential for repair of ultraviolet-damaged DNA. *Proc. Natl Acad. Sci. USA*, **78**, 2742–2746.

Nalbantoglu, J., Gonclaves, O. and Meuth, M. (1983) Structure of mutant alleles at the *aprt* locus of Chinese hamster ovary cells. *J. Mol. Biol.*, **167**, 575–594.

Nasim, A. and Auerbach, C. (1967) The origin of complete and mosaic mutants from mutagenic treatment of single cells. *Mutation Res.*, **4**, 1–14.

Nes, I. F. (1980) Purification and properties of a mouse cell DNA repair endonuclease, which recognises lesions in DNA induced by ultraviolet light, depurination, X-rays and OsO₄ treatment. *Eur. J. Biochem.*, **112**, 161–168.

Nevers, P. and Spatz, H. C. (1975) *Escherichia coli* mutants *uvrD* and *uvrE* deficient in gene conversion of λ-heteroduplexes. *Mol. Gen. Genet.*, **139**, 233–243.

Newbold, R. F. and Overell, R. W. (1983) Fibroblast immortality is a prerequisite for transformation by EJ c-Ha-*ras* oncogene. *Nature*, **304**, 648–651.

Newbold, R. F., Warren, W., Medcalf, A. S. C. and Amos, J. (1980) Mutagenicity of carcinogenic methylating agents is associated with a specific DNA modification. *Nature*, **283**, 596–599.

Nishioka, H. and Doudney, C. O. (1969) Different modes of loss of photoreversibility of mutation and lethal damage in ultraviolet-light resistant and sensitive bacteria. *Mutation Res.*, **8**, 215–228.

Nishioka, H. and Doudney, C. O. (1970) Different modes of loss of photoreversibility and suppressor mutations to tryptophan independence in an auxotrophic strain of *Escherichia coli*. *Mutation Res.*, **9**, 349–358.

Orr-Weaver, T. and Szostak, J. W. (1985) Fungal recombination. *Microbiol. Rev.*, **49**, 33–58.

Paffenholz, V. (1978) Correlation between DNA repair of embryonic fibroblasts and different lifespan of 3 inbred mouse strains. *Mech. Ageing Dev.*, **7**, 131–150.

Painter, R. B., Clarkson, J. M. and Young, B. R. (1973) Ultraviolet induced repair replication in ageing diploid human cells (WIr38). *Radiat. Res.*, **56**, 560–564.

Paterson, M. C., Lohman, P. H. M. and Slutyer, M. L. (1973) Use of a UV endonuclease from *Micrococcus luteus* to monitor the progress of DNA repair in UV-irradiated human cells. *Mutation Res.*, **19**, 245–256.

Pegg, A. E. (1984) Methylation of the O⁶ position of guanine in DNA is the most likely initiating event in carcinogenesis by methylating agents. *Cancer Invest.*, **2**, 223–231.

Peleg, L., Raz, E. and BenIshai, R. (1976) Changing capacity for DNA excision repair in mouse embryonic cells in vitro. *Exp. Cell Res.*, **104**, 301–307.

Peterson, T. A., Prakash, L., Prakash, S., Osley, M. A. and Reed, S. T. (1985) Regulation of *CDC9*, the *Saccharomyces cerevisiae* gene that encodes DNA ligase. *Mol. Cell. Biol.*, **5**, 226–235.

Picksley, S. M., Ahfield, P. V. and Lloyd, R. G. (1984) Repair of DNA double-strand breaks in *Escherichia coli* K12 requires a functional *recN* product. *Mol. Gen. Genet.*, **195**, 267–274.

Pollard, E. and Randall, E. P. (1973) Studies on the inducible inhibitor of radiation-induced DNA degradation of *E. coli*. *Radiat. Res.*, **55**, 265–279.

Pollard, E. C., Fluke, D. and Kazanis, D. (1981) Induced radioresistance: An aspect of induced repair. *Mol. Gen. Genet.*, **184**, 421–429.

Prakash, L. (1981) Characterization of postreplication repair in *Saccharomyces cerevisiae* and effects of *rad6*, *rad18*, *rev3* and *rad52* mutations. *Mol. Gen. Genet.*, **184**, 471–478.

Pukkila, P. J. (1978) The recognition of mismatched base pairs in DNA by DNAse from *Ustilago maydis*. *Mol. Gen. Genet.*, **161**, 245–250.

Pukkila, P. J., Peterson, J., Herman, G. (1983) Effects of high levels of DNA adenine methylation on methyl-directed mismatch repair in *Escherichia coli*. *Genetics*, **104**, 571–582.

Radany, E. H. and Friedberg, E. C. (1980) A pyrimidine dimer-DNA glycosylase activity associated with the *v* gene product of bacteriophage T4. *Nature*, **286**, 182–185.

Radding, C. M. (1982) Homologous pairing and strand exchange in genetic recombination. *Ann. Rev. Genet.*, **16**, 405–437.

Radman, M., Wagner, R. E., Glickman, B. W. and Meselson, M. (1980) DNA methylation, mismatch correction and genetic stability. In *Progress in Environmental Mutagenesis* (ed. M. Alacevic), Elsevier-North Holland Biomedical Press, Amsterdam, pp. 121–130.

Razzaque, A., Chakrabarti, S., Joffee, S. and Seidman, M. (1984) Mutagenesis of a shuttle vector plasmid in mammalian cells. *Mol. Cell. Biol.*, **4**, 435–441.

Regan, J. D. and Setlow, R. B. (1974a) Two forms of repair in DNA of human cells damaged by chemical carcinogens and mutagens. *Cancer Res.*, **34**, 3318–3325.

Regan, J. D. and Setlow, R. B. (1974b) DNA repair in human progeroid cells. *Biochem. Biophys. Res. Commun*, **59**, 858–864.

Regan, J. D., Trosko, J. E. and Carrier, W. L. (1968) Evidence for excision of ultraviolet induced pyrimidine dimers from the DNA of human cells *in vitro*. *Biophys. J.*, **8**, 319–325.

Regan, J. D., Carrier, W. L., Samet, C. and Olla, B. L. (1982) Photoreactivation in two closely related marine fishes having different longevities. *Mech. Ageing Dev.*, **18**, 59–66.

Resnick, M. A. (1978) The importance of DNA double-strand break repair in yeast. In *DNA Repair Mechanisms* (eds P. C. Hanawalt, E. C. Friedberg and C. F. Fox), Academic Press, New York, pp. 417–420.

Resnick, M. A. and Moore, P. D. (1979) Molecular recombination and the repair of DNA double strand breaks in CHO cells. *Nucl. Acid Res.*, **6**, 3145–3160.

Reynolds, R. J. and Friedberg, E. C. (1981) Molecular mechanisms of pyrimidine dimer excision in *Saccharomyces cerevisiae*: Incision of ultraviolet-irradiated deoxyribonucleic acid in vivo. *J. Bacteriol.*, **146**, 692–704.

Reynolds, R. J., Love, J. D. and Friedberg, E. C. (1981) Molecular mechanisms of pyrimidine dimer excision in *Saccharomyces cerevisiae*: Excision of dimers in cell extracts. *J. Bacteriol.*, **147**, 705–708.

Riazzudin, S. and Lindahl, T. (1978) Properties of 3-methyladenine-DNA glycosylase from *Escherichia coli*. *Biochemistry*, **17**, 2110–2118.

Riggs, A. D. and Jones, P. A. (1983) 5-methylcytosine, gene regulation and cancer. *Adv. Cancer Res.*, **40**, 1–30.

Roberts, J. W. and Devoret, R. (1983) Lysogenic induction. In *Lambda II* (eds P. W. Hendrix, J. W. Roberts, F. W. Stahl and R. A. Weisberg), Cold Spring Harbor Laboratory, Cold Spring Harbor.

Robbins, J. H. (1983) Hypersensitivity to DNA-damaging agents in primary degenerations of excitable tissue. In *Cellular Responses to DNA Damage* (eds E. C. Friedberg and B. A. Bridges), Alan R. Liss Inc., New York, pp. 671–700.

Robbins, J. H., Moshel, A. N., Lutzner, M. A., Ganges, M. B. and Dupuy, J. M. (1983) A new patient with both xeroderma pigmentosum and Cockayne's syndrome is in a new xeroderma pigmentosum complementation group. *J. Invest. Dermatol.*, **80**, 331–340.

Robins, P. and Cairns, J. (1979) The numerology of the adaptive response to alkylating agents. *Nature*, **280**, 74–76.

Rogers, S. G. and Weiss, B. (1980) Exonuclease III of *Escherichia coli* K-12 is an AP endonuclease. *Methods Enzymol.*, **65**, 201–211.

Rolfe, M. (1985a) UV-inducible protein in *Saccharomyces cerevisiae*. *Curr. Gen.*, **9**, 529–532.

Rolfe, M. (1985b) UV-inducible transcripts in *Saccharomyces cerevisiae*. *Curr. Gen.*, **9**, 533–538.

Rommelaere, J. and Miller-Faures, A. (1975) Detection by density equilibrium centrifugation of recombinant-like DNA molecules in somatic mammalian cells. *J. Mol. Biol.*, **98**, 195–218.

Ruby, S. W. and Szostak, J. W. (1985) Specific *Saccharomyces cerevisiae* genes are expressed in response to DNA damaging agents. *Mol. Cell. Biol.*, **5**, 75–84.

Ruley, E. (1983) Adenovirus early region 1A enables viral and cellular transforming genes to transform primary cells in culture. *Nature*, **304**, 602–606.

Rydberg, B. and Lindahl, T. (1982) Nonenzymatic methylation of DNA by the intracellular methyl group donor S-adenosyl-L-methionine is a potentially mutagenic reaction. *EMBO J.*, **1**, 211–216.

Rydberg, G. (1978) Bromouracil mutagenesis and mismatch repair in mutator strains of *Escherichia coli*. *Mutation Res.*, **52**, 11–24.

Sacher, G. A. and Hart, R. W. (1978) Longevity, ageing and comparative cellular and molecular biology of the housemouse, *Mus musculus*, and the white-footed mouse, *Peromyscus leucopus*. In *Birth Defects: Original Article Series*, Vol. 14 (eds D. Bergsma and D. Harrison), Alan R. Liss Inc., New York, pp. 71–96.

Sage, E. and Haseltine, W. A. (1984) High ratio of alkali-sensitive lesions to total DNA modification induced by benzo (a) pyrene diol epoxide. *J. Biol. Chem.*, **259**, 11 098–11 102.

Samson, L. and Cairns, J. (1977) A new pathway for DNA repair in *Escherichia coli*. *Nature*, **267**, 281–282.

Samson, L. and Schwartz, J. F. (1980) Evidence for an adaptive DNA repair pathway in CHO and human skin fibroblast cell lines. *Nature*, **287**, 801–863.

Sancar, A. and Rupp, W. D. (1983) A novel repair enzyme UVRABC excision nuclease of *Escherichia coli* cuts a DNA strand on both sides of the damaged region. *Cell*, **33**, 249–260.

Sancar, A. and Sancar, G. B. (1984) *Escherichia coli* DNA photolyase is a flavoprotein. *J. Mol. Biol.*, **172**, 223–227.

Sancar, A., Franklin, K. A. and Sancar, G. B. (1984) *Escherichia coli* DNA photolyase stimulates uvrABC excision nuclease in vitro. *Proc. Natl Acad. Sci. USA*, **81**, 7397–7401.

Sancar, A., Smith, F. W. and Sancar, G. B. (1984) Purification of *Escherichia coli* DNA photolyase. *J. Biol. Chem.*, **259**, 6028–6032.

Sarasin, A. and Benoit, A. (1980) Induction of an error prone mode of DNA repair in UV-irradiated monkey kidney cells. *Mutation Res.*, **70**, 71–81.

Sarasin, A. and Hanawalt, P. C. (1980) Replication of ultraviolet-irradiated simian virus 40 in monkey kidney cells. *J. Mol. Biol.*, **138**, 299–319.

arasin, A., Gaillard, C. and Benoit, A. (1981) Molecular mechanism of error-prone DNA replication induced in UV-irradiated or acetoxyacetylaminofluorene treated monkey cells. *J. Supra. Struc. Supp.*, **5**, 203.

chaaper, R. M. and Glickman, B. W. (1982) Mutability of bacteriophage M13 by ultraviolet light: role of pyrimidine dimers. *Mol. Gen. Genet.*, **185**, 404–407.

chaaper, R. M. and Loeb, L. A. (1981) Depurination causes mutations in SOS induced cells. *Proc. Natl Acad. Sci. USA*, **78**, 1773–1777.

chaaper, R. M., Glickman, B. W. and Loeb, L. A. (1982) Role of depurination in mutagenesis by chemical carcinogens. *Cancer Res.*, **42**, 3480–3485.

chaaper, R. M., Kunkel, T. A. and Loeb, L. A. (1983) Infidelity of DNA synthesis associated with bypass of apurinic sites. *Proc. Natl Acad. Sci. USA*, **80**, 487–491.

chendel, P. F. and Robins, P. (1978) Repair of O^6-methylguanine in adapted *Escherichia coli. Proc. Natl Acad. Sci. USA*, **75**, 6017–6020.

chendel, P. F., Edington, B. V., McCarthy, J. G. and Todd, M. L. (1983) Repair of alkylation damage in *E. coli*. In *Cellular Responses to DNA Damage* (eds E. C. Friedberg and B. A. Bridges), Alan R. Liss Inc., New York, pp. 227–240.

chorpe, M., Mallick, U., Rahmsdorf, H. J. and Herrlich, P. (1984) UV-induced extracellular factors from human fibroblasts communicates the UV responses to non-irradiated cells. *Cell*, **37**, 801–868.

edgwick, S. G. and Goodwin, P. A. (1985) Differences in mutagenic and recombinational DNA repair in enterobacteria. *Proc. Natl Acad. Sci. USA*, **82**, 4172–4176.

eeberg, E., Nissen-Meyer, J. and Strike, P. (1976) Incision of ultraviolet-irradiated DNA by extracts of *E. coli* requires three different gene products. *Nature*, **263**, 524–526.

ekiya, T., Fushini, M., Hori, H., Hirohashi, S., Nishimura, S. and Sugimura, T. (1984) Molecular cloning and the total nucleotide sequence of the human c-Ha-*ras*-1 gene activated in a melanoma from a Japanese patient. *Proc. Natl Acad. Sci. USA*, **81**, 4771–4775.

elsky, C. A., Henson, P., Weichselbaum, R. R. and Little, J. B. (1979) Defective reactivation of ultraviolet light irradiated Herpes virus by a Bloom's syndrome fibroblast strain. *Cancer Res.*, **39**, 3392–3396.

etlow, R. B. (1978) Repair deficient human disorders and cancer. *Nature*, **271**, 713–717.

etlow, R. B., Regan, J. D., German, J. and Carrier, W. L. (1969) Evidence that xeroderma pigmentosum cells do not perform the first step in repair of ultraviolet damage to their DNA. *Proc. Natl Acad. Sci. USA*, **64**, 1035–1041.

harp, J. D., Capecchi, N. E. and Capecchi, M. R. (1973) Altered enzymes in drug resistant variants of mammalian tissue culture cells. *Proc. Natl Acad. Sci. USA*, **70**, 3145–3149.

hinura, Y., Ise, T., Kato, T. and Glickman, B. W. (1983) umuC Mediated misrepair mutagenesis in *Escherichia coli*: extent and specificity of SOS mutagenesis. *Mutation Res.*, **111**, 515–519.

iede, W. and Eckardt, F. (1984) Inducibility of error-prone repair in yeast? *Mutation Res.*, **129**, 3–11.

inden, R. R. and Cole, R. S. (1978) Topography and kinetics of genetic recombination in *Escherichia coli* treated with psoralen and light. *Proc. Natl Acad.*

Sci. USA, **75**, 2373–2377.

Sinzinis, B. I., Smirnov, G. B. and Saenko, A. A. (1973). Repair deficiency in *Escherichia coli* UV sensitive mutator strain *uvr502*. *Biochem. Biophys. Res. Comm.*, **53**, 309–316.

Smith, C. A. and Hanawalt, P. C. (1976) Repair replication in cultured normal and transformed human fibroblasts. *Biochim. Biophys. Acta*, **447**, 121–132.

Smith, C. A. and Hanawalt, P. C. (1978) Phage T4 endonuclease V stimulates DNA repair replication in isolated nuclei from ultraviolet-irradiated human cells, including xeroderma pigmentosum fibroblasts. *Proc. Natl Acad. Sci. USA*, **75**, 2598–2602.

Spatz, H. C. and Trautner, T. A. (1970) One way to do experiments on gene conversion? *Mol. Gen. Genet.*, **109**, 84–106.

Steinborn, G. (1978) Uvm mutants of *Escherichia coli* K12 deficient in UV mutagenesis. I. Isolation of uvm mutants and their phenotypical characterization of DNA repair and mutagenesis. *Mol. Gen. Genet.*, **165**, 87–93.

Stockdale, F. E. (1971) DNA synthesis in differentiating skeletal muscle cells: Initiation by ultraviolet light. *Science*, **171**, 1145–1147.

Strauss, B., Rabkin, S., Sagher, D. and Moore, P. (1982) The role of DNA polymerase in base substitution mutagenesis on non-instructional templates. *Biochimie*, **64**, 829–838.

Streisinger, G., Okada, Y., Emrich, J., Newton, J., Tsugita, A., Terzaghi, E. and Inouye, M. (1966) Frameshift mutations and the genetic code. *Cold Spring Harbor Symp. Quant. Biol.*, **31**, 77–84.

Sukumar, S., Notario, V., Martin-Zanca, D. and Barbacid, M. (1983) Induction of mammary carcinomas in rats by nitrosomethylurea involves malignant activation of H-*ras*-1 locus by single point mutations. *Nature*, **306**, 658–661.

Sutherland, B. M. (1974) Photoreactivating enzyme from human leukocytes. *Nature*, **248**, 109–112.

Sutherland, B. M. (1981) Photoreactivating enzymes. In *The Enzymes* (ed. P. D. Boyer), Academic Press, New York, pp. 482–515.

Sutherland, B. M., Rice, M. and Wagner, E. K. (1975) Xeroderma pigmentosum cells contain low levels of photoreactivating enzyme. *Proc. Natl Acad. Sci. USA*, **72**, 103–107.

Swift, S., Sholman, L., Perry, M. and Chase, C. (1976) Malignant neoplasms in the families of patients with Ataxia telangiectasia. *Cancer Res.*, **36**, 209–215.

Szostak, J. W., Orr-Weaver, T. L. and Rothstein, R. J. (1983) The double-strand-break repair model for recombination. *Cell*, **33**, 25–35.

Tanaka, K., Sekiguchi, M. and Okada, Y. (1975) Restoration of ultraviolet-induced unscheduled DNA synthesis of xeroderma pigmentosum cells by the concomitant treatment with bacteriophage T4 endonuclease V and HVJ (Sendai virus). *Proc. Natl Acad. Sci. USA*, **72**, 4071–4075.

Taylor, A. M. R., Harnden, D. G., Arlett, C. F., Harcourt, S. A. and Lehmann, A. R. (1975) Ataxia telangiectasia: a human mutation with abnormal radiation sensitivity. *Nature*, **258**, 427–429.

Teo, I., Sedgwick, B., Demple, B., Li, B. and Lindahl, T. (1984) Induction of resistance to alkylating agents in *E. coli*: the ada^{+} gene product serves both as a regulatory protein and as an enzyme for repair of mutagenic damage. *EMBO J.*, **3**,

2151–2157.

Thacker, J., Debenham, P. G., Stretch, A. and Webb, M. B. T. (1983) The use of a cloned bacterial gene to study mutation in mammalian cells. *Mutation Res.*, **111**, 9–23.

Thompson, K. V. A. and Holliday, R. (1983) Genetic effects on the longevity of cultured human fibroblasts II. DNA repair deficient syndromes. *Gerontology*, **29**, 83–88.

Thompson, L. H. and Carrano, A. V. (1983) Analysis of mammalian cell mutagenesis and DNA repair using *in vitro* selected CHO cell mutants. In *Cellular Responses to DNA Damage* (eds E. C. Friedberg and B. A. Bridges), Alan R. Liss Inc., New York, pp. 125–143.

Tindall, K. R., Stankowski, L. F., Machanoff, R. and Hsie, A. W. (1984) Detection of deletion mutations in pSV2*gpt*-transformed cells. *Mol. Cell. Biol.*, **4**, 1411–1415.

Todd, P. A. and Glickman, B. W. (1982) Mutational specificity of UV light in *Escherichia coli*: Indications for a role of DNA secondary structure. *Proc. Natl Acad. Sci. USA*, **79**, 4123–4127.

Toman, Z., Dambly, C. and Radman, M. (1980) Induction of a stable heritable epigenetic change by mutagenic carcinogens – a new test system. *Mutation Res.*, **74**, 242–243.

Treton, J. A. and Curtois, Y. (1982) Correlation between DNA excision repair and mammalian lifespan in lens epithelial cells. *Cell Biol. Int. Reports*, **6**, 253–260.

Tye, B. K., Chien, J., Lehman, I. R., Duncan, B. K. and Warner, H. R. (1978) Uracil incorporation: A source of pulse-labelled DNA fragments in the replication of the *Escherichia coli* chromosome. *Proc. Natl Acad. Sci. USA*, **75**, 233–237.

Unrau, P. (1975) The excision of pyrimidine dimers from the DNA of mutant and wild-type strains of *Ustilago*. *Mutation Res.*, **29**, 53–65.

Unrau, P., Wheatcroft, R. and Cox, B. S. (1971) Excision of pyrimidine dimers from DNA of ultraviolet irradiated yeast. *Mol. Gen. Genet.*, **113**, 359–362.

Upcroft, P., Carter, B. and Kidson, C. (1980) Analysis of recombination in mammalian cells using SV40 genome segments having homologous overlapping termini. *Nucl. Acids Res.*, **8**, 2725–2736.

Van Sluis, C. A., Mattern, I. E. and Patterson, M. C. (1974) Properties of *uvrE* mutants of *Escherichia coli* K12 I. Effects of UV irradiation on DNA metabolism. *Mutation Res.*, **25**, 273–279.

Villani, G., Boiteux, S. and Radman, M. (1978) Mechanism of ultraviolet-induced mutagenesis: Extent and fidelity of *in vitro* DNA synthesis on irradiated templates. *Proc. Natl Acad. Sci. USA*, **75**, 3037–3041.

Volkert, M. R., George, D. L. and Witkin, E. M. (1976) Partial suppression of the LexA phenotype by mutations (*rnm*) which restore ultraviolet resistance but not ultraviolet mutability to *Escherichia coli* B/r *uvrA lexA*. *Mutation Res.*, **36**, 17–28.

Volkert, M. R. and Nguyen, D. C. (1984) Induction of specific *Escherichia coli* genes by sublethal treatments with alkylating agents. *Proc. Natl Acad. Sci. USA*, **81**, 4110–4114.

Wade, M. H. and Chu, E. H. Y. (1979) Effects of DNA damaging agents on cultured fibroblasts derived from patients with Cockayne's syndrome. *Mutation Res.*, **59**, 49–60.

Wagner, R. and Meselson, M. (1976) Repair tracts in mismatched DNA heteroduplexes. *Proc. Natl Acad. Sci. USA*, **73**, 4135–4139.

Wahl, G. M., Hughes, S. H. and Capecchi, I. R. (1974) Immunological characterization of hypoxanthine-guanine phosphoribosyl transferase mutants of mouse L cells. Evidence for mutations at different loci in the HGPRT gene. *J. Cell. Physiol.*, **85**, 307–320.

Waldstein, E. A., Cao, E. H. and Setlow, R. B. (1982) Adaptive resynthesis of O^6-methylguanine-accepting protein can explain the differences between mammalian cells proficient and deficient in methyl excision repair. *Proc. Natl Acad. Sci. USA*, **79**, 5117–5121.

Walker, G. C. (1984) Mutagenesis and inducible responses to deoxyribonucleic acid damage in *Escherichia coli*. *Microbiol. Rev.*, **48**, 60–93.

Weigle, J. J. (1953) Induction of mutation in a bacterial virus. *Proc. Natl Acad. Sci. USA*, **39**, 628–636.

West, S. C. and Howard-Flanders, P. (1984) Duplex-duplex interactions catalysed by RecA protein allow strand exchanges to pass double-strand breaks in DNA. *Cell*, **37**, 683–691.

West, S. C., Countryman, J. K. and Howard-Flanders, P. (1983) Enzymatic formation of biparental figure eight molecules from plasmid DNA and their resolution in *E. coli*. *Cell*, **32**, 817–829.

White, R. and Fox, M. S. (1974) On the molecular basis of high negative interference. *Proc. Natl Acad. Sci. USA*, **71**, 1544–1548.

Wilcox, D. R. and Prakash, L. (1981) Incision and postincision steps of pyrimidine dimer removal in excision-defective mutants of *Saccharomyces cerevisiae*. *J. Bacteriol.*, **148**, 618–623.

Wildenberg, J. and Meselson, M. (1976) Mismatch repair in heteroduplex DNA. *Proc. Natl Acad. Sci. USA*, **72**, 2202–2206.

Wilson, J. H. (1977) Genetic analysis of host range mutant viruses suggests an uncoating defect in simian virus 40-resistant monkey cells. *Proc. Natl Acad. Sci. USA*, **74**, 3503–3507.

Wilson, V. L. and Jones, P. A. (1983) Inhibition of DNA methylation by chemical carcinogens *in vitro*. *Cell*, **32**, 239–246.

Witkin, E. M. (1967) Mutation proof and mutation prone modes of survival in derivatives of *Escherichia coli* B differing in sensitivity to ultraviolet light. *Brookhaven Symp. Biol.*, **20**, 17–55.

Witkin, E. M. (1976) Ultraviolet mutagenesis and inducible DNA repair in *Escherichia coli*. *Bacteriol. Rev.*, **40**, 869–907.

Witkin, E. M. and George, D. L. (1973) Ultraviolet mutagenesis in *polA* and *uvrA polA* derivatives of *Escherichia coli* B/r: Evidence for an inducible error-prone repair system. *Genetics Suppl.*, **73**, 91–108.

Witkin, E. M. and Kogoma, T. (1984) Involvement of the activated form of RecA protein in SOS mutagenesis and stable replication in *Escherichia coli*. *Proc. Natl Acad. Sci. USA*, **81**, 7539–7543.

Wood, R. D. and Hutchinson, F. (1984) Non-targeted mutagenesis of unirradiated lambda phage in *Escherichia coli* host cells irradiated with ultraviolet light. *J. Mol. Biol.*, **173**, 293–305.

Wood, R. D., Skopek, T. R. and Hutchinson, F. (1984) Changes in DNA base sequence induced by targeted mutagenesis of lambda phage by ultraviolet light. *J.*

Mol. Biol., **173**, 273–291.

Woodhead, A. D., Setlow, R. B. and Grist, E. (1980) DNA repair and longevity in three species of cold-blooded vertebrates. *Exp. Gerontol.*, **15**, 301–304.

Wright, S. J. L. and Hill, E. C. (1968) The development of radiation-resistant cultures of *Escherichia coli* by a process of 'growth-irradiation cycles'. *J. Gen. Microbiol.*, **51**, 97–106.

Yamamoto, K., Satake, M. and Shinagawa, H. (1984) A multicopy *phr*-plasmid increases the ultraviolet resistance of a *recA* strain of *Escherichia coli*. *Mutation Res.*, **131**, 11–18.

Yarosh, D. B., Foote, R. S., Mitra, S. and Day, R. S. (1983) Repair of O^6-methylguanine in DNA by demethylation is lacking in mer⁻ human-tumour cell strains. *Carcinogenesis*, **4**, 199–205.

Young, C. S. H. and Fisher, P. B. (1980) Adenovirus recombination in normal and repair-deficient human fibroblasts. *Virology*, **100**, 179–184.

Youngs, D. A. and Smith, K. C. (1977) The involvement of polynucleotide ligase in the repair of UV induced DNA damage in *Escherichia coli* K-12 cells. *Mol. Gen. Genet.*, **152**, 37–41.

Zurcher, C., van Zwieten, M. J., Solleveld, H. A. and Hollander, C. F. (1982) Ageing research. In *The Mouse in Biomedical Research, IV* (eds H. L. Foster, J. D. Small and J. G. Fox), Academic Press, New York, pp. 11–35.

10 Kinetic and probabilistic thinking in accuracy

J. NINIO

10.1 Introduction

The cell contains thousands of molecular species in permanent motion. The nucleic acids breathe. The proteins fold and unfold. Groups of atoms swing or rotate. Charges jump from one place to the other in the molecule. The environment threatens the cell in many ways. Yet the cell is able to reproduce itself accurately and to remain essentially unchanged throughout millions of generations. When DNA, the key molecule of heredity, is manufactured by the assembly of millions or billions of its elementary subunits, there may be less than one error in ten billion per added subunit. What are then the strategies used by the cell to achieve such a high degree of precision with rather unreliable components?

Traditional textbooks simply ignore the problem. Specificity in molecular interactions is presented as a matter of geometric complementarity between the shapes of the interacting molecules. If an enzyme catalyses the reactions between the molecules A and B to form a product $A-B$, then it is generally pictured as a kind of ball with two cavities, the shapes of which match exactly the shapes of the substrates A and B. 'Locks and keys', 'matrices and moulds' were the common metaphors to explain how an enzyme recognizes its substrates. The same picture applied to antibody–antigen recognition. When a healthy mammal is invaded by a foreign molecule, it is able to design and manufacture within a few days (so it was thought) antibodies that bind very specifically to the intruder and distinguish it from all other molecules of the self. If Nature can do this in a few days for antibodies, why bother about enzymic systems that have been refined over hundreds of millions of years?

In the beginning of the 1970s, paradoxical observations on the specificity of particular processes started to accumulate in several areas of molecular biology. The way specificity was altered in various situations could not be explained by lock and key theories (reviewed in Ninio and Chapeville, 1980). A new framework for understanding specificity was thus developed: the kinetic theory of accuracy. In parallel, the immunologists had to admit the fundamental looseness of antibody specificity. Despite the commercial interests vested in the theory of the high specificity of monoclonal antibodies, these pure molecular species were found to be able to bind many substrates in addition to the antigens against which they were raised. Actually, a good deal of the complexity of the immune system as we understand it today comes from the need to circumvent the problem of the high degeneracy of antibody–antigen interactions.

Long ago, Gestalt psychologists made experiments in which they trained chickens to take the food presented on dark-grey paper, and reject the food presented simultaneously on medium-grey. Afterwards, when the chickens had to choose between food on medium-grey and light-grey paper, they selected the former. Thus, the animals did not recognize in an absolute manner a particular shade of grey, but had developed a criterion for choosing between two competing offers. Similarly, an enzyme does not recognize its substrate in an absolute manner, but has a strategy for utilizing the normal substrate rather than a number of other related molecules present in the cell.

For most biologists, the recognition of a substrate by an enzyme and its further processing to transform it into product are two successive disconnected steps in enzyme action. First the enzyme looks around to select its complementary substrate. If it is not abundantly present, the enzyme eventually accepts as a second choice a non-complementary ligand. Having selected the substrate it then goes on to perform its catalytic function. Contrary to this view, the kinetic theory of accuracy teaches that the events that follow the binding of the substrate are essential to enzyme discrimination. The recognition strategy is in part written down in the reaction scheme of the enzyme. To understand this, it is not enough to look at reaction schemes and compare them. Actually the reaction mechanisms as they are written today, with their collections of arrows connecting discrete enzyme states, are a symbolic representation of physical reality. This symbolism is highly practical for the automatic writing of rate equations, but it conceals aspects of the reactions that are crucial for the accuracy theory. A great deal of confusion in the field has arisen from attempts to discuss enzyme discrimination directly from the symbolic reaction schemes, and not from the underlying physical processes.

Thus, I shall begin in the next section by making explicit a number of physical principles behind the current enzymological formalism. Then I shall discuss three well-studied kinetic effects: (1) the sequestration effect that occurs when the substrates are depleted by their interaction with the enzyme, (2) kinetic modulation which is the way a non-specific change at a point in a reaction can predictably alter the overall specificity of the process, and (3) the hypothetical mechanisms of kinetic amplification of specificity. After giving practical recipes for calculating the specificity of a reaction scheme, I shall outline some of the directions in which the field may move in the future.

10.2 Hidden principles behind the kinetic formalism

10.2.1 SOME BASIC ASSUMPTIONS

Take a DNA polymerase that moves along a template at a rate of about 5000 nucleotides per second. Since two base-pairs are separated by about 3 Å, one may say that the rate of progression of the enzymes is $1.5 \mu m \, s^{-1}$. Suppose we want to describe the motion of the enzyme by a chemical equation A $\rightarrow$ B and a single kinetic parameter k for the rate of conversion of A to B, where A represents the polymerase bound to the template at a certain position, and B the polymerase bound to the next position (nucleotide) on the template. In all rigour this cannot be done, for in chemical kinetics a rate is never the velocity of a motion from one place to another. A rate is always a frequency of conversion from one discrete state to another discrete state. The prototype of an elementary chemical step is radioactive decay, such as $^{14}C \rightarrow {}^{14}N$. If k is the rate constant for this reaction, then during any short interval of time dt, there will be a probability kdt for any ^{14}C atom to be converted into ^{14}N. Several aspects of this statement must be noted:

(1) The process is assumed to be probabilistic.
(2) The duration of the rearrangement that leads to a change of state does not matter. The kinetic constant k just describes the probability (or, if you prefer, the frequency) of the transition.
(3) The law of probabilities that applies to the process is absolutely determined. It is taken that there is a constant probability per unit time for an atom in state A to switch to state B. Thus, even if the process is very slow (we should say instead very infrequent), a given atom initially in state A has the usual kdt probability of being changed into B during the interval from time 0 to time dt.

These three assumptions apply to the whole field of chemical kinetics. When writing down a kinetic scheme, every single arrow connecting two states in the scheme is assumed to represent a process obeying the three constraints enumerated above. Otherwise, the rate equations that one computes by routine application of the mathematical techniques of enzymology (e.g. the Briggs–Haldane steady-state treatment) would all be wrong. Of course, one is allowed to postulate that a certain step in a reaction does not follow the same probabilistic behaviour as that of radioactive decay. But then the step cannot be represented with a single arrow in the scheme. It might necessitate a rather complex description, involving many 'intermediate' states and a whole network of arrows to connect them, or it may just be impossible to represent the step in this manner. Consider again the motion of the DNA polymerase from one nucleotide to the next. Since any elementary step in a kinetic scheme obeys a law of uniform probability, any event represented by one arrow may occur (although with a low probability) in as short an interval of time as one wishes. Thus, even if there are many steps in the reaction leading from substrate to product, some molecules of product may be generated in an infinitely small amount of time. If the chemical formalism applies to DNA replication, a whole chromosome may be replicated faster than the speed of light, even with the slowest of all polymerases!

One may object that, after all, we are not interested in detailed probabilistic distributions, and are satisfied with a formalism that predicts correctly the average time taken by each elementary step in a reaction. The objection may hold in usual enzymological work, but is invalid in the field of accuracy, for the kinetic theory of accuracy shows that error rates cannot be predicted from average rates: the probability distribution of reaction times is of utmost importance.

Another example will show the distance between kinetic schemes and physical reality. Take a very simple enzyme, E_1, that binds four substrates, A, B, C and D, as they come to make a product P. Take a second enzyme, E_2, that obeys rather stringent specifications. It cannot bind B, C or D unless A is already there. In order to bind C, both A and B must be already on the enzyme, and the presence of all these three substrates is necessary for the binding of D. In the notations of chemical kinetics, the mechanism of E_2 is a very simple sequential one: $E \rightleftharpoons EA \rightleftharpoons EAB \rightleftharpoons EABC \rightleftharpoons EABCD \rightleftharpoons P$. On the other hand, E_1 would appear to have a very complicated reaction mechanism, since it has no less than sixteen states, and we need thirty-two double arrows to connect them. Thus, counting the number of steps in a reaction is an artificial operation that relates to the properties of the kinetic formalism, but not necessarily to the real physics of the situation. To discuss accuracy, we must first analyse the physical aspects, then transcribe the schemes in chemical notation.

10.2.2 TIMES, PROBABILITIES, AND KINETIC CONSTANTS

The remainder of this section is devoted to a more quantitative discussion of some of the consequences of the kinetic formalism. Take an elementary reaction $A \rightarrow B$ with an elementary kinetic constant ($=$ frequency of transition) k. If we take at time 0 a population of molecules of A at concentration $[A]$ what will be the concentration at a later time? Since the probability of an A molecule being converted to B during an interval of time dt is kdt, the average behaviour of the population is described by the law of exponential decay:

$$\frac{d[A]}{dt} = -kdt$$

thus
$$[A]_t = [A]_0 e^{-kt} \tag{10.1}$$

How long does it take, on average, to change A into B? If we start with a population of A molecules at time 0 and record, for each molecule, the time at which it was converted into B and then take the average of all such times, we find $t = 1/k$. The average transition time is just the reciprocal of the kinetic constant for the transition. This statement is true for an exponentially decaying population. It would not be true of events occurring at the same frequency, but with a different probability law. If you go to a bus station at a time chosen at random, and know that the buses arrive at a frequency k of ten per hour, how long will you wait on average? If the buses come regularly spaced in time, at intervals of 6 min, then your average waiting time will be 3 min, hence $1/2k$. However, if the buses reach the station with the same probability law as for the docking of a substrate to an enzyme, then to the same frequency of ten per hour there will correspond an average waiting time of $1/k = 6$ min.

Consider instead a chemical example. We have a two-step reaction $A \rightarrow B \rightarrow C$ with rate constants k_1 and k_2 for the first and second step. The average transition time from A to B is $t_1 = 1/k_1$ and from B to C it is $t_2 = 1/k_2$ so that the average transition time from A to C is $1/k_1 + 1/k_2$. Now, if we contract the reaction scheme into $A \rightarrow C$ what kinetic constant must be attached to the reaction? The answer is k_1! There is a frequency k_1 for switching from A to B, and all molecules of B are ultimately converted into C so that, under steady-state conditions, $k_1[A]dt$ molecules of C will be produced in an interval of time dt, whatever the value of k_2. For a compound step, the average conversion time is not the reciprocal of the rate constant. Furthermore, the probability distribution of conversion times is no longer given by the exponential decay curve. If a transformation occurs in n steps of comparable frequencies then, as n increases, the distribution of transformation times becomes more and more gaussian (Fig. 10.1).

We turn now our attention to a different question. Consider a molecule A which can be transformed either into A_1 with a rate k_1, or into A_2 with a rate k_2:

$$A \begin{array}{c} \overset{k_1}{\nearrow} A_1 \\ \underset{k_2}{\searrow} A_2 \end{array} \qquad (10.2)$$

How will A_1 and A_2 turn up in time, and in what ratio? The answer is simple: for every short interval of time dt, $k_1[A]dt$ molecules of A_1 and $k_2[A]dt$ molecules of A_2 are formed on average. The population of A molecules declines as though there was a single transformation of frequency $k_1 + k_2$, the average transformation time being $1/(k_1 + k_2)$. All the time, A_1 and A_2 are produced in proportional amounts, with a concentration ratio k_1/k_2.

Now let us take one molecule of A and follow its fate. At a certain time it will be converted into either A_1 or A_2. What is the probability that it has been converted into A_1? The answer is $p = k_1/(k_1 + k_2)$. The formula can be

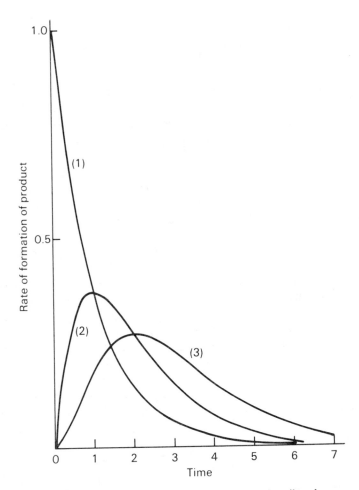

Figure 10.1 The curve labelled (1), an exponential decay, describes the conversion of a substrate into product, in the case of a one-step reaction $S \rightarrow P$, with an average conversion time of one unit. The rate of formation of P is maximum at time zero, then decreases steadily. For two or more steps, the curves are qualitatively different (curves labelled (2) and (3) for two-step and three-step reactions, respectively). Such qualitative differences are of paramount importance, as accuracy theory shows that specificity depends not only upon the average transformation times, but also, and very critically, upon the distribution of such times.

generalized to any number of transformation modes. If A can be converted into $A_1, A_2, \ldots A_n$ by single step reactions with frequencies $k_1, k_2, \ldots k_n$ then the average conversion time is $t = 1/(k_1 + k_2 + \ldots + k_n)$ and the probability of forming A_1 is $p = k_1 t$. The formula for p is still valid if the state A is anywhere within a complex reaction scheme and $A_1, A_2, \ldots A_n$ are some of the states into which A can be directly converted.

10.2.3 RATES FROM TIMES AND PROBABILITIES

Let us now make the link between times and probabilities on one hand, and some more familiar but less physical entities such as k_{cat} and K_m. Consider an enzyme E which has only one state to which the substrate S may bind. Let us call t_w the average time that the free enzyme waits until a molecule of substrate binds to its active site. Let k_1 be the rate constant relative to the binding step $E + S \rightarrow ES$. The waiting time t_w is just $1/k_1[S]$. Once the complex ES is formed, there are two possibilities: (1) the reaction goes to completion and a product P is formed (we assume that there is no backreaction), or (2) the reaction aborts, S is released and E is free again. We do not put any additional constraint on the mechanism of the reaction, it may have an unlimited number of steps, cycles, branch points. The substrate may leave at a number of different places in the scheme. The only stringent requirements here are that there is only one entry point for the substrate, and no back flow of the product. Then we can associate two quantities to the reaction, which are totally independent of enzyme, substrate or product concentration. First, there is an average reaction time t_R. Here, it is defined as the average time that elapses between the instant the enzyme–substrate complex forms and the instant when the enzyme is free again and available for substrate binding. Second, a probability p that the reaction goes to product, rather than aborting.

Let us take one molecule of enzyme and follow its fate as time passes. It is alternately free, waiting for the substrate (for an average duration t_w), and processing the substrate (average time t_R). One cycle then takes an average time $t_w + t_R$ (van Slyke and Cullen, 1914). But since there is a probability p, not necessarily equal to one, that the cycle is productive, the average time taken by the enzyme to transform a molecule of substrate into product is $1/p$ times the average duration of a cycle:

$$\theta = (t_w + t_R)/p. \tag{10.3}$$

The speed of the reaction, per molecule of enzyme, is $1/\theta$ so that at a total concentration $[E]$ of enzyme the rate of formation of the product is:

$$v = \frac{[E]}{\theta} = \frac{[E]p}{t_w + t_R} = \frac{[E]p}{(1/k_1[S]) + t_R}. \tag{10.4}$$

The identification of this formula with the Michaelis–Menten* equation gives $k_{cat} = p/t_R$ and $K_m = 1/(k_1 t_R)$.

We have shown here, effortlessly, that any enzyme mechanism that presents only one entry site for the substrate and excludes product back-flow, however complex it may be otherwise, will end up by giving Michaelis–Menten kinetics. This shows the power of probabilistic reasoning in enzymology. This power will become even more obvious when we need to compute error rates.

10.3 The sequestration effect

There are many situations in the cell where an enzyme or a complex R has to choose between two ligands or sites A and B, and where it is found that the specificity of the choice, at constant A/B ratio, depends upon the global concentration of A and B, relative to that of R. Where, in practice, an RNA polymerase initiates transcription depends upon how the molecules of enzyme are distributed between weak and strong promoters, and this distribution varies with the RNA polymerase concentration. Similarly, the relative frequencies of translation of various mRNAs change with ribosome concentration. The effect was first observed and understood in the context of the choice between two competing tRNAs during the translation of a particular codon (Pestka, Marshall and Nirenberg, 1965; Grunberg-Manago and Dondon, 1965). It has a serious incidence on modern *in vitro* assays of protein synthesis (Jelenc and Kurland, 1979) and appropriate corrections must be made in the interpretation of the results (Carrier and Buckingham, 1984).

When poly(U) is translated in the presence of a full assortment of tRNAs and one measures isoleucine and phenylalanine incorporation into polypeptide, the error rate (Ile/Phe) increases as the global concentration of tRNA is decreased (Pestka, Marshall and Nirenberg, 1965; Grunberg-Manago and Dondon, 1965). On first inspection, if for a concentration $[A]$ of available Phe-tRNA^{Phe} and $[B]$ of available Leu-tRNA^{Leu}, the error rate is α on poly(U) programmed ribosomes, then the error rate should remain the same if we multiply or divide both concentrations of tRNA by the same amount. More generally, if we multiply $[A]$ by m and $[B]$ by n, we expect to find an error rate of $\alpha n/m$. Now, the crucial point, clearly acknowledged by Pestka, Marshall and Nirenberg is that the ratio A/B of the available tRNAs

---

*Editors' note: The author originally used the designation 'Henri–Michaelis' for this equation and its associated scheme of kinetics. Whilst we agree that the seminal contribution of Henri to the development of enzyme kinetics (see Segal, H. L. (1959) In *The Enzymes*, Vol. 1, 2nd edn (eds P. D. Boyer, H. Lardy and K. Myrbach), Academic Press, New York, pp. 1–48) is too often overlooked, we have substituted the more usual designation 'Michaelis–Menten' here and elsewhere in the chapter to be consistent throughout the book.

may not be equal to the ratio of the originally acylated tRNAs. If A is consumed more rapidly than B, the ratio of the remaining species B/A will increase constantly, resulting in an increase of errors. This effect is somewhat trivial and should disappear if the tRNAs are properly reacylated immediately after rejection from the ribosome. But even then, a residual effect is predicted at very low concentrations of substrates (in practice, when the concentration of tRNA becomes comparable to that of active ribosomes). For then, the amounts of tRNAs processed by the ribosomes are not negligible compared to the amounts of total available tRNAs. Hence, the ratio A/B of the available tRNA species may be different from the input A/B ratio. This is the sequestration effect. Since low concentrations of substrates result in low rates of polymerization, we may describe the speed and accuracy aspects of the sequestration effects as follows:

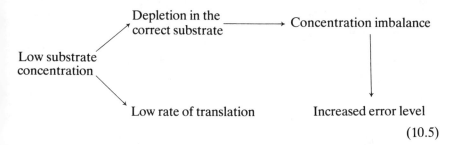

$$(10.5)$$

Assume now that we have poly$(A-U-A)_n$ and poly(U) programmed ribosomes in the same test tube. Both A and B are now appreciably sequestrated and the concentration imbalance may even be cancelled. Thus, for appropriate concentrations of the two mRNAs, and a low concentration of total tRNA, the error level may be as low as that obtained for high concentrations of tRNA.

If we transpose the above discussion to the case of two competing tRNA species A and B and two corresponding aminoacyl-tRNA ligases E_A and E_B, we expect the following:

(1) High A and B in the presence of E_A $\rightarrow$ normal error rate
(2) High A and B in the presence of E_A and E_B $\rightarrow$ normal error rate
(3) Low A and B in the presence of E_A $\rightarrow$ high error rate (sequestration effect)
(4) Low A and B in the presence of E_A and E_B $\rightarrow$ error rate comparable to (1) and (2) (sequestration effect counteracted)

This situation was first analysed by Yarus (1972a) who reached at that time different conclusions.

10.4 Kinetic modulation

10.4.1 DISCRIMINATION IN A MICHAELIS–MENTEN SCHEME

Consider a reaction scheme of the kind defined above, namely, the enzyme has only one state for substrate admission, there is no backflow of the product, but the mechanism may be otherwise of unlimited complexity. As we have seen (Equation (10.4)), the kinetics are of the Michaelis–Menten type. Assume the enzyme acts on either of two substrates A and B. There is a kinetic constant k^A for the association of A with the enzyme and after binding, a probability p^A that a molecule of product derived from A be formed. Similarly, the reaction with B is characterized by the parameters k^B and p^B.

Let us follow the fate, over a long period of time, of one enzyme molecule which is free at time 0, when both A and B are present. If we just count the binding events, we find that EA and EB are formed proportionately to $k^A[A]$ and $k^B[B]$. This does not mean that the concentrations of EA and EB are proportional to these expressions, for EA and EB, once they are formed, may decay at different rates. Once a molecule of substrate A or B is bound, there is a certain probability p^A or p^B that a product will be formed. The products derived from A and B will thus be formed in the ratio:

$$\frac{v^A}{v^B} = \frac{k_1^A[A]}{k_1^B[B]} \times \frac{p^A}{p^B} \tag{10.6}$$

Since the concentrations of A and B intervene in a trivial manner, it is convenient to define the 'discrimination' of the enzyme as the ratio of the products formed from A or B divided by the ratio of the concentrations of the two substrates:

$$D = \frac{v^A}{v^B} \bigg/ \frac{[A]}{[B]} = \frac{k_1^A p^A}{k_1^B p^B} \tag{10.7}$$

The conditions of validity of this expression (Ninio, 1974, 1975b) were given at the beginning of this section.

Orthodox enzymologists (such as Dixon and Webb, 1958; Fersht, 1977), solving the steady-state equations for the strict (i.e. with only one ES and no EP state) Michaelis–Menten scheme, obtain.

$$\frac{v^A}{v^B} = \frac{[A]}{[B]} \times \frac{k_{cat}^A/K_m^A}{k_{cat}^B/K_m^B} \tag{10.8}$$

Others will prefer to apply the classical formula for competitive inhibition and rewrite (Equation (10.8)) in an ornamented form, thus:

$$\frac{v^A}{v^B} = \frac{v_{max}^A}{v_{max}^B} \times \frac{[S^A]}{[S^B]} \times \frac{K_m^B\{1 + ([S^A]/K_m^A)\} + [S^B]}{K_m^A\{1 + ([S^B]/K_m^B)\} + [S^A]} \tag{10.9}$$

which is an authentic example taken from the biochemical literature.

Remembering (Section 10.2.3) that $k_{cat} = p/t_R$ and $K_m = 1/(k_1 t_R)$ makes the equivalence between Equation (10.6) and Equation (10.8) obvious. Still, Equation (10.6) is definitely preferable for two reasons. First, it is more transparent and follows directly from primary physical principles, without the need to solve sets of equations and rearrange algebraic expressions. In contrast, Equation (10.8) is obtained by solving steady-state equations explicitly and rearranging the results. Furthermore, the k_{cat}s and K_ms in Equation (10.8) are taken as experimental rather than theoretical constants and have little physical meaning. Second, when complex schemes are considered, the k_{cat}s and K_ms become extremely complicated expressions; in fact, most of the complexity comes from t_R. Practically, when one divides k_{cat} by K_m most of the terms cancel away, and one is left with the terms that made up p. Computing p directly is very easy (see the recipes in Section 10.6). To compute the relevant probabilities in a complex scheme, one can break the scheme into subschemes, analyse the parts separately, and combine them to obtain the final result (Section 10.6). There is no known method for achieving a similar breakdown of complexity with k_{cat}s amd K_ms.

10.4.2 MODULATION OF SPECIFICITY BY NON-SPECIFIC FACTORS

For a strict Michaelis–Menten scheme, when the enzyme–substrate complex is formed, it can either go to product (rate constant k_2) or dissociate into free enzyme and substrate (rate constant k_{-1}) so that the probability p of product formation is just $k_2/(k_2 + k_{-1})$. Then,

$$\frac{v^A}{v^B} = \frac{k_1^A[S^A]}{k_1^B[S^B]} \frac{k_2^A/(k_2^A + k_{-1}^A)}{k_2^B/(k_2^B + k_{-1}^B)} \tag{10.10}$$

Let us now examine that part of the discrimination attributable to p^A/p^B and take two numerical examples.

Example 1 $k_{-1}^A = 1$ $k_{-1}^B = 10$ $k_2^A = k_2^B = 1000$

Example 2 $k_{-1}^A = 1$ $k_{-1}^B = 10$ $k_2^A = k_2^B = \quad 1$

We have assumed here that after the formation of the enzyme–substrate complex, the dissociation of B is ten times as fast as the dissociation of A. In Example 1, we have taken a rather high rate constant for the $ES \to P$ reaction. Whether A or B is on the enzyme, there will be a high probability of product formation. In Example 2, the EA complex has an equal chance of dissociating or going to product, but most often, the complex EB will abort, since the dissociation of EB is faster than the $EB \to P$ reaction. More precisely, using the formula $p = k_2/(k_2 + k_{-1})$, we find:

Example 1 $p^A = 0.999$ $p^B = 0.99$ $p^A/p^B = 1$

Example 2 $p^A = 0.5$ $p^B = 0.09$ $p^A/p^B > 5$

If we take a very small k_2, p^A/p^B will increase further towards the limit $k^B_{-1}/k^A_{-1} = 10$.

This, in its simplest form, is the kinetic modulation effect. The enzyme has a certain potential for discriminating between A and B by virtue of the different kinetic constants k^A_{-1} and k^B_{-1}. This potential can be expressed more or less, depending upon the value of some non-specific kinetic constant (k_2) of the reaction scheme. By changing this critical non-specific constant, the overall discrimination of the enzyme is altered. The same effect is obtained if, instead of multiplying the k_2s by a certain amount, we divide the k_{-1}s by the same amount. Speed (in the catalytic act) or strength (in binding) go together with low discrimination.

10.4.3 KINETIC MODULATION IN TRANSLATION

We have a ribosome translating an mRNA at a given codon. There are two competing acylated tRNAs A and B. A is a strong binder, and sticks well to the codon. Let us call θ^A its sticking time, the average time it remains bound if no peptide bond can be formed. B is a weak binder, with a short sticking time θ^B. Let us call τ the average time the ribosome takes to make a peptide bond. If τ is very small compared to the sticking times (i.e. peptide bond formation is fast), the tRNA A or B nearly always sticks long enough to the ribosome for elongation to occur. If τ is large compared to the sticking times the tRNAs often fall off, but the longer a tRNA stays associated to the codon, the higher the chances of the peptide chain being added to it. In the first case, the ribosome does not discriminate between A and B. In the second, it discriminates well, at the expense of being less efficient (Ninio, 1973, 1974; Schwartz and Lysikov, 1974). We can easily identify θ^A and θ^B of this description with $1/k^A_{-1}$ and $1/k^B_{-1}$ of the previous treatment, and τ with $1/k_2$.

A low θ/τ (or a low k_2/k_{-1}) implies a high proportion of abortive binding and therefore a reduced rate of protein synthesis:

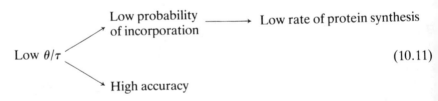

$$(10.11)$$

The effect on synthesis rates should be the most visible at the lowest concentrations of tRNA.

The kinetic ideas were unnatural in the field of protein synthesis for several reasons:

(1) The notion was that the ribosome acted through a series of disconnected steps. First it recognized, then it cleaved GTP, later it made the peptide

bond, etc. Countless studies with inhibitors of protein synthesis contributed strongly to this misconception. There could be no room for kinetic modulation which is the interplay between kinetic constants acting at different places in a reaction scheme. Spirin and his collaborators dislodged this view of ribosomology by showing that the events at a 'late' step could affect the outcome of a selection process assumed to have been completed much earlier (review in Spirin, 1978).

(2) That weak binding could be correlated with good discrimination was counter-intuitive. Even in recent work, it is often stated as being obvious that strong codon–anticodon interactions are required for a high accuracy in translation.

(3) The natural identification of complementary pairing with good binding, and non-complementary pairing with poor binding. Thus, a nonsense suppressor tRNA with an anticodon complementary to a termination codon was assumed to bind well to ribosomes and to behave, in various circumstances, more like a normal, cognate tRNA than a noncognate one. The ribosomal screen hypothesis (Gorini, 1971) was an attempt to explain the observation that nonsense suppression increased or decreased in correlation with protein synthesis errors, since just the reverse was expected on the basis of the complementarity concept. With the kinetic ideas we only need to assume that nonsense suppressors are weak binders and we can, by the way, estimate their binding constants as compared to those of normal tRNAs (Ninio, 1974).

(4) The most disturbing implication of the kinetic model was perhaps in the notion that after binding (i.e. recognition) the tRNA could dissociate from the ribosome, even if it was the correct cognate tRNA. It is now known that sometimes in *E. coli* the peptidyl-tRNA, carrying a whole polypeptide chain, falls from the ribosome, thereby wasting all the energy spent in synthesizing the chain (Caplan and Menninger, 1979).

Bacterial ribosomes cleaved in their RNA by cloacin E13 are slower and more accurate than wild type ribosomes (Twilt, Overbeek and Van Duin, 1979). Some eukaryotic ribosomes contain hidden breaks in their RNA (Leipoldt and Engel, 1983). A nuclease, indirectly activated by interferon, is able to cut ribosomal RNA (Wreschner *et al.*, 1981). Suppose that cleaved eukaryotic ribosomes are also more accurate than uncleaved ones, so that read-through of termination codons becomes very improbable. Then, because read-through products (or other 'erroneous' products of translation) may be essential to a number of viruses, the development of such viruses would be hindered on the more accurate ribosomes. Alternatively, the slowing down of translation at some 'hungry' codons would enhance the vulnerability of the viral RNA to nucleolytic attack. This may be one of the modes of action of interferon (Ninio, 1981).

10.4.5 KINETIC MODULATION IN DNA REPLICATION

Suppose we have a DNA polymerase molecule which has just incorporated the nth nucleotide in a chain, S_n. It can now either elongate the primer with the next nucleotide (correct S_{n+1}^A or incorrect S_{n+1}^B) or remove it (by pyrophosphorolysis, or by excision when there is an associated exonuclease activity). Then, there is the choice:

$$\ldots S_{n-1} \xleftarrow{k_2^n} \ldots S_{n-1}S_n \begin{array}{c} \xrightarrow{k_1^A} S^A \longrightarrow S_{n-1}S_n S_{n+1}^A \\ \\ \xrightarrow{k_1^B} S^B \longrightarrow S_{n-1}S_n S_{n+1}^B \end{array} \qquad (10.12)$$

The probability that S_n will not be removed is:

$$p^n = (k_1^A[S_{n+1}^A] + k_1^B[S_{n+1}^B])/(k_1^A[S_{n+1}^A] + k_1^B[S_{n+1}^B] + k_2^n) \qquad (10.13)$$

or, assuming that k_1^B is sufficiently small compared to k_1^A:

$$p^n = k_1^A[S_{n+1}^A]/(k_2^n + k_1^A[S_{n+1}^A]) \qquad (10.14)$$

Assume now that k_2 (the rate constant for removing the last nucleotide) is higher for an incorrect than for a correct base. Then, good use of the discrimination potential due to the different k_2 values requires $k_1^A[S_{n+1}^A]$ to be as low as possible. Therefore, error rates at position n are predicted to depend upon the concentration of the substrate that is to be incorporated at the next position, and to increase with that concentration (Ninio, 1975b; Galas and Branscomb, 1978; Fersht, 1979; Clayton et al., 1979; Kunkel et al., 1981).

Another form of kinetic modulation, not yet observed experimentally, could play a key role in the specificity of initiation in transcription and replication. The intracellular concentration of pyrophosphate in E. coli is rather high: 0.5 mM, according to Kukko and Heinonen (1982). In some DNA regions of high A : T content, the insertion of nucleotides and their removal by pyrophosphorolysis could be equally probable. Thus, a DNA or RNA polymerase having a high affinity for such sites would idle there without producing any real initiation. However, if external conditions change so that the ratio of pyrophosphate to nucleoside triphosphates decreases, the polymerase will be able to move forward. Thus, a switch from G·C-rich to A·T-rich promoters can be imagined without the need to invoke an altered binding specificity of the polymerase.

10.4.6 THE NEED TO SUPPRESS PRODUCT BACK FLOW

So far we have not considered the reverse of the reaction $ES \to P$, that is, $P \to ES$. In fact, if the reverse reaction is allowed, and the reaction goes to equilibrium, the ratio of correct to incorrect product will be determined by the equilibrium constants, and will no longer be related to the enzyme's discriminative abilities. But in the cells, once a product is formed, it is consumed at a subsequent stage of metabolism so that the reaction is at least partially irreversible. For instance, even if the acylation of tRNA works reversibly, there is a net uptake of acylated tRNA on the ribosomes. By considering $P \rightarrow P_{inc}$ the pumping of the product into the next metabolic step, we complicate the initial reaction scheme:

$$E + S \underset{k_{-1}}{\overset{k_1}{\rightleftharpoons}} ES \underset{k_{-2}}{\overset{k_2}{\rightleftharpoons}} P \xrightarrow{k_3} P_{inc} \qquad (10.15)$$

Once ES is formed, the probability R that it will go to P_{inc} rather than dissociate is (see Section 4.6):

$$1/R = 1 + \frac{k_{-1}}{k_2}\left(1 + \frac{k_{-2}}{k_3}\right) \qquad (10.16)$$

Hence $R = k_2'/(k_2' + k_{-1})$ with $k_2' = k_2 k_3/(k_3 + k_{-2})$. Since R has the same form as in the strict Michaelis–Menten case, k_2' replacing k_2, we can conclude similarly that discrimination will depend upon the ratio k_2'/k_{-1}. The lower this ratio, the better the discrimination. For fixed values of k_2 and k_{-2}, discrimination is maximized when k_3 (the pumping rate) is the largest. At the limit, k_3 becoming infinite, $k_2' = k_2$ and the discrimination in the full scheme (10.15) becomes equal to what it would be without the $P \rightarrow ES$ reaction. Thus, with respect to $P \rightarrow P_{inc}$ we have the relationship:

High rate of pumping⟨ Better discrimination
                      High rate of synthesis   (10.17)

Good discrimination requires then that k_2 be low and k_3 high.

The general importance for accuracy of pumping reactions that displace substrates and products from equilibrium was stressed by Kurland (1978) and analysed in detail by Ehrenberg and Blomberg (1980) (see also Chapter 11). This line of work has clarified the problem of the cost of accuracy. Whatever the detailed arrangement of a kinetic scheme, there is a minimum thermodynamic cost which is the cost of operating the pumps that maintain substrates and products out of equilibrium. In addition there may be a small extra cost due to the non-optimality of the scheme (Ehrenberg and Blomberg, 1980; Savageau and Lapointe, 1981; Kremen, 1982).

When driving, if one pushes on the accelerator pedal the car goes faster and consumes more fuel. Yet, on the whole, it consumes less fuel on the freeway, going fast, than in town where it runs more slowly. Similarly, the kinetic ideas show that speed and accuracy are related, but with many different modalities, and all situations are observed experimentally (Laughrea, 1981).

10.5 Kinetic amplification

10.5.1 THREE ANALOGIES

If there is just one mistake in a billion successive operations that are performed by a computer, all the calculations may have been done in vain. Is it possible to increase the reliability of the calculations beyond the reliability of the individual electronic components used in the computer? Early discussions of the problem led to a number of solutions (von Neumann, 1956; McCulloch, 1960; Winograd and Cowan, 1963; see also Moore and Shannon, 1956, for discussion of a similar problem), all of them being based on redundant calculations made in parallel and the use of a majority-taking device.

Similarly, can we increase the reliability of molecular processes beyond the reliability of the constituent components? We have an enzyme which discriminates between substrates A and B by virtue of two different kinetic constants, for instance the sticking times $t^A = 1/k^A$ and $t^B = 1/k^B$. All other kinetic constants are equal for both substrates. Is it possible to find a reaction scheme such that the ratio P^A/P^B of the two products may exceed the ratio t^A/t^B of the kinetic constants (at equal concentrations of substrates)? Four schemes that meet this requirement have been proposed under various headings: kinetic proofreading (Hopfield, 1974), delayed reaction (Ninio, 1975a, b), delayed escape of the product (Ninio, 1975b; Hopfield, 1978), and energy relay (Hopfield, 1980). None of them uses the kind of redundancy that is the key to computer reliability. There has been considerable confusion over these schemes. In particular, the difference between classical proofreading and kinetic amplification of enzyme discrimination is not always clear in the minds of the biochemists. Thus, before discussing reaction schemes, I shall give two more analogies.

Consider a coin machine (Fig. 10.2) with a discriminating box D1. A good coin going through D1 is always accepted. A bad coin is rejected with a probability 0.9 and accepted with a probability 0.1. How to improve the performance of the machine? One way of doing so would be to add a second discrimination box D2, similar but not necessarily identical to D1, across the *accept* outlet of D1. Thus a bad coin will have a probability of, say, 0.1×0.1 of being finally accepted. There is a way of achieving the same final result without using a second discrimination box. Consider a device R called the recycler on the outlet pathway of D1 (Fig. 10.2) which sends the good

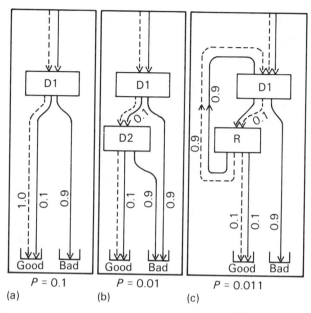

Figure 10.2   The coin-machine metaphor. Consider first the simple machine, (a), with a discriminator D1 sending the good coins (interrupted lines) to the 'good' counter with a probability of 1, and the bad coins (continuous lines) either to the 'bad' counter with a probability of 0.9 or to the good counter with a probability of $P = 0.1$. A way to improve the selectivity of the machine would be to add a second discriminator D2 on the path from D1 to the counter of good coins, as shown in (b). The probability that a bad coin will reach the good counter is now $P = 0.01$. This is analogous to classical proofreading. A more subtle strategy is shown in (c). D2 has been replaced by R, a recycler, which just sets apart 90% of the coins that were on the way from D1 to the good counter, and sends them back indiscriminately into D1. Now a bad coin will reach the good counter with a probability $P = 0.011$. Such a mechanism bears an analogy to the kinetic amplification schemes.

coins and the bad coins back to D1 indiscriminately with a probability of 0.9, and accepts the coins indiscriminately with a probability of 0.1. Now in this case, the final probability of being accepted is, for a bad coin: $(0.1 \times 0.1) + (0.9 \times 0.1) \times (0.1 \times 0.1) + \ldots = 0.011$.

The classical proofreading idea involves two discriminative capacities. The first is related to the initial choice between substrates. The second lies in the specificity of the destruction of correct and incorrect products. Thus, the coin machine with two discriminative boxes in succession can serve as an analogy to classical proofreading. On the other hand, kinetic amplification mechanisms are analogous to the coin machine with the recycler in that they make use of only one discriminative capacity. This unique discrimination may appear at two or several different places in a reaction scheme, due to the idiosyncrasies of the symbolic representation of schemes (Section 10.2). The coin machine analogy gives an intuitive notion of what one is aiming at, but, when considered in detail, it is not an adequate transposition of the kinetic amplification schemes.

A better analogy was put forward by Guéron (1978) in the form of a parable. It goes as follows. A fisherman attempts to catch fish by immersing in a lake a rack containing open glass jars. A fly is painted as a bait at the bottom of each jar. Now, it is assumed that both black and red fish, attracted by the painted flies, enter the rack then the jars at the same rate. However, the black fish quickly perceive the absence of real prey and leave the jars within one second, whereas the less astute red fish stay about a hundred seconds in the jars before quitting. Thus, assuming equal numbers of black and red fish in the lake, when the fisherman collects the contents of his rack, he will have, say, a hundred red fish for one black fish. How can discrimination against black fish be increased? At the time the rack is raised above the water, there is a hundred to one ratio in favour of reds. Assume the rack is held high above the water level for about ten seconds. Some fish will fall back into the lake – a few of the reds and most of the blacks. The losses are not compensated by arrivals, since fish will not jump from the lake up into the rack. Hence, the rack's content has been enriched in red relative to black fish, at the expense of losing some of the red fish. How is the enhancement of selectivity obtained? Not by changing in any manner the unique discriminating principle (the departure of the fish is in no way affected), but by inhibiting, at a certain stage, the entry of both black and red fish and postponing until after that stage the collecting of the rack's content.

10.5.2 THE FOUR KNOWN KINETIC AMPLIFICATION MECHANISMS

In the simplified description of enzyme action in Section 10.4, we had two substrates A and B which formed complexes with the enzyme, with sticking times θ^A and θ^B. Considering that the catalytic activity of the enzyme (quantified by the conversion time $\tau = 1/k_2$) is equally efficient with A and B, we saw that there was a limit to the θ contribution to discrimination, the limit being θ^A/θ^B. The enzyme, which is capable of synthetic activity all the time that a substrate is present is, in a sense, comparing the two sticking times to the conversion time. Suppose $\theta^A = 15$ and $\theta^B = 5$. The limit of the θ contribution to discrimination is $15/5 = 3$. Assume now that catalysis cannot occur immediately after binding of the substrates, but is postponed by four time units, during which the substrates are still allowed to leave. In this case, the enzyme will compare the two sticking times minus the waiting time to the conversion time. This would give a discrimination limit of $(15-4)/(5-4) = 11$.

Now we can restate the principle behind kinetic amplification in a more precise form. In ordinary kinetics, the enzyme compares sticking times to conversion times, with the crucial assumption that both underlying processes (dissociation of ES and conversion of S into P) are random and occur with uniform probabilities. All kinetic amplification mechanisms work by making the probability distribution of one process non-uniform, and somewhat more gaussian. The kinetic strategy is simple, the difficulty is to embody the abstract idea of time delay into plausible enzyme mechanisms. I shall now describe the four known kinetic amplification mechanisms. They are named here according to their titles in the original publications. The diversity of the names should not mask the fact that all four mechanisms make use of a common principle: the introduction of a time delay at a critical point in the reaction scheme. In contradistinction, time delays are useful but not necessary in classical proofreading mechanisms.

(a) Delayed reaction scheme

Consider that in order to form the product P^A from the substrate S^A, a second substrate (ATP, for instance) is needed. The free enzyme and S^A associate, but as long as ATP is not there there is no conversion of S into P. Then, the waiting time for ATP constitutes a (random) time delay for the catalytic reaction. The events on the enzyme may be pictured as follows:

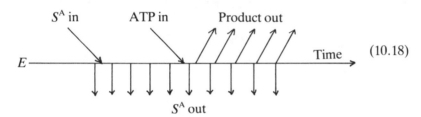

(10.18)

Now let us transcribe this picture into chemical notation. Our first attempt is:

$$E \rightleftharpoons ES^A \rightleftharpoons ES^A \cdot ATP \longrightarrow P \qquad (10.19)$$

Unfortunately, this is not equivalent to Reaction (10.18) above, but a transcription of the Reaction (10.20) below:

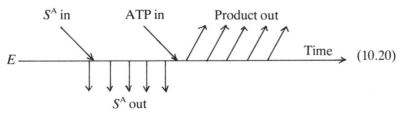

(10.20)

According to Reaction (10.19), S^A may leave the enzyme from the ES^A state, but not from the $ES^A \cdot ATP$ state. Thus, we must add a branch to the scheme to allow the departure of S^A when ATP is present:

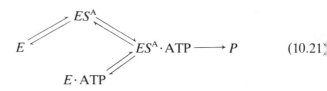

$$\text{(10.21)}$$

There is a problem, however, because if S^A binds to the $E \cdot ATP$ complex and follows the subpathway $E \cdot ATP \rightleftharpoons ES^A \cdot ATP \rightarrow P$ there will be no amplification of discrimination in this pathway. For the scheme to work therefore we need to forbid the $E \cdot ATP \rightarrow ES^A \cdot ATP$ reaction. We can reach this goal practically by having a very quick $E \cdot ATP \rightarrow E$ transformation, mediated, for instance, by a non-specific hydrolysis of ATP on the enzyme when it is in the $E \cdot ATP$ form (Ninio, 1975a, b).

(b) Kinetic proofreading scheme
Another version of the same fundamental scheme (Reaction (10.18)) is Hopfield's 1974 kinetic proofreading scheme:

$$\begin{array}{ccc} ATP & & AMP \\ \searrow & & \nearrow \\ E \rightleftharpoons ES^A & \rightleftharpoons E^* S^A & \longrightarrow P \\ & & \updownarrow \\ & & E \end{array} \qquad \text{(10.22)}$$

Here, the complex ES^A is converted into a high energy complex $E^* S^A$ thanks to the hydrolysis of a molecule of ATP. The subpathway $E \rightleftharpoons E^* S^A \rightarrow P$ is very improbable since $E^* S^A$ is a high energy state and it must be reached without the help of ATP hydrolysis. During all the time it takes to switch from the ES^A to the $E^* S^A$ state, the substrate is allowed to dissociate, but no product can be formed. The time delay is thus created by the $ES^A \rightarrow E^* S^A$ transition.

(c) Energy-relay scheme
In a third variant of the same fundamental scheme called the 'energy-relay' model (Hopfield, 1980), there are two high energy states. The energy is derived from the $S \rightarrow P$ conversion (when it is energetically favoured) and used to amplify discrimination at the next cycle of enzyme action. If the reaction aborts, the enzyme is left without its supply of free energy and so there must be an auxiliary system to charge the enzyme again. Again, the time

delay is generated by that part of the cycle in which the substrate may leave, but no product is formed.

(d) Delayed-escape scheme

So far, we have discussed examples in which the substrates differed by their sticking times. We now present a scheme in which it is *not binding specificity*, but product hydrolysis specificity that is amplified (Ninio, 1975b; Hopfield, 1978).

Consider an enzyme with a classical proofreading activity. It can destroy products P^A or P^B with rate constants k^A and k^B. As soon as the product is formed, it may leave the enzyme with a rate constant k (or a departure time $\tau = 1/k$). Thus, the probability of leaving the enzyme rather than being destroyed by the proofreading activity is $k/(k + k^A)$ for P^A and $k/(k + k^B)$ for P^B. If we take the ratio of the escape probabilities, we find $(k + k^B)/(k + k^A)$. Assuming $k^B > k^A$, the ratio cannot be larger than k^B/k^A. So here again we have a natural limit to the discrimination that can be extracted from the different rate constants k^A and k^B. We can go beyond this limit, however, by changing the probability distribution of the departure time τ.

In the absence of dissociation, the probability that a molecule of product has not yet been hydrolysed at time t is $e^{-k^A t}$ for P^A and $e^{-k^B t}$ for P^B. The ratio of the two probabilities, $e^{-(k^B - k^A)t}$, can be made much larger than k^B/k^A provided t is large enough. What we need then is a retention of the product, so that its escape from the enzyme is delayed. We have to make the departure of the product (and not its destruction) dependent upon some event that may occur only after the product is formed.

(e) Further schemes

Can we transpose the amplification schemes to a forward reaction? Assuming that the only difference in kinetics between the substrates lies in the k_2s of Reaction (10.21), is there a way to discriminate beyond the k_2 ratio? The answer, within a certain well-defined conceptual framework, is negative. There is no manipulation of the probability distribution of conversion times that can result in kinetic amplification (Ninio, 1977).

The reader is strongly encouraged, to gain a better view of the subject, to refer to the original publications (Hopfield, 1974; Ninio, 1975a, b, 1977; Blomberg, 1977; Hopfield, 1980) since many secondary discussions are fraught with confusions and conceptual errors.

10.5.3 THE CASE FOR OR AGAINST KINETIC AMPLIFICATION MECHANISMS

Classical proofreading, which is the destruction of undesirable products, is known to occur in several systems, but whether or not kinetic amplification is really used is still in debate. I shall briefly review the evidence in three domains (aminoacyl-tRNA ligases, DNA polymerases, ribosomes) trying to draw, in each domain, the demarcation line between those arguments that relate to classical proofreading, and those that are more specific to kinetic amplification.

(a) Aminoacyl-tRNA ligases

Early reports that aminoacyl-tRNA ligases could hydrolyse wrong intermediates (Norris and Berg, 1964) or products (Eldred and Schimmel, 1972; Yarus, 1972b) were soon followed by claims that the hydrolytic activities were non-specific and unrelated to error correction. The confusion arose because a link was missing. If an aminoacyl-tRNA ligase makes a wrong product it must destroy it at once or else, if the wrongly acylated tRNA is allowed to quit the enzyme, it may go to the ribosome and cause a translation error (Ninio, 1975b). Proof that the enzyme destroys precisely those wrong intermediates or products that it has itself generated came from three different groups (Hopfield et al., 1976; Fersht and Kaethner, 1976; von der Haar and Cramer, 1976). The degradation of wrong aminoacyl-adenylates or wrong aminoacyl-tRNAs is now well documented (reviewed by Fersht in Chapter 4). In particular, Fersht and Kaethner (1976) showed that after acylation the tRNA tail bearing the amino acid moves to a second site and is subject there to hydrolytic attack.

In general, no effort was made to check the specific predictions of the kinetic amplification schemes, due perhaps to psychological factors or to problems of technical feasibility. One of the four mechanisms (delayed escape of the product) must be comparatively easy to check or eliminate. Considering that the enzyme does manufacture and proofread wrongly acylated tRNAs, the theory shows (Section 10.5.2) that discrimination at the proofreading level will be amplified if the departure of the product is delayed, and does not occur as a single step dissociation (Ninio, 1975b). There are indeed grounds to believe that after aminoacylation, the aminoacyl tRNA stays on the enzyme, and does not leave unless a new cycle of aminoacylation has started on the same enzyme, with another tRNA. This would effectively create a time delay in the departure of the product.

Concerning error correction or prevention at the level of the aminoacyl-adenylate. Hopfield's kinetic proofreading scheme will be extremely difficult to distinguish from classical proofreading. The chances of clarification are somewhat greater with my delayed reaction scheme. This requires a gratuitous ATPase activity, one that can be exerted if ATP binds before the amino acid. Biochemists will be tempted to discard such an ATPase as being an artefact. But Nature also may conspire to hide it! For the destruction of some cellular compounds, if it benefits accuracy under normal conditions, may endanger the cell when the substrates are in short supply. Thus, one expects the cell to protect itself by making the destructive activities of the enzyme conditionally dependent upon signals saying that everything is normal. In particular, the destructive activity may have the correct amino acid or the correct aminoacyl-tRNA as activator.

(b) DNA polymerases

The situation is about as clear with DNA polymerases. Classical proofreading – in this case, the removal through hydrolytic excision – of the last nucleotide incorporated was first documented in 1972 (Brutlag and Kornberg, 1972; Muyzyczka, Poland and Bessman, 1972). Is there evidence for kinetic amplification going beyond classical proofreading?

When applied to DNA polymerases, Hopfield's kinetic proofreading scheme implies that the incoming dNTP is first cleaved into dNMP and pyrophosphate and later linked to the primer terminus. Hopfield (1974) proposed to test the model by attempting the direct incorporation of dNMPs by the polymerases. The difficulty resides in distinguishing dNMP incorporated via such a route from the dNMP incorporated through the reverse of the exonuclease activity.

Using dideoxy-NMPs instead of the standard dNMPs, Doubleday, Lecomte and Radman (1983) obtained a dramatic result. The four dideoxy-NMPs were incorporated into DNA at preferred positions along the template and with a complete lack of specificity with respect to the opposite base on the template. When the same authors used dideoxy-NTPs, they observed dideoxy-NMP incorporation at the primer's terminus but no generation of free dideoxy-NMP, which rather goes against the kinetic proofreading scheme.

I find the results of Doubleday, Lecomte and Radman (1983) very suggestive of a ribosome-like mechanism, involving two distinct sites. The incoming dNTP would pair to the base to be read, unstacked and sticking out at site 1. Next, the dNTP would be cleaved into dNMP and PPi. Then, the base-pair would be disrupted and the dNMP (and eventually, the pyrophosphate) would be transferred to a site 2. Finally, the dNMP would be linked to the primer while, competitively, the previous base could be removed by hydrolytic excision or pyrophosphate attack. Such a mechanism would account for the total lack of specificity in dideoxy-NMP incorporation and would also help to resolve a long-standing theoretical difficulty. If an error has been made by the enzyme, due to a tautomery it will be stabilized by base-pairing and difficult to detect and correct afterwards. If, however, the disruption of the base-pair is an obligatory step in the incorporation mechanism, then when the base-pair is formed again and proofread the bases will usually be in their normal tautomeric states.

Another line of work to understand the mechanism of DNA polymerases was to measure the ratio of dNMP incorporation versus dNMP release as a function of substrate concentration (Bernardi and Ninio, 1978; Bernardi *et al.*, 1979; Clayton *et al.*, 1979). Simple kinetic schemes of the DNA polymerases predict the incorporated/released dNMP ratio to increase linearly with [dNTP]. On the other hand, two kinetic amplification mechanisms (kinetic proofreading and delayed escape of the product) predict this ratio to go asymptotically to a limit when [dNTP] increases indefinitely. Both behaviours were observed with *E. coli* DNA polymerase I, depending upon the template (Bernardi and Ninio, 1978; Bernardi *et al.*, 1979; Saghi and Dorizzi, 1982). The second behaviour, compatible with kinetic amplification, was observed with phage T4 DNA polymerase (Clayton *et al.*, 1979).

Actually, the kinetics of *E. coli* DNA polymerase I are complicated by the existence of a mnemonic effect; when the DNA polymerase dissociates from the template, it progressively unlearns how to excise, so that upon reinitiation the exonuclease activity may be inefficient and warms up little by little as the enzyme moves along the template (Papanicolaou, Dorizzi and Ninio, 1984). After correction for this effect it is expected that all kinetics will be similar and compatible with kinetic amplification.

It is clear in my view that there are at least two sources of dNMPs in the DNA polymerase reactions. One is due to the authentic exonuclease (classical proofreading). This is the dNMP produced while the DNA polymerase is waiting for the next dNTP. The extra dNMP (seen for instance at infinite dNTP concentration) is attributable to kinetic amplification of the true exonuclease (delayed escape mechanism) or to a cleavage of dNTP prior to incorporation (kinetic proofreading mechanism).

In an earlier work, we reported that both DNA and RNA polymerases were able to degrade dNTPs into dNDPs and phosphate (Ninio *et al.*, 1975). Because of a counting artefact, the quantitative relationships given in the paper were not correct, so that the relationship between these activities and kinetic amplification is doubtful. With the progress in the chemical synthesis of DNAs of defined sequences, we hope to clarify the point in the near future.

(c) Ribosomes
Ribosomes are far more complex than DNA polymerases or aminoacyl-tRNA ligases, yet they provide the best opportunity for the unambiguous demonstration of kinetic amplification mechanisms. Classical proofreading on the ribosome would be a hydrolysis of the last peptide bond, promoted by an incorrect codon–anticodon interaction. This may occur in a very extreme form: the tRNA bearing the whole polypeptide chain being synthesized may fall off, and the polypeptide may be completely digested (Caplan and Menninger, 1979; Menninger, 1983). Such a costly event must be rare and cannot occur every time a weak codon–anticodon association is formed. On the other hand, it is hard to see how an immediate removal of just the last amino acid incorporated, influenced by the strength of the codon–anticodon interaction, might be arranged physically. This clears the way for kinetic amplification.

If tRNA always binds to the ribosome as a ternary complex with elongation factor EF-Tu and GTP, Hopfield's kinetic proofreading scheme, but not my delayed reaction scheme, may apply. The incorporation of an amino acid is accompanied by the hydrolysis of two molecules of GTP. The first, coming from the ternary complex, is a potential signature of kinetic proofreading. The second, which is mediated by elongation factor EF–G, occurs afterwards and is related to the pumping of the product. The technical difficulty then is to evaluate the first GTP hydrolysis separately. This has been accomplished in two different ways.

In the first strategy (Thompson and Stone, 1977), poly(U)-primed ribosomes are assayed at low temperature, in the absence of EF-G. A number of kinetic constants have been determined under these conditions, for both cognate and noncognate tRNAs. The use of a slowly hydrolysable analogue of GTP has allowed a good sharpening of the picture (Thompson and Karim, 1982) which was found consistent with the kinetic proofreading scheme.

Another strategy, which is now producing a mass of decisive results, is to manipulate the concentrations of the various elongation factors so that the availability of free EF-Tu becomes the limiting factor in the speed of protein synthesis. Now, if the kinetic proofreading scheme is correct, the incorporation of one amino acid into polypeptide requires the dissipation of f ternary complexes, f being close to one for a cognate tRNA and clearly larger than one for a noncognate tRNA. The way the rates of incorporation vary with the concentrations of various factors give an indication of the f values. The first estimates were somewhat indirect (Ruusala, Ehrenberg and Kurland, 1982). The results are now substantiated by a more transparent experimental strategy (Bohman et al., 1984). While some of the assumptions of the treatments – in the absence of more background information – are debatable, there is no doubt in my mind that the method has a bright future. On the whole, the results show internal consistency, they agree with the kinetic proofreading scheme and clarify the mysteries of low and high fidelity ribosomes.

If transposed to the ribosome, my delayed reaction scheme requires (which is not generally the case) independent binding of tRNA on one side, and EF-Tu-GTP on the other, and a rapid hydrolysis of GTP whenever EF-Tu-GTP binds first. Thus, the scheme strongly predicts a ribosome-dependent hydrolysis of GTP in the EF-Tu-GTP complex. Such a hydrolysis, if observed (see Fig. 1 in Thompson et al., 1981) will undoubtedly be treated by most experimentalists as background noise. I believe that the question has not been completely settled. In particular, under conditions of amino acid starvation in stringent cells, the metabolite ppGpp is produced which switches the synthetic machinery to a higher accuracy regime (reviewed by Gallant and Foley, 1980). Apparently, ppGpp diminishes the affinity of elongation factors for tRNA (Pingoud et al., 1983; Rojas et al., 1984) so that, under the viscosity conditions that prevail within bacteria, it may well have the effect of favouring a pathway of independent binding of tRNA and EF-Tu-GTP to ribosomes. This would be a high accuracy pathway, which would use two time delays; the first, when the tRNA is on the ribosome, waiting for EF-Tu-GTP (delayed reaction scheme) and the second, when both the tRNA and EF-Tu-GTP are there and the ribosome is waiting for GTP hydrolysis (kinetic proofreading scheme). The crucial experiment, yet to be done, is to measure GTP hydrolysis versus amino acid incorporation, as the concentration of acylated tRNA varies, under stringent and relaxed conditions.

10.6 Recipes for calculation

Several non-standard procedures have been applied to compute error rates for given enzyme mechanisms (Ninio, 1974, 1975b; Blomberg, 1977; Gala and Branscomb, 1978; Bernardi *et al.*, 1979; Malygin and Yashina, 1980; Ehrenberg and Blomberg, 1980). The method I shall describe here is subtle but simple and straightforward once the key idea is assimilated.

To begin with, we take the simplest example:

$$E + S^A \rightleftharpoons ES^A \longrightarrow P^A$$

$$E + S^B \rightleftharpoons ES^B \longrightarrow P^B$$

(10.23)

There are a number of binary choices to which there correspond elementary probabilities:

$$ES^A \xleftarrow{p^A} E \xrightarrow{p^B} ES^B, \text{ with } p^A + p^B = 1$$

$$E + S^A \xleftarrow{1-q^A} ES^A \xrightarrow{q^A} P^A$$

$$E + S^B \xleftarrow{1-q^B} ES^B \xrightarrow{q^B} P^B$$

(10.24)

We are interested in the overall probability of obtaining P^A or P^B from E:

$$P^A \xleftarrow{R^A} E \xrightarrow{R^B} P^B, \text{ with } R^A + R^B = 1$$

(10.25)

How to compute R^A? By summing up the probabilities of all pathways leading from E to P^A, that is:

(1) $E \longrightarrow ES^A$ and $ES^A \longrightarrow P^A$ $\qquad\qquad :p^A q^A$

(2) $E \longrightarrow ES^A$ and $ES^A \longrightarrow E$ and $E \longrightarrow P^A$ $\quad :p^A(1-q^A)R^A$ (10.26)

(3) $E \longrightarrow ES^B$ and $ES^B \longrightarrow E$ and $E \longrightarrow P^A$ $\quad :p^B(1-q^B)R^A$

The key point is that once we are back to E we still have a probability R^A of forming the product P^A, by definition of R^A, and by virtue of the fact that such probabilities are independent of the past history of E. Hence,

$$R^A = p^A q^A + \boxed{p^A(1-q^A)}\, R^A + \boxed{p^B(1-q^B)}\, R^B$$
$$R^B = p^B q^B + \boxed{p^A(1-q^A)}\, R^B + \boxed{p^B(1-q^B)}\, R^B \qquad (10.27)$$

From there,

$$R^A = p^A q^A / (p^A q^A + p^B q^B) \qquad (10.28)$$

$$R^A / R^B = p^A q^A / p^B q^B \qquad (10.29)$$

Note that if we are only interested in the rato R^A/R^B, it is not necessary to take into account cases (2) and (3) in which there is a return to the initial state. Therefore, we can simplify the treatment further by computing 'reduced probabilities' r^A and r^B that take into account only those pathways that do not pass through the initial state; thus:

$$r^A = p^A q^A;\; r^B = p^B q^B;\; R^A/R^B = r^A/r^B = p^A q^A / p^B q^B \qquad (10.30)$$

The expressions for the ps and the qs are straightforward (Section 10.2.2). Thus, the error rate can be written by simple inspection of the scheme, once the method is understood.

We may also view the above simplification (replacing R^A/R^B by r^A/r^B) as mere arithmetic trick. Knowing that

$$\frac{X}{Y} = \frac{a}{b} \quad \text{implies} \quad \frac{X}{Y} = \frac{a-mX}{b-mY} \qquad (10.31)$$

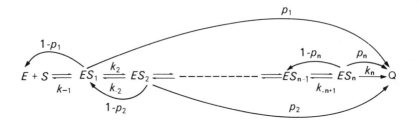

$$\frac{1}{p_1} = 1 + \frac{k_{-1}}{k_2}\left(\frac{1}{p_2}\right) = 1 + \frac{k_{-1}}{k_2}\left[1 + \frac{k_{-2}}{k_3}\left(\frac{1}{p_3}\right)\right] = \ldots; \quad \frac{1}{p_n} = \frac{k_n + k_{-n+1}}{k_n}$$

(a)

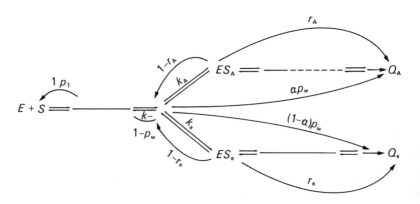

Compute $1/p_1$ as a function of $1/p_M$ as in Case 1. Then,

$$\frac{\alpha}{1-\alpha} = \frac{k_A}{k_B}\frac{r_A}{r_B} \quad \text{and} \quad \frac{1}{p_M} = 1 + \frac{k_-}{k_A r_A + k_B r_B}$$

(b)

Figure 10.3 Recipes for the calculation of error rates. Ratios of rates can be expressed as ratios of probabilities. Parts (a), (b) and (c) show how the relevant probabilities can be computed in some simple situations. (a) Case 1, (b) Case 2, (c) Case 3.

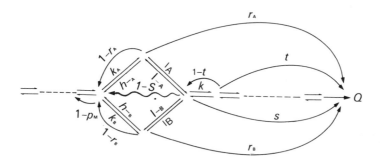

Compute $1/p_1 = f(1/p_M)$ (Case 1) then $1/p_M = f(1/r_A, 1/r_B)$ (Case 2)

Then: $r_A = l_A s/(l_A + h_{-A})$; $r_B = l_B s/(l_B + h_{-B})$ Finally:

$$\frac{1}{s} = 1 + \frac{E}{kt} \quad \text{with} \quad E = \frac{h_{-A}}{l_A + h_{-A}} l_{-A} + \frac{h_{-B}}{l_B + h_{-B}} l_{-B}$$

(c)

m being any number, Equation (10.29) is readily deduced from Equation (10.27). From $R^B = 1 - R^A$, we deduce:

$$\frac{1}{R^A} = 1 + \frac{p^B q^B}{p^A q^A} \tag{10.32}$$

This form is particularly suitable for recurrent calculations. With all these essential tricks in mind, one can deal easily with more complex situations (Fig. 10.3).

A special difficulty arises when dealing with DNA polymerases that carry a 3' to 5' exonuclease activity. The enzyme may move back and forth along the template, undoing what it had done earlier (Goodman *et al.*, 1974). The expression for the overall frequency of errors as a function of the elementary rate constants was given for simple situations by Ninio, 1975b; Galas and Branscomb, 1978; Bernardi *et al.*, 1979. The effect of pyrophosphorolysis on error rates has been given further attention (Herbomel and Ninio, 1980). The 'peelback' problem has since been solved in cases of increasing generality: when the elementary rates of incorporation of a base depend upon the nature of the preceding base (Malygin and Yashina, 1980) and when both rates of incorporation and excision depend upon the preceding base (Durup, 1982).

10.7 Outlook

10.7.1 PHYSICAL ENZYMOLOGY

There is one environmental condition to which little attention has been paid so far, and which may turn out to be vitally important in accuracy studies: *viscosity*. The viscosity of the intracellular medium, in the case of prokaryotes, may be twenty-five to a hundred times higher than that of liquid water. However, because the cell's interior is extremely heterogeneous, the diffusion rates of the various molecular species are not simply related to the overall viscosity. A small molecule will travel in the channels of liquid water, and will perceive an apparent viscosity at most twice as high as that of ordinary water (Cooke and Kuntz, 1974). The larger the molecule, the more severe the restrictions to its motion due to the collisions with other large molecules. Thus, the diffusion rate of tRNA inside the cell is perhaps one-fifth its rate in liquid water (Gouy and Grantham, 1980). Accuracy depends critically upon ratios of rates. The passage from the highly viscous conditions of the cell's interior to the liquid conditions of *in vitro* experiments may affect quite differently the various kinetic constants of a reaction. A polymerase will wait longer for the incoming nucleotide, but if it dissociates from the template, it might fail to diffuse away, and might reassociate readily, the more so as imprinting effects on the template may last longer. Side products of the reaction like pyrophosphate or deoxynucleoside monophosphate might hang around and be in abnormally high local concentrations, with respect to usual *in vitro* assays. Is this the reason why RNA polymerase carries a companion activity which eliminates pyrophosphate by an unusual method (Volloch, Rits and Tumerman, 1979)? Could it be that some differences in design between the polymerases or the ribosomes from eukaryotes and prokaryotes follow from the different intracellular viscosities of the two classes of organisms? Perhaps in the near future cells can be grown under conditions in which the internal amount of water can be varied, and 'viscosity-sensitive' mutants can be isolated and studied.

In agreement with formal enzyme kinetics, the kinetic theory of accuracy has assumed so far that every chemical reaction can be decomposed into elementary steps that occur with uniform temporal probability. As a consequence, any enzymic reaction, whatever its complexity, can theoretically be completed (with a vanishingly small probability) in an infinitely small amount of time. A piece of double-stranded DNA would be able to go, with the help of a topoisomerase, across another piece of double-stranded DNA instantaneously. But motion cannot occur at infinite speed. The exponential decay rule, valid as a good approximation for small molecular rearrangements, may not be the appropriate building block for constructing schemes that describe the interactions of large macromolecules. In a theoretical work on DNA helicases, Mizraji and Linn (1984) describe the motion of the enzyme with the help of a viscoelastic model. The enzyme moves forward by an expansion and contraction cycle, mediated by ATP hydrolysis. The movement is opposed by frictional forces. As far as average rates are concerned, this physical model can be equated to a classical kinetic scheme (Mizraji and Linn, 1984). Conceivably, physical models of enzyme action will be developed in the future, to the point where they will cease to be compatible with the present formalism of chemical kinetics. Then, the theoretical treatments of accuracy will have to be modified to take into account a less formal, more physical, understanding of enzyme mechanisms.

10.7.2 POPULATIONAL SPECIFICITY

So far, the kinetic theory of accuracy has been mainly concerned with DNA replication and protein synthesis. There are other domains of biology where challenging recognition problems are encountered, and which require, I think, a new conceptual framework.

One of the unsolved problems in eukaryote biology is the logic of regulation by protein phosphorylation and dephosphorylation (or other post-trans-lational modifications). In eukaryotes, most proteins are there all the time, but their level of activity is controlled by their state of chemical modification, say phosphorylation. Genes themselves are locked or unlocked according to the degree of chemical modification of the bound chromosomal proteins. Now, if the modifying enzymes, phosphorylases and kinases are taken individually, they display a rather loose specificity. Although both classes of enzymes may use in principle kinetic amplification, I believe that the real trick is elsewhere. We might be in the presence of a new kind of specificity, 'populational specificity'. The simultaneous presence of two antagonistic series of reactions (addition and removal of phosphate groups) may result in a pattern of phosphorylation that could be much more specific than expected from the study of any individual phosphorylase or kinase (Ninio and Chapeville, 1980). Actually, some ideas on populational specificity are emerging, but in another field where things are more complex, yet better understood: immunology. Two analogies taken in the field of visual per-ception will guide our more technical discussions.

An object may be seen under many different angles and give rise to widely different appearances. If you look for an object in its usual place, say on your desk, even if the place is untidy and only a tiny part of the object is visible, then it will be spotted at once. Knowing the place well and the objects it usually contains, these will be distinguished one from the other according to a small set of attributes. But the attributes that suffice for identification in a given well-known context may be of little use in another context. If the same object is in an unfamiliar place, then it may be in front of your nose and you may be unable to detect it.

In order to facilitate discrimination between self and foreign molecules, the immunological system has set up a constant context to look upon them. The context is provided by macromolecules which probably act as universal molecular glues. These are the products of the MHC (major histocompatibility complex) genes. When bound to an MHC protein, a molecule makes only a restricted part of itself accessible to the exterior. This is the part upon which the immune system of an individual bases its distinction between self and non-self. When self and non-self molecules are viewed within a different context (i.e. bound to different MHC proteins) severe errors arise. The criteria for self/non-self discrimination are developed at an early stage of maturation of the immune system, and involve the generation and diversification of 'receptors', the T-cell receptors. These are the true eyes of the immune system. They signal the presence of foreign molecules to another part of the immune system, which obeys another logic. Here again, we can take an analogy in vision.

The objects around us generally appear well delineated by sharp contours. They are within boundaries that separate them from each other and from the background. Yet, if we were to measure physically the change in the intensity of light across the edge of an object, we would find a smooth sigmoid, rather than a neat edge. The contours that we perceive are in fact the result of a perceptual processing that starts at the retinal level. The technique used for that purpose, 'lateral inhibition', was postulated by Ernst Mach as early as 1865 (see Ratliff, 1972). Consider a network of cells on the retina. Each cell is stimulated according to the intensity of light it receives. However, an excited cell A inhibits a neighbouring cell B, the inhibition increasing with the excitation of A and decreasing with the distance A–B. This type of relationship permits edge detection and hence improves the signal/noise ratio.

The lateral inhibition trick can be reversed and transposed to immunology. The production of antibodies is under the responsibility of cell families arranged in networks (Jerne, 1974). Unlike nerve cells, they do not form spatially stable networks with physical connections, but these cells can influence each other when they meet by releasing factors that promote or inhibit growth and differentiation. A given B lymphocyte of the immunological network will receive stimulation from an external source, the antigen (the analogue of light), and its activity will be subject to 'lateral' facilitation or inhibition by other cells of the network. Physical proximity on the retina will be transposed here as 'having related antigenic binding sites'. However, as in DNA replication where a nucleotide does not recruit its like directly, but is replicated by taking the complementary nucleotide, and then the complement of the complement, immunological recruiting of antibodies is indirect. Cells producing an antibody Ab1 will stimulate cells producing antibodies anti-Ab1 and these in turn will stimulate cells producing antibodies anti-(anti-Ab1). Among these will be found antibodies binding the same antigen as Ab1 (Urbain, Wuilmart and Cazenave, 1981). Thus, we have a kind of lateral facilitation instead of lateral inhibition. This will favour the centre of gravity of a distribution, rather than its edges. Thus one may expect through this method a sharpening of the immunological response to foreign antigen, with the selective production of the antibodies having the highest affinity for the antigen (Grossman, 1984). But strong binding goes together with low specificity (Section 10.4) and this is also true in immunology. The most effective antibodies are also the most likely to bind to normal cellular components and are a potential danger to the organism.

The immunological system gets around this difficulty by having two networks: the network of B-cells, which produces weapons that are effective but which can explode in one's hands, and the signalling system with the MHC molecules on the surface of the 'antigen presenting cells' and the family of T cell detectors. In the presence of a foreign molecule, the detecting system says 'fire!'. The danger of self-destruction is real, but temporary. As soon as the enemy is expelled, the firing order is cancelled. If, after a certain time, the enemy is still around, the immunological system switches to a second strategy: suppression. Better live with the enemy than die fighting him.

Thus the immunological system offers today a wide range of phenomena that illustrate the difficulty of achieving accuracy in biological systems. A network to distinguish foreign from self molecules can be built, but only by restricting the part of the molecules upon which discrimination is based. Hence, some foreign molecules will be considered as self. Antibodies can be manufactured, having a high affinity for almost any foreign molecules, but at the price of being loosely specific. The division of labour between the specific part (MHC + T cells) and the active part (B cells) implies that the immunological system must either find a quick response, or renounce fighting (suppression). An MHC-bearing cell will normally capture an antigen at the surface of a B cell, then present the antigen to a T cell. Ideally, one would like the T cell to deliver its signal specifically to the B cell from which the antigen was captured. In order to work, such a triangular device requires that all affinities and diffusion constants are within suitable ranges. This has not yet been worked out quantitatively.

I wonder if the logic behind the immunological system could not be the key to regulation in eukaryotes. A regulatory protein, say a phosphorylase, would receive stimulations or inhibitions from two channels: the external channel (saying that the concentration of a metabolite is too low or too high) and lateral channels (the phosphorylase itself is the target of several phosphorylases and kinases). Thus, a formal similarity to the B cell network logic is not excluded. This perhaps can be worked out quantitatively much more easily than it can be done in the immune system, and it could become one of the future goals of researchers now working in the field of accuracy. Whether or not an alarm system, much more rudimentary than the MHCs and T cell receptors, is operative at the molecular level, within eukaryotic cells, I cannot predict. Let us keep an eye out for manifestations of such a possibility.

References

Bernardi, F. and Ninio, J. (1978) The accuracy of DNA replication. *Biochimie*, **60**, 1083–1095.

Bernardi, F., Saghi, M., Dorizzi, M. and Ninio, J. (1979) A new approach to DNA polymerase kinetics. *J. Mol. Biol.*, **129**, 93–112.

Blomberg, C. (1977) A kinetic recognition process for tRNA at the ribosome. *J. Theor. Biol.*, **66**, 307–325.

Bohman, K., Ruusala, T., Jelenc, P. C. and Kurland, C. G. (1984) Kinetic impairment of restrictive streptomycin-resistant ribosomes. *Mol. Gen. Genet.*, **198**, 90–99.

Brutlag, D. and Kornberg, A. (1972) Enzymatic synthesis of deoxyribonucleic acid. XXXVI. A proofreading function for the $3' \rightarrow 5'$ exonuclease activity in deoxyribonucleic acid polymerases. *J. Biol. Chem.*, **247**, 241–248.

Caplan, A. B. and Menninger, J. R. (1979) Tests of the ribosomal editing hypothesis: Amino acid starvation differentially enhances the dissociation of peptidyl-tRNAs from the ribosome. *J. Mol. Biol.*, **134**, 621–637.

Carrier, M. J. and Buckingham, R. H. (1984) An effect of codon context on the mistranslation of UGU codons *in vivo*. *J. Mol. Biol.*, **175**, 29–38.

Clayton, L. K., Goodman, M. F., Branscomb, E. W. and Galas, D. J. (1979) Error induction and correction by mutant and wild type T4 DNA polymerases. Kinetic error discrimination mechanisms. *J. Biol. Chem.*, **254**, 1902–1912.

Cooke, R. and Kuntz, I. D. (1974) The properties of water in biological systems. *Ann. Rev. Biophys. Bioeng.*, **3**, 95–126.

Dixon, M. and Webb, E. C. (1958) *Enzymes*, Longmans, London.

Doubleday, O. P., Lecomte, Ph. J. and Radman, M. (1983) A mechanism for nucleotide selection associated with the pyrophosphate exchange activity of DNA polymerases. In *Cellular Responses to DNA Damage* (eds E. C. Friedberg and B. A. Bridges), Alan R. Liss, New York, pp. 489–499.

Durup, J. (1982) On the relations between error rates in DNA replication and elementary chemical rate constants. *J. Theor. Biol.*, **94**, 607–632.

Ehrenberg, M. and Blomberg, C. (1980) Thermodynamic constraints on kinetic proofreading in biosynthetic pathways. *Biophys. J.*, **31**, 333–358.

Eldred, E. W. and Schimmel, P. R. (1972) Rapid deacylation by isoleucyl transfer ribonucleic acid synthetase of isoleucine-specific transfer ribonucleic acid aminoacylated with valine. *J. Biol. Chem.*, **247**, 2961–2964.

Fersht, A. R. (1977) *Enzyme Structure and Mechanism*, Freeman and Co., San Francisco.

Fersht, A. R. (1979) Fidelity of replication of phage ϕX174 DNA by DNA polymerase III holenzyme: Spontaneous mutation by misincorporation. *Proc. Natl Acad. Sci. USA*, **76**, 4946–4950.

Fersht, A. R. and Kaethner, M. M. (1976) Enzyme hyperspecificity. Rejection of threonine by the valyl-tRNA synthetase by misacylation and hydrolytic editing. *Biochemistry*, **15**, 3342–3346.

Galas, D. J. and Branscomb, E. W. (1978) Enzymatics determinants of DNA polymerase accuracy. Theory of coliphage T4 polymerase mechanisms. *J. Mol. Biol.*, **124**, 653–687.

Gallant, J. and Foley, D. (1980) On the causes and prevention of mistranslation. In:

Ribosomes Structure, Function and Genetics (eds G. Chambliss *et al.*), University Park Press, Baltimore, pp. 615–638.

Goodman, M. F., Gore, W. C., Muzyczka, N. and Bessman, M. J. (1974) Studies on the biochemical basis of spontaneous mutation. III. Rate model for DNA polymerase-effected nucleotide misincorporation. *J. Mol. Biol.*, **88**, 423–435.

Gorini, L. (1971) Ribosomal discrimination of tRNAs. *Nature New Biol.*, **234**, 261–264.

Gouy, M. and Grantham, R. (1980) Polypeptide elongation and tRNA cycling in *Escherichia coli*: a dynamic approach. *FEBS Lett.*, **115**, 151–155.

Grossman, Z. (1984) Recognition of self and regulation of specificity at the level of cell populations. *Immunol. Rev.*, **79**, 119–138.

Grunberg-Manago, M. and Dondon, J. (1965) Influence of pH and S-RNA concentration on coding ambiguities. *Biochem. Biophys. Res. Commun.*, **18**, 517–522.

Guéron, M. (1978) Enhanced selectivity of enzymes of kinetic proofreading. *American Scientist*, **66**, 202–208.

Herbomel, P. and Ninio, J. (1980) Fidélité d'une réaction de polymérisation selon la proximité de l'équilibre. *C.R. Acad. Sc. Paris, Série D*, **291**, 881–884.

Hopfield, J. J. (1974) Kinetic proofreading: a new mechanism for reducing errors in biosynthetic processes requiring high specificity. *Proc. Natl Acad. Sci. USA*, **71**, 4135–4139.

Hopfield, J. J. (1978) Origin of the genetic code: A testable hypothesis based on tRNA structure, sequence, and kinetic proofreading. *Proc. Natl Acad. Sci. USA*, **75**, 4334–4338.

Hopfield, J. J. (1980) The energy relay: A proofreading scheme based on dynamic cooperativity and lacking all characteristic symptoms of kinetic proofreading in DNA replication and protein synthesis. *Proc. Natl Acad. Sci. USA*, **77**, 5248–5252.

Hopfield, J. J., Yamane, T., Yue, Y. and Coutts, S. M. (1976) Direct experimental evidence for kinetic proofreading in amino acylation of tRNAIle. *Proc. Natl Acad. Sci. USA*, **73**, 1164–1168.

Jelenc, P. C. and Kurland, C. G. (1979) Nucleoside triphosphate regeneration decreases the frequency of translation errors. *Proc. Natl Acad. Sci. USA*, **76**, 3174–3178.

Jerne, N. K. (1974) Towards a network theory of the immune system. *Ann. Immunol. (Inst. Pasteur)*, **125c**, 373–389.

Kremen, A. (1982) Information-theoretic significance of Gibbs energy supply to editing mechanisms. *Biophys. J.*, **40**, 149–154.

Kukko, E. and Heinonen, J. (1982) The intracellular concentration of pyrophosphate in the batch culture of *Escherichia coli*. *Eur. J. Biochem.*, **127**, 347–349.

Kunkel, T. A., Schaaper, R. M., Beckman, R. A. and Loeb, L. A. (1981) On the fidelity of DNA replication. Effect of the next nucleotide on proofreading. *J. Biol. Chem.*, **256**, 9883–9889.

Kurland, C. G. (1978) The role of guanine nucleotides in protein biosynthesis. *Biophys. J.*, **22**, 373–392.

Laughrea, M. (1981) Speed-accuracy relationships during *in vitro* and *in vivo* protein biosynthesis. *Biochimie*, **63**, 145–168.

Leipoldt, M. and Engel, W. (1983) Hidden breaks in ribosomal RNA of phylogenetically tetraploid fish and their possible role in the diploidization process.

Biochem. Genet., **21**, 819–841.

McCulloch, W. S. (1960) The reliability of biological systems. In *Self-organizing Systems* (eds M. C. Yovits and S. Cameron), Pergamon Press, New York, pp. 264–281.

Malygin, E. G. and Yashina, L. N. (1980) Kinetic description of the error-correcting mechanism of bifunctional DNA polymerases. (In Russian.) *Doklad. Akad. Nauk. SSSR*, **250**, 246–250.

Menninger, J. R. (1983) Computer simulation of ribosome editing. *J. Mol. Biol.*, **171**, 383–399.

Mizraji, E. and Linn, J. (1984) Chemicomechanical transduction performed by enzymes activated by polymers. *J. Theor. Biol.*, **108**, 173–189.

Moore, E. F. and Shannon, C. E. (1956) Reliable circuits using less reliable relays. *J. Franklin Inst.*, **262**, 191–208.

Muzyczka, N., Poland, R. L. and Bessman, M. J. (1972) Studies on the biochemical basis of spontaneous mutation. 1. A comparison of the deoxyribonucleic acid polymerases of mutator, antimutator and wild type strains of bacteriophage T4. *J. Biol. Chem.*, **247**, 7116–7122.

Ninio, J. (1973) Recognition in nucleic acids and the anticodon families. *Progress in Nucleic Acid Res. Mol. Biol.*, **13**, 301–337.

Ninio, J. (1974) A semi-quantitative treatment of missense and nonsense suppression in the *strA* and *ram* ribosomal mutants of *Escherichia coli*. Evaluation of some molecular parameters of translation *in vivo*. *J. Mol. Biol.*, **84**, 297–313.

Ninio, J. (1975a) La précision dans la traduction génétique. In *Ecole de Roscoff 1974. L'Évolution des Macromolécules Biologiques* (ed. C. Sadron), C.N.R.S., Paris, pp. 51–68.

Ninio, J. (1975b) Kinetic amplification of enzyme discrimination. *Biochimie*, **57**, 587–595.

Ninio, J. (1977) Are further kinetic amplification schemes possible? *Biochimie*, **59**, 759–760.

Ninio, J. (1981) Molecular evolution: a walk in the sequence space. In *Food, Nutrition and Evolution* (eds D. N. Walcher and N. Kretchmer), Masson, New York, pp. 9–13.

Ninio, J. and Chapeville, F. (1980) Recognition: the kinetic concepts. In *Chemical Recognition in Biology* (eds F. Chapeville and A. L. Haenni), Springer-Verlag, Heidelberg, pp. 78–85.

Ninio, J., Bernardi, F., Brun, G., Assairi, L., Lauber, M. and Chapeville, F. (1975) On the mechanism of nucleotide incorporation into DNA and RNA. *FEBS Lett.*, **57**, 139–144.

Norris, A. T. and Berg, P. (1964) Mechanism of aminoacyl RNA synthesis: studies with isolated aminoacyl adenylate complexes of isoleucyl RNA synthetase. *Proc. Natl Acad. Sci. USA*, **52**, 330–337.

Papanicolaou, C., Dorizzi, M. and Ninio, J. (1984) A memory effect in DNA replication. *Biochimie*, **66**, 115–119.

Pestka, S., Marshall, R. and Nirenberg, M. (1965) RNA codewords and protein synthesis. V. Effect of streptomycin on the formation of ribosome-sRNA complexes. *Proc. Natl Acad. Sci. USA*, **53**, 639–646.

Pingoud, A., Gast, F.-U., Block, W. and Peters, F. (1983) The elongation factor Tu from *Escherichia coli*, aminoacyl-tRNA, and guanosine tetraphosphate form a ternary complex which is bound by programmed ribosomes. *J. Biol. Chem.*, **258**, 14 200–14 205.

Ratliff, F. (1972) Contour and contrast. *Sci. Am.*, **226** (6), 91–101.

Rojas, A.-M., Ehrenberg, M., Anderson, S. and Kurland, C. G. (1984) ppGpp inhibition of elongation factors Tu, G and Ts during polypeptide synthesis. *Mol. Gen. Genet.*, **197**, 36–45.

Ruusala, T., Ehrenberg, M. and Kurland, C. G. (1982) Is there proofreading during polypeptide synthesis? *EMBO J.*, **1**, 741–745.

Saghi, M. and Dorizzi, M. (1982) Polymerization/excision kinetics of *Escherichia coli* DNA polymerase I. Stability in kinetic behaviour and variation of the rate constants with temperature and pH. *Eur. J. Biochem.*, **123**, 191–199.

Savageau, M. A. and Lapointe, D. S. (1981) Optimization of kinetic proofreading: a general method for derivation of the constraint relations and an exploration of a specific case. *J. Theor. Biol.*, **93**, 157–177.

Schwartz, V. S. and Lysikov, V. N. (1974) Physical mechanisms of ribosomal screen. (In Russian.) *Doklady Akad Nauk SSSR*, **217**, 1446–1448.

Spirin, A. S. (1978) Energetics of the ribosome. *Progr. Nucl. Acid Res. Mol. Biol.*, **21**, 39–62.

Thompson, R. C., Dix, D. R., Gershon, R. B. and Karim, A. M. (1981) A GTPase reaction accompanying the rejection of Leu-tRNA$_2$ by UUU-programmed ribosomes. Proofreading of the codon–anticodon interaction by ribosomes. *J. Biol. Chem.*, **256**, 81–86.

Thompson, R. C. and Karim, A. M. (1982) The accuracy of protein biosynthesis is limited by its speed: high fidelity selection by ribosomes of aminoacyl-tRNA ternary complexes containing GTP (γS). *Proc. Natl Acad. Sci. USA*, **79**, 4922–4926.

Thompson, R. C. and Stone, P. J. (1977) Proofreading of the codon–anticodon interaction on ribosomes. *Proc. Natl Acad. Sci. USA*, **74**, 198–202.

Twilt, J. C., Overbeek, G. P. and van Duin, J. (1979) Translational fidelity and specificity of ribosomes cleaved by cloacin DF13. *Eur. J. Biochem.*, **94**, 477–484.

Urbain, J., Wuilmart, C. and Cazenave, P.-A. (1981) Idiotypic regulation in immune networks. In: *Contemporary Topics in Molecular Immunology* (eds F. P. Inwan and W. J. Mandy), Vol. 8. Plenum Press, New York, pp. 113–148.

van Slyke, D. D. and Cullen, G. E. (1914) The mode of action of urease and of enzymes in general. *J. Biol. Chem.*, **19**, 141–180.

Volloch, V. Z., Rits, S. and Tumerman, L. (1979) Pyrophosphate-condensing activity linked to nucleic acid synthesis. *Nucl. Acids Res.*, **6**, 1521–1534.

von der Haar, F. and Cramer, F. (1976) Hydrolytic action of aminoacyl- tRNA synthetases from baker's yeast: 'chemical proofreading' preventing acylation of tRNA[ile] with misactivated valine. *Biochemistry*, **15**, 4131–4138.

von Neumann, J. (1956) Probabilistic logics and the synthesis of reliable organisms from unreliable components. In *Automata Studies* (eds C. E. Shannon and J. McCarthy), Princeton University Press, pp. 43–98.

Winograd, S. and Cowan, J. D. (1963) *Reliable Computation in the Presence of Noise*, MIT Press, Cambridge, Massachusetts.

Wreschner, D. H., James, T. C., Silverman, R. H. and Kerr, I. N. (1981) Ribosomal RNA cleavage, nuclease activation and 2-5A (ppp (A2'p)$_n$ A) in interferon-treated cells. *Nucl. Acids Res.*, **9**, 1571–1581.

Yarus, M. (1972a). Increased specificity in the aminoacylation reaction due to the use of parallel systems of ligands. *Nature New Biol.*, **239**, 106–108.

Yarus, M. (1972b) Phenylalanyl-tRNA synthetase and isoleucyl-tRNA[Phe]: a possible verification mechanism for aminoacyl-tRNA. *Proc. Natl Acad. Sci. USA*, **69**, 1915–1919.

11 Kinetic costs of accuracy in translation

M. EHRENBERG, C. G. KURLAND
and C. BLOMBERG

11.1 Introduction

In this chapter we focus attention on enzymic selections in which kinetic proofreading (Hopfield, 1974; Ninio, 1975) contributes to the accuracy. We discuss, in particular, ribosomal translation, which is the most costly (Ingraham, Maaløe and Neidhardt, 1983), as well as the noisiest step (Bouadloun, Donner and Kurland, 1983) on the path from gene to protein. In a short, historical, section we discuss the shortcomings of equilibrium thermodynamics as well as of information theory to describe enzymic selections. This discussion provides a background to the proofreading concept. It is followed by discussions on the thermodynamic flows and driving forces that are necessary for enzymic selections in general (Kurland, 1978), as well as for kinetic proofreading in particular (Blomberg, Ehrenberg and Kurland, 1980; Blomberg, 1983a, b). Our discussion about the costs of accuracy in translation starts with an experimental section where the data obtained from a set of bacterial mutants are reviewed. These measurements illustrate in a dramatic way that 'high accuracy' ribosomes have a reduced kinetic efficiency *in vitro* as well as impaired elongation rate and growth rate *in vivo*. These results motivate a novel analysis of the kinetic costs of accuracy in growing bacteria (Kurland and Ehrenberg, 1984; Ehrenberg and Kurland, 1984), which is contrasted with previous cost analyses based on the dissipative losses of enzymic selections (Bennet, 1979; Savageau and Freter, 1979a, b; Freter and Savageau, 1980; Blomberg, Ehrenberg and Kurland, 1980).

We begin our discussion of costs of accuracy with remarks on some general conditions for rapid growth in bacteria and continue by analysing optimal conditions of efficiency. These results are then used to define the kinetic costs of accuracy at the initial selection step as well as in the proofreading step of the ribosome.

Our discussion is informal throughout this chapter. To obtain more rigorous derivations for the flows and forces of enzymatic selections the reader is referred to other studies (Blomberg, Ehrenberg and Kurland, 1980; Blomberg, 1983a, b). Aspects of optimal translation are treated by Kurland and Ehrenberg (1984) and a number of formal derivations are explained by Ehrenberg and Kurland (1984). In addition, subjects closely related to the topics of the present chapter are treated in Chapter 6.

11.2 Kinetic proofreading revisited

Hopfield (1974) and Ninio (1975) have described how an enzymic selection can be more accurate than the intrinsic selectivity (Blomberg, Ehrenberg and Kurland, 1980; Fersht, 1977a) of a single step. They start their analysis by suggesting that the intrinsic selectivity of enzymes is limited and that this limitation becomes crucial when the cognate substrate and its competitor are chemically very similar. The intrinsic selectivity is determined by the maximal difference in molar interaction standard free energy $(\Delta\Delta\mu_0)$ between the cognate substrate molecule and one of its competitors in the reaction pathway from substrate to product on the enzyme (Hopfield, 1974). In order for an enzymic pathway to be more accurate than its maximum single step selectivity, two conditions must be fulfilled, according to Ninio and Hopfield. First, the reaction route must be branched. The competing substrates enter the enzyme along the main branch. On the way to product formation there exist branch points where the incorrect substrate molecules are discarded with higher frequency than the correct ones. In this way the intrinsic selectivity of the enzyme can be used several times and the ratio between correct and incorrect flows increased for each step. However, a second condition is absolutely essential for such a scheme to give the desired enhancement of accuracy. This condition can be derived from the principle of detailed balance (de Groot and Mazur, 1969) which states that at equilibrium the chemical flows around every closed loop in a reaction scheme are zero.

The rate constants in a proofreading scheme, therefore, cannot be varied independently since every discard branch in the diagram corresponds to such a closed loop. The law of detailed balance necessitates a special driving force for the discard branches since they must be able to release a net flow of substrate molecules out from the main pathway. Hopfield (1974) and Ninio (1975) suggested that hydrolysis of a nucleoside triphosphate accompanies every discard event in proofreading. The displacement from equilibrium of the nucleoside triphosphate from its hydrolytic products could then give the requested driving force for the discard reaction. Furthermore, this would immediately suggest how to identify and measure proofreading experimentally (Hopfield, 1974; Ninio, 1975).

Since every discard reaction in translation will be accompanied by the hydrolysis of a GTP molecule, the number of GTP molecules hydrolysed by

EF-Tu per incorrect peptide bond, divided by the corresponding number per correct peptide bond, will be the factor by which proofreading enhances the accuracy of the ribosomes.

This powerful and simple connection has inspired a number of *in vitro* investigations of the accuracy of the aminoacylation reaction (Fersht and Kaethner, 1976; Fersht, 1977b; Hopfield *et al.*, 1976; Yamane and Hopfield, 1977; von der Haar and Cramer, 1976; Fersht and Dingwall, 1979), as well as of the accuracy of translation (Thompson and Stone, 1977; Jelenc and Kurland, 1979; Thompson, Dix and Eccleston, 1980; Thompson *et al.*, 1981; Thompson and Karim, 1982; Yates, 1979; Ruusala, Ehrenberg and Kurland, 1982a; Andersson and Kurland, 1983; Ruusala *et al.*, 1984; Ruusala and Kurland, 1984; Bohman *et al.*, 1984; Jelenc and Kurland, 1984).

A new link between the dissipative losses of enzymic selections and their accuracy guided those experiments. They showed that the understanding of code translation requires, not information theory, but a conceptual framework in which there exist flows of matter and energy as well as driving forces for these flows. They showed that it is not sufficient to study *at equilibrium* the structures and sequences of proteins and nucleic acids and their interaction standard free energies. For example, Pauling (1957) made estimates of maximal differences in interaction standard free energies ($\Delta\Delta\mu_0$) between different pairs of amino acids. From these estimates he predicted that amino acid substitutions in proteins occur with error frequencies in the percent range or even higher. Fersht (1981) showed that Pauling had underestimated the real intrinsic selectivity of the aminoacyl tRNA-synthetases. However, Fersht's figures would also lead to predictions of very high error levels for amino acid substitutions in proteins. In contrast, measured *in vivo* error frequencies in proteins (Loftfield, 1963; Loftfield and Vanderjagt, 1972; Edelmann and Gallant, 1977; Bouadloun, Donner and Kurland, 1983) showed that Pauling's predictions, based on equilibrium considerations, are wrong by several orders of magnitude. This discrepancy between Pauling's predictions and the real *in vivo* error frequencies demonstrated the shortcomings of previous equilibrium-based views and contributed to the formulation of the kinetic proofreading concept (Ninio, 1975; Hopfield, 1974).

Kurland (1978) emphasized the fact that when there is equilibrium between substrates and products in an enzymic reaction the selectivity of that pathway is lost. The displacement of the product of the reaction from equilibrium with its substrate must therefore always influence the enzyme's accuracy. He also showed formally that the displacement from equilibrium of the participating nucleoside triphosphate with its hydrolytic products makes proofreading possible in branched kinetic schemes. In his description, chemical flows and thermodynamic driving forces were used explicitly so that here enzymic selections were connected with irreversible thermodynamics. A closer look at these relations between displacements and accuracy is our next topic.

11.3 Displacements in enzymic selections

Let E denote an enzyme or a molecular complex which selects the correct substrate (S^c) from its incorrect competitors (S^w). Let G be a cosubstrate for the reaction (ATP in aminoacylation, GTP in translation). The pathway is constructed so that the transition $G \to M$, where M is the hydrolytic product of the cosubstrate (G), is coupled to the formation of the selected products (P^c or P^w):

$$E + S + G \underset{k_{-1}}{\overset{k_1}{\rightleftharpoons}} (ESG) \underset{k_{-2}}{\overset{k_2}{\rightleftharpoons}} E + P + M. \qquad (11.1)$$

The concentration of product, $[P]$, is, in equilibrium, related to the concentration of substrate, $[S]$, by the difference between the *standard* molar free energies of S and P. A similar relationship holds for $[G]$ and $[M]$:

$$\left(\frac{[P]}{[S]}\right)_{eq} = \exp(\Delta\mu_0^{PS}/RT) = K_{PS} \qquad (11.2)$$

$$\left(\frac{[M]}{[G]}\right)_{eq} = \exp(\Delta\mu_0^{MG}/RT) = K_{MG}.$$

If the kinetic pathway between S and P is open, the reaction will stop when the first relation in Reaction (11.1) is fulfilled. Similarly, with an open pathway between G and M the reaction will stop when the second relation in Equation (11.2) is fulfilled. In Reaction (11.1), however, these two transformations are coupled and there will therefore be a coupled equilibrium relation for the reaction. In this state the two relations in Equation (11.2) are in general not valid separately:

$$\left(\frac{[P][M]}{[S][G]}\right)_{eq} = K_{PS} K_{MG} = K. \qquad (11.3)$$

Here $[P]$ may be shifted far above detailed equilibrium with $[S]$ provided that $[M]$ is shifted below equilibrium with $[G]$ by the same factor. The total free energy loss in going from G to M is, in such a coupled equilibrium, exactly compensated for by the total free energy gain in going from S to P so that the free energy of $S + G$ equals the free energy of $P + M$. Reaction (11.1) describes, in a simple way, the transduction of free energy from the cosubstrate G to the selected product P. This transduction of free energy can be related to the accuracy if chemical flows are introduced.

In the steady state, where products, P, are continuously removed and nucleoside triphosphates, G, regenerated from M the product of the ratios of concentrations in Equation (11.3) is smaller than the equilibrium constant, K. The factor by which the left member of Equation (11.3) is shifted from the equilibrium constant is the displacement, δ:

$$\frac{[P][M]}{[S][G]} = K\delta. \tag{11.4}$$

When $\delta = 1$ in Equation (11.4) the equilibrium situation (Equation (11.3)) is obtained. When $\delta < 1$, there is a net chemical flow from S to P. One mole of substrate corresponds to the total free energy W_S, where

$$W_S = RT\ln([G][S]) + \mu_{0G} + \mu_{0S} + C. \tag{11.5}$$

One mole of product corresponds to the total free energy W_p, where

$$W_P = RT\ln([M][P]) + \mu_{0M} + \mu_{0P} + C. \tag{11.6}$$

These two expressions are equal only at equilibrium. In the steady state there are always free energy losses $\Delta W = W_S - W_p$. The displacement, δ, defined in Equation (11.4), is related to the molar free energy loss, ΔW, by an exponential:

$$\delta = \exp(-\Delta W/RT). \tag{11.7}$$

ΔW is obtained from δ by the inverse of this expression:

$$\Delta W = -RT\ln(\delta). \tag{11.8}$$

As long as the rate constants of the enzymic reaction do not change with the degree to which the enzyme is saturated or with the displacement from equilibrium of the pathway, we may ascribe to the enzyme an R factor ($R = k_{cat}/K_m$) which summarizes its catalytic capacity to transform a particular substrate to product (Fersht, 1977a; Ehrenberg and Blomberg, 1980). The R factor can be used to write the flow in the forward direction from substrate to product as

$$J_F = R[E][S][G].$$

The enzyme cannot change the equilibrium distribution of the reactants (see Fersht, 1977a) and the back flow (J_B) must therefore be

$$J_B = \frac{R}{K}[E][P][M].$$

With the displacement, δ, in Equation (11.4) the net forward flows $J^{c,w} = J_F^{c,w} - J_B^{c,w}$ can be written

$$J^c = [E][S^c][G]R^c(1 - \delta^c),$$

$$J^w = [E][S^w][G]R^w(1 - \delta^w) \tag{11.9}$$

for correct (S^c) and incorrect (S^w) substrates.

The R factor is related to the association rate constant k_1 for the encounter between a substrate and the enzyme by a weighting factor, α, which is the probability that a product molecule forms and leaves the enzyme, given that an enzyme–substrate complex has been formed (see Ninio, 1975):

$$R = k_1\alpha. \tag{11.10}$$

For Reaction (11.1) we have

$$R = \frac{k_1 k_2}{k_{-1} + k_2}. \tag{11.11}$$

Thus, the probability α is given by

$$\alpha = \frac{1}{1 + k_{-1}/k_2} = \frac{1}{1 + a_1}. \tag{11.12}$$

To calculate the error level in this enzymic selection, standard conditions are assumed. These are: (1) equal concentrations of correct and incorrect substrates; (2) the standard free energy difference between substrate and product is the same for the two competitors; (3) there is a non-selective removal of products. Then the accuracy, A, defined as the ratio of the two flows in Equation (11.9), is given by

$$A = d\left(1 - \frac{d-1}{d} \times \delta^c\right). \tag{11.13}$$

The factor d is the ratio R^c/R^w and defines the intrinsic selectivity of the enzyme. From Equation (11.13) it follows that when the displacement, δ^c, is 1.0, the accuracy, A, is 1.0 as well, so that the selectivity of the pathway is lost in this (equilibrium) limit. In the other extreme, when δ^c approaches zero and the dissipation goes to infinity, the accuracy tends to its maximum value, d. The level of the product displacement is thus important for ordinary enzymic selections. However, thermodynamic driving forces play an even more prominent role in kinetic proofreading and this will be our next topic.

11.4 Displacements in kinetic proofreading

By the introduction of discard branches, Reaction (11.1) may be transformed into a reaction with kinetic proofreading.

$$E + S + G \underset{k_{-1}}{\overset{k_1}{\rightleftharpoons}} (ESG)_1 \underset{k_{-2}}{\overset{k_2}{\rightleftharpoons}} (ESG)_2 \ldots (ESG)_n \underset{k_{-(n+1)}}{\overset{k_{n+1}}{\rightleftharpoons}} E + P + M$$

$$q_{-1} \Bigg\Vert q_{+1} \qquad q_{-2} \Bigg\Vert q_{+2} \qquad q_{-n} \Bigg\Vert q_n \qquad\qquad (11.14)$$

$$E + S + M \qquad\quad E + S + M \qquad E + S + M$$

The selectivity of this new diagram is specified by its intrinsic selectivity as well as by the thermodynamic driving forces that are associated with its branches. 'Partial' R factors for the main pathway can be defined as:

$$R_i = C_{i-1} k_i$$
$$C_0 = 1. \qquad\qquad (11.15)$$

C_i is the equilibrium constant between a state $(ESG)_i$ and the free state $E + S + G$. Rate factors for the discard steps are given by the parameters Q_i:

$$Q_i = C_i q_i. \qquad\qquad (11.16)$$

Intrinsic selectivities in the forward direction of the diagram are defined as R_i^c / R_i^w and the selectivities along the discard steps as Q_i^w / Q_i^c.

When all steps have the same selectivity every discard step can maximally contribute with the same factor $d = (R_i^c / R_i^w)(Q_i^w / Q_i^c)$ to the accuracy of the reaction (Blomberg, Ehrenberg and Kurland, 1980; Blomberg, 1983b).

When the intrinsic selectivity of the mechanism originates exclusively from a higher affinity (C_i^c) for the correct substrate to the enzyme than for its competitor (C_i^w) then $Q_i^w / Q_i^c = 1$ and $d = R_i^c / R_i^w$. In the 'double sieve' mechanism (Fersht, 1977a; and see Chapter 4) the incorrect substrate is assumed to be smaller than the correct one. Here, the discard reaction is much more selective than the forward selection along the main branch of the diagram. In this case Q_i^w / Q_i^c is larger than R_i^c / R_i^w.

There are two external thermodynamic driving forces in Reaction (11.14). The first comes from the displacement, δ, of the products of the reaction, P, M, from equilibrium with its substrates S, G, and this drives the product formation flow. The second is the drive associated with the discard reactions. When a substrate, S, is discarded at a proofreading branch point and returned to the free state, the reaction cycle is accompanied by the hydrolysis of the cosubstrate, G, to its product, M. The displacement $1/\epsilon$ of the cosubstrate from equilibrium with its product provides a driving force for the dissociation of substrate, S:

$$\frac{[M]}{[G]} = K_{MG}\epsilon. \tag{11.17}$$

The molar free energy associated with the discard cycle is the difference between the total free energy of G (W_G) and the total free energy of M (W_M):

$$\Delta W_{GM} = W_G - W_M = -RT\ln(\epsilon). \tag{11.18}$$

When M is in equilibrium with G, then $\epsilon = 1$ in Equations (11.17) and (11.18), and there is no thermodynamic driving force for the discard cycles ($\Delta W = 0$). When $\epsilon = 1$, detailed balance requires that there is a net *inflow* over the discard branches. From this it follows that there can be no error reduction by proofreading when the discard displacement, ϵ, equals one, since proofreading requires a net *outflow* over the discard branches (Blomberg, Ehrenberg and Kurland, 1980; Blomberg, 1983b). When the discard displacement is smaller than one, the situation is different. First, the free energy loss over the discard steps (Equation (11.18)) is not zero but has a positive value, and the smaller ϵ is, the larger is the molar free energy loss per discard event according to Equation (11.18). Second, the detailed balance constraint has now been shifted from its ground level ($\epsilon = 1$) to a much more 'relaxed' level ($\epsilon \ll 1$). When ϵ is small, the inflow over the discard steps is much smaller than the outflow over those steps. There can be a net flow out from the consecutive branch points of the diagram until one of its states $(ESG)_i$ has been shifted below equilibrium with the free state $E + S + G$ by the discard factor. At that step, where this happens, there is no driving force for the discard reaction since here the enzyme–substrate complex $(ESG)_i$ has the same total free energy as the product of the discard reaction, $E + S + M$. In other words, here $(ESG)_i$ is in equilibrium with $E + S + M$.

This means that the proofreading of incorrect substrates can only be carried to a point where the incorrect enzyme–substrate complex has been displaced by a factor ϵ from equilibrium with the free state. Accordingly, the accuracy of any proofreading mechanism cannot exceed d/ϵ, where d is the intrinsic discrimination in the forward direction. This argument leads to a generalization of Equation (11.13):

$$A < \frac{d}{\epsilon}\left(1 - \frac{d-1}{d} \times \delta^c\right) \tag{11.19}$$

The driving force for the discard reaction thus sets a theoretical upper limit for the accuracy, A, of all kinetic proofreading schemes. When the cosubstrate is ATP, as in the charging of transfer RNA, ϵ is 10^{-10}, and for GTP it is 10^{-8} (Kurland, 1978). The actual *in vivo* error level in protein synthesis, between 3×10^{-4} and 2×10^{-3} (Edelmann and Gallant, 1977; Loftfield and Vanderjagt, 1972; Bouadloun, Donner and Kurland, 1983), is far above what the limit of Equation (11.19) would allow even if small values of d were assumed. Only in DNA replication, which has error frequencies as low as 10^{-10} (Loeb and Kunkel, 1982), may accuracy approach the limit (Equation (11.19)). It is possible that the requirements of gene copying have made the small *in vivo* values of the discard displacement necessary.

In a proofreading reaction such as (11.14), every discard branch can increase the selectivity of the reaction, but by not more than a factor d. Therefore, a second inequality for the accuracy is also valid:

$$A < d^{n+1}. \tag{11.20}$$

The smaller of the two expressions (11.19) and (11.20) determines the upper limit for the accuracy that can be obtained by proofreading. However, in order for a discard branch to increase the accuracy by the factor d, the discard reaction must dominate over the forward reaction not only for incorrect, but also for correct substrates. An additional requirement is that the product displacement, δ^c, is kept at very low values and that the discard displacement is small. Therefore, the free energy loss per mole of product must approach very high values when the selection is brought near either of its upper limits (11.19, 11.20). Now we identify the inflow of substrate in Reaction (11.14) as J_0 and the product flow as J_n. The difference between the flows $(J_0 - J_n)$ then corresponds to the discard flows summed over all branches of the diagram. The free energy loss of the reaction per mole of product consists of two parts. The first is related to the discard reactions and the second to the driving force for the product formation. Thus,

$$\Delta W = -RT \frac{[(J_0^c - J_n^c) + (J_0^w - J_n^w)] \ln(\epsilon) + J_n^c \ln(\delta^c) + J_n^w \ln(\delta^w)}{J_n^c + J_n^w}.$$

The accuracy is the ratio between the correct and incorrect product formation flows, i.e. $A = J_n^c / J_n^w$. If we use the standard relation $A = \delta^c / \delta^w$, this can be written as

$$\Delta W = -RT \left[\left(\frac{J_0^c}{J_n^c} - 1 \right) \frac{A}{A+1} + \left(\frac{J_0^w}{J_n^w} - 1 \right) \frac{1}{A+1} \right] \ln(\epsilon)$$

$$-RT \left[\ln(\delta^c) - \frac{\ln(A)}{A+1} \right]. \tag{11.21}$$

The first part of this expression contains the factor $-RT\ln(\epsilon)$, which is the molar free energy loss of the discard reaction. Thus, the first bracket in Equation (11.21) contains the free energy loss over the discard steps per mole of product. The second part in this equation is the free energy loss common to all enzymic selections, irrespective of whether or not they have error correction. This is the dissipation corresponding to the transformation of substrate to product.

When the discard displacement decreases, the molar loss in Equation (11.21) increases. Furthermore, the approach to the limit (11.20) necessarily implies that the discard reactions become more and more dominant, which also implies that the molar losses of free energy increase. In other words, when the accuracy becomes higher the mechanism always dissipates more free energy per product. If, on the other hand, the discard displacement tends to zero and the number of error-correcting steps tends to infinity there is no obvious upper limit to the accuracy even for moderate values of the intrinsic selectivity, d. This brings us back to an earlier argument. By a detailed calculation one could, in principle, predict the intrinsic selectivity of, for example, an aminoacyl-tRNA synthetase with respect to its binding of two competing amino acids. However, we have seen here that the connection between this parameter and the accuracy of the tRNA synthetase, tacitly assumed by Pauling, simply does not exist. In fact, the whole kinetic scheme in Reaction (11.14), as well as the displacements associated with it, have to be known before the proper dependence of the accuracy on the intrinsic selectivity can be calculated. Furthermore, since every a priori determined accuracy can be surpassed in principle simply by an increase in the number of discard steps and the magnitude of driving forces, this suggests that the accuracy levels in vivo of replication, transcription and translation are not constrained by a theoretical limit of attainable accuracy. Instead, the accuracy must be optimized with respect to some other parameter(s). This notion of optimality gets support from the fact that there exist ribosomal mutants with lower as well as higher accuracy than wild type. The proofreading properties of such mutants will be discussed in the next section.

11.5 Kinetic proofreading in translation

The first attempts to detect kinetic proofreading in E. coli ribosomes by Thompson and Stone (1977) were soon followed by other similar experiments (Yates, 1979; Thompson, Dix and Eccleston, 1980; Thompson et al., 1981; Thompson and Karim, 1982). All results of those experiments, which were performed in the absence of EF-G, pointed in the same direction: between one and two GTP molecules are hydrolysed per peptide bond when poly(U)-programmed ribosomes enzymically bind a correct tRNA^{Phe} to the A-site. In contrast, when an aminoacyl-tRNA with a nonmatching anticodon binds to the A-site there is a much larger number of GTP molecules hydrolysed per peptide bond. For the isoacceptor $tRNA_2^{Leu}$ about twenty-five times more GTPs were hydrolysed per peptide bond than for the correct $tRNA^{Phe}$ (Thompson *et al.*, 1981).

The design of those experiments leads to interpretation ambiguities (Kurland, 1978; Kurland and Ehrenberg, 1984) and we have therefore made a different experimental approach for the investigation of proofreading in translation. An *in vitro* system for steady-state translation was used at 37 °C with all three elongation factors present (Jelenc and Kurland, 1979; Pettersson and Kurland, 1980; Wagner *et al.*, 1982). This system has buffer conditions which mimic the composition of ions *in vivo* (Jelenc and Kurland, 1979). It is described in detail in Chapter 6. The most important property of this buffer system is that the error frequency is similar to the amino acid substitution rate *in vivo* (Loftfield and Vanderjagt, 1972; Edelmann and Gallant, 1977; Bouadloun, Donner and Kurland, 1983), while its maximal elongation rate of five to ten amino acids per second (Wagner *et al.*, 1982) is only somewhat smaller than the elongation rates *in vivo* (Maaløe, 1979; Churchward, Bremer and Young, 1982).

A nucleoside triphosphate energy regenerating system (Jelenc and Kurland, 1979) brings the thermodynamic driving forces to high values as in the living bacterium. The close similarities between this *in vitro* system and the characteristics of *E. coli* indicate that the results obtained regarding its proofreading properties are of biological relevance.

The detection of kinetic proofreading in translation is based on a counting of the number of cycles of EF-Tu that are necessary to make an incorrect peptide bond (f_w) and the number of cycles to make a correct peptide bond (f_c) (Ruusala, Ehrenberg and Kurland, 1982a). The ratio between these two numbers, f_w/f_c, is the factor, F, by which proofreading increases the accuracy of translation. The accuracy of ternary complex selection is composed of an initial selection factor I before GTP hydrolysis and the factor F from proofreading:

$$A = IF. \tag{11.22}$$

Two requirements must be met in these experiments which are designed to measure the efficiency of ternary complex usage. First, the rate of protein synthesis must be limited by the availability of ternary complex. Second, the cycle of EF-Tu must be rate limited by the release rate, k_d, of GDP so that the amount of EF-Tu (Tu_0) equals the amount of TuGDP. This latter condition can easily be fulfilled for both correct and incorrect tRNAs by excluding EF-Ts and thus taking advantage of the very slow, spontaneous, release rate of the guanine nucleotide from EF-Tu (Ruusala, Ehrenberg and Kurland, 1982b). When both these conditions are met the simultaneous flows into polypeptide of correct (j_c) and incorrect (j_w) amino acids are linearly related as follows:

$$f_w j_w + f_c j_c = \text{Tu}_0 k_d. \tag{11.23}$$

$\text{Tu}_0 k_d$ is the total number of EF-Tu cycles that are completed per unit time. This number is the sum of the cycles per unit time related to incorrect peptide bond formation, $f_w j_w$ and those related to correct elongation, $f_c j_c$. When there is no proofreading, f_c as well as f_w is equal to one. If, in contrast, there is proofreading, then f_w is much larger than one and also much larger than f_c. A rearrangement of terms in Equation (11.23) gives

$$j_w = -j_c/F + \text{Tu}_0 k_d/f_w. \tag{11.24}$$

By plotting j_w as a function of j_c, where the latter flow can be varied, for example by changing the charging level of correct tRNA (Ruusala, Ehrenberg and Kurland, 1982a; Andersson and Kurland, 1983; Bohman *et al.*, 1984), the *F* factor can be calculated from the slope of the straight line.

Alternatively, separate determinations of f_c and f_w can be made by keeping, respectively, j_w or j_c equal to zero and by varying the input amount of EF-Tu (Ruusala and Kurland, 1984; Bohman *et al.*, 1984). From Equation (11.24), with $j_w = 0$

$$j_c = \text{Tu}_0(k_d/f_c). \tag{11.25}$$

From Equation (11.24) with $j_c = 0$

$$j_w = \text{Tu}_0(k_d/f_w). \tag{11.26}$$

The detailed technical aspects of these proofreading assays are described in a forthcoming article (M. Ehrenberg, C. G. Kurland and T. Ruusala, 1985). Here, as well as in Bohman *et al.* (1984) the two ways (Equations (11.24)–(11.26)) of measuring the *F* factor are in good agreement. The data show that *E. coli* ribosomes indeed seem to proofread ternary complexes (Ruusala, Ehrenberg and Kurland, 1982a), at least *in vitro*. The isoacceptor tRNA$_4^{\text{Leu}}$ has a normalized error frequency of 6×10^{-4} when it competes with tRNA^{Phe} in the poly(U)-programmed *in vitro* system. This error level is obtained by an initial selection, *I*, of about forty, multiplied by a proofreading selection, *F*, which also contributes to the accuracy by a factor of forty. Measurement on the tRNA$_2^{\text{Leu}}$ isoacceptor shows that this tRNA has an error frequency of 1.1×10^{-4} when it competes with tRNA^{Phe}. Here, the initial selectivity is 90 and the proofreading factor is 100. For these two isoacceptors tRNA$_4^{\text{Leu}}$ and tRNA$_2^{\text{Leu}}$, the ribosome increases its accuracy by approximately the same factor in each of the two selection steps. For each Phe–peptide bond that is made, between 1.05 and 1.2 cycles of EF-Tu are required (Ruusala, Ehrenberg and Kurland, 1982a; Bohman $et\ al.$, 1984).

If the proofreading occurs in a single step, the intrinsic selectivity, d, as defined above, can be calculated (Bohman $et\ al.$, 1984). For $tRNA_4^{Leu}$, which normally reads UUG, d is approximately 400 and for $tRNA_2^{Leu}$, which normally reads CUU, d is approximately 1000. These estimates depend critically on the difference between the measured value of f_c and 1.0. The values of d are estimated to be correct within a factor of two up or down (lower limit of f_c = 1.05, upper limit of f_c = 1.2). Therefore, the ribosomes use only between five and twenty percent of their maximal discrimination capacity at the proofreading branch under our conditions. Now, if d is the same at both the proofreading and initial selection steps, the actual accuracy in translation is only one percent or less of its maximum. Thompson and Karim (1982) attempted to measure d for $tRNA_2^{Leu}$ in competition with $tRNA^{Phe}$ on poly(U). Their experiment, which we have discussed in detail elsewhere (Kurland and Ehrenberg, 1984), gave a d value of about 40000; a figure forty-fold above the maximum estimate given here. However, this large d-value is derived by a set of arguments containing logical errors (Thompson and Dix, 1982). A more conservative estimate, which says that d in initial selection is larger than 2000, can be obtained from the data given by Thompson and Karim (1982). This value is similar to the d values obtained for the proofreading step. This underusage of the full discrimination capacity of the programmed ribosome emphasizes the point made above, that translation accuracy is optimized rather than maximized. It also turns out that this underusage of the accuracy can make translation stable against error propagation as discussed below (see also Kurland and Ehrenberg, 1984; Ehrenberg and Kurland, 1984).

Indeed, we have investigated how the translation machinery responds when the ribosomal accuracy is shifted away from its optimum. Such error shifts are associated with several classes of ribosomal mutants, where the mutation can increase as well as decrease the accuracy. The error levels can also be increased by a variety of drugs.

Changes in the ribosomal protein S4 can lead to greatly enhanced nonsense suppression (Gorini, 1971). The rates and error levels (Andersson $et\ al.$, 1982) of three such ribosomal ambiguity mutants (ram) were studied as well as their proofreading characteristics (Andersson and Kurland, 1983). The three ram phenotypes elongate $in\ vivo$ (Andersson $et\ al.$, 1982) with rates indistinguishable from those of wild type. $In\ vitro$, the maximal turnover rates, k_{cat}, for all mutants are the same as the maximal rate for the wild type. The increased nonsense suppression $in\ vivo$ has its counterpart $in\ vitro$ in a five- to ten-fold increase in the missense error levels (Andersson $et\ al.$, 1982). These error enhancements are all due to changes in the selectivity of the proofreading step (Andersson and Kurland, 1983). Thus, although there is a five-to ten-fold error increase as a result of the reduced proofreading factor in the mutants, those large changes do not lead to detectable changes in the translation rate as compared to wild type, either *in vivo* or *in vitro*.

The reason for this lack of response, which contradicts predictions of Galas and Branscomb (1976), becomes evident when one considers that the proofreading parameter which is relevant for cognate peptide elongation is not the F factor, which changes dramatically with the S4 modification, but the number of dissipated EF-Tu cycles per *correct* peptide bond, f_c. This parameter drops from approximately 1.1 in the wild type to values indistinguishable from 1.0 in the *ram* mutants. Thus, not more than a ten percent increase in k_{cat}/K_m for the *ram* mutant and an even smaller increase in k_{cat} would be expected as a result of the decreased proofreading (Ehrenberg and Kurland, 1984; see also below). Thus, the gain in translation rate appears to be very small when proofreading is decreased from its wild type level, whether one considers the R factor or the maximum turnover rate. The failure to detect an increased translation rate of *ram* mutants may also come from a reduced kinetic efficiency of the ribosomes due to the fact that they contain between five and ten times more amino acid errors than wild type ribosomes (Ehrenberg and Kurland, 1984). These questions will be given a more thorough discussion below when we have inspected mutants with increased accuracy and proofreading and after we have considered some of the supporting conditions for maximal growth rates. It seems clear already at this point, however, that *ram* ribosomes are inferior to wild type ribosomes in that they are more error prone and at the same time they do not offer any advantage with respect to their translation rate.

We have investigated five streptomycin resistant (*SmR*) ribosomal phenotypes with respect to their accuracy and elongation rates *in vivo* as well as *in vitro* (Bohman et al., 1984). The mutants were either restrictive, with strongly reduced nonsense suppression *in vivo*, or non-restrictive with unchanged (or slightly increased) nonsense suppression. Those *SmR* mutants which were error restricted also had a smaller elongation rate *in vivo*. The two most restrictive strains had elongation rates, calculated from β-galactosidase completion, of 7.8 and 9.3 amino acids per second per ribosome which is significantly lower than the fourteen amino acids per second per ribosome obtained for wild type. The unrestricted ribosomes, in contrast, elongate *in vivo* at the same rate as wild type ribosomes.

The nonsense restricted mutant ribosomes had considerably lower missense errors *in vitro* than ribosomes from wild type bacteria. Most of the error reduction is associated with an enhanced proofreading activity of the restrictive *SmR* ribosomes. This behaviour was seen particularly clearly when the competition between the isoacceptor $tRNA_2^{Leu}$ and $tRNA^{Phe}$ was observed. Here, the variations in the error levels of the different ribosomal phenotypes followed precisely the variations of the proofreading factors, F.

These changes in error levels, p_E, caused by changes in the proofreading factors of the different mutants, were related to changes in the numbers of EF-Tu cycles that are required to form correct peptide bonds. The more restrictive a mutant is with respect to nonsense suppression (*in vivo*) or with respect to missense errors (*in vitro*), the more EF-Tu cycles are used per Phe–peptide bond *in vitro*.

Along with those changes in f_c, we observe a corresponding reduction in the k_{cat}/K_m values of the restrictive mutants and a concomitant increase in the K_m value for ternary complex interaction with the ribosomes. The maximum turnover rate was, in contrast, virtually unaffected by the mutation, even for the most restrictive ribosomal phenotype. It could be concluded from this, first, that the reduced *in vivo* β-galactosidase elongation rate in restrictive *SmR* bacteria is not due to a reduction in the maximal translation rate. The explanation for this rate reduction may be, instead, the increase in K_m value for ternary complex, or the increased excess hydrolysis of guanine nucleotides due to the increased proofreading in the mutant. If the first explanation was true, that would imply that the ribosomes are not running very close to their maximal rate *in vivo*, in contrast to previous views (Maaløe, 1979). Instead, the ternary complex concentrations are so small that a two-fold increase in K_m value can give a significant effect on the elongation rate.

In recent *in vivo* experiments, where elongation rates were measured for different proteins, Pedersen (1984) showed that the elongation rate for one and the same protein varies with the growth medium so that it is higher in a rich medium than in a poor one. He also noted that the elongation rate is different for different proteins: high copy number proteins are translated significantly faster than low copy number proteins (see below). These results are consistent with the notion that ribosomes are unsaturated with ternary complex *in vivo* so that small variations in K_m can also have significant effects on the translation rate.

The second explanation of the reduced elongation rate *in vivo* is that the GTP-regenerating kinases or EF-Ts cannot keep up with the increased consumption of guanosine triphosphates due to the increased proofreading in the mutants. Therefore, an increase in f_c could decrease the ternary complex/Tu. GDP ratio down to a level which significantly decreases the rate of protein synthesis.

These correlations between proofreading activities detected *in vitro* and elongation, as well as growth rate *in vivo*, were even more dramatic for another bacterial mutant (Ruusala *et al.*, 1984). This is an intermediate between the classical streptomycin-resistant (*SmR*) and streptomycin-dependent (*SmD*) phenotypes, which are associated with a change in the ribosomal protein S12. It does not strictly depend on Sm but its growth rate was stimulated by the presence of Sm in the medium (Zengel *et al.*, 1977); we call this the *SmP* phenotype.

In parallel with this growth stimulation of the SmP mutant by Sm there is an increase in the *in vivo* ribosomal elongation rate (Zengel *et al.*, 1977; Ruusala *et al.*, 1984). From five amino acids per second per ribosome in the absence of Sm, the rate increased to almost ten amino acids per second per ribosome in the presence of the drug, compared with a wild type translation of fifteen amino acids per second under the same conditions. The reduced elongation rate in the absence of Sm correlated with a drastically reduced nonsense suppression frequency of the SmP mutant. The UGA suppression frequency was context-dependent and varied between values twenty and forty times smaller than the corresponding frequency in the wild type. When Sm was added, however, the SmP mutant nonsense suppression frequency became larger than the suppression frequency of the wild type in the absence of Sm, by almost an order of magnitude. In summary, *in vivo*, in the absence of Sm, this phenotype is characterized by small translation rates, low error levels, and a significantly reduced growth rate. When Sm is added, the translation rate is stimulated and the growth rate is increased, even though the accuracy of the translation machinery has now decreased considerably.

The missense error level of the SmP ribosomes, measured *in vitro* as a leucine error in poly(Phe) synthesis on poly(U), was somewhat lower than that of the wild type (Ruusala *et al.*, 1984). This error reduction is associated with enhanced proofreading. The number of EF-Tu cycles per correct peptide bond was 1.7 for the SmP mutant and 1.1 for the wild type. The corresponding numbers of the isoacceptor $tRNA_2^{Leu}$ were 180 and 45 for mutant and wild type ribosomes, respectively. The enhanced proofreading of SmP ribosomes was correlated with a reduced kinetic efficiency as for the restricted SmR mutants described above. The K_m value for correct ternary complexes was almost two-fold above the wild type level for the mutant, and the k_{cat}/K_m value (R factor) was reduced more than two-fold. When Sm was added to the SmP ribosomes, their missense error level went up to values more than one order of magnitude above the missense error level of the wild-type ribosome in the absence of Sm. The effects of Sm on accuracy for this mutant were exclusively a result of perturbation of the proofreading step. Here, as well as in other cases (Ruusala and Kurland, 1984), the drug almost completely eliminated the discard probability at the proofreading branch. Thus, f_c went from 1.7 down to 1.0 and f_w dropped from 180 to 3.5, while the initial selectivity, I, was unchanged. Sm improved the kinetic efficiency of the SmP ribosomes with respect to their K_m value, which was reduced two-fold, and with respect to their R factor, which was increased by a factor of two. However, the maximum turnover rate of the SmP ribosomes does not increase when they interact with the drug. In contrast, a small reduction in k_{cat} was observed when Sm was added to the system.

Thus, the stimulatory effects on growth rate of Sm can be explained from our *in vitro* data if we assume that translation is normally under K_m limitation so that it is the R factor for the ternary complex–ribosome interaction rather than the translation k_{cat} that determines the elongation rate. Alternatively, it is conceivable that the extra GTP hydrolysis of the *SmP* mutant leads to a reduced elongation rate *in vivo*. As a consequence here, as well as for the *SmR* mutant discussed above (Bohman *et al.*, 1984), the regeneration of ternary complex from EF-Tu.GDP is not fast enough to support an unimpeded elongation rate. Therefore, a high f_c value could lead to a depletion of ternary complex levels and thus bring down the *in vivo* efficiency of the ribosomes. Such an effect could be the result of a limited rate of conversion of GDP to GTP by the kinases of the bacterium as suggested above, or alternatively, the limitation could be the consequence of restrictive amounts of EF-Ts present. If the cycling rate of EF-Tu goes up when f_c increases, then an insufficient concentration of EF-Ts leads to increased levels of EF-Tu. GDP and to a depletion of ternary complexes.

In any case, we wish to suggest that both the *SmR* phenotype and the *SmP* phenotype illustrate the notion that excessively accurate translation is bad for the bacterium. In order to discuss optimal translation error levels, we must consider first some general requirements for optimal growth rates of cells.

11.6 Efficiency of biochemical pathways

In this section, we analyse how a condition of maximal growth rate influences the biochemical pathways in growing bacteria. One aspect of these pathways may be outlined as follows (Ingraham, Maaløe and Neidhardt, 1983). A reaction sequence in a growing cell transforms substrate molecules via a number of intermediate steps to product. The molecules that enter the pathway of substrate will, in their turn, be the final product of another, preceding pathway. The product of the pathway will, in a similar way, enter a subsequent reaction chain as initial substrate. These general features of coupled biochemical pathways lead to metabolite pool constraints for exponentially growing bacteria (Ehrenberg and Kurland, 1984; Kurland and Ehrenberg, 1984). The introduction of a couple of formal parameters will facilitate the discussion of those constraints and will, in particular, allow a new definition of costs of biochemical reactions. This definition will make it possible to compare different and previously unrelated costs of accuracy in translation (Ehrenberg and Kurland, 1984; Kurland and Ehrenberg, 1984) as well as in other selections. Let ρ_1 be the protein density of a particular pathway (1) of the growing bacterium defined so that

$$\rho_1 = \sum_i [E_{1i}]N_{1i} \tag{11.27}$$

where $[E_{1i}]$ is the concentration and N_{1i} is the number of amino acids of an enzyme i in the pathway 1. Furthermore, let j_{p1} be the flow of product from this pathway into subsequent pathways. A measure of the load of the pathway in relation to the total flow (j_a) of amino acids into polypeptide is defined as the ratio (α_1) between j_{pl} and j_a:

$$j_{pl} = \alpha_1 j_a. \tag{11.28}$$

The total amino acid density (ρ_0) of the bacterium is the sum of the partial densities (ρ_1) of all its pathways:

$$\rho_0 = \sum_1 \rho_1 \tag{11.29}$$

where we assume that the protein density (ρ_0) is a constant. This assumption (Ehrenberg and Kurland, 1984) is further discussed in forthcoming work (Kurland and Ehrenberg, in preparation) along with other constraints on the growth rate (k). The exponential growth rate can, under conditions of balanced growth, be defined as

$$k = \frac{1}{m} \frac{dm}{dt} \tag{11.30}$$

where m is the amount of amino acids in protein per unit volume of the bacterium. Since j_a is the total flow of amino acids into protein per unit time and unit volume the following identification can be made

$$\frac{dm}{dt} = j_a V. \tag{11.31}$$

From the definition of ρ_0

$$m = \rho_0 V, \tag{11.32}$$

so that by combining Equations (11.30)–(11.32),

$$k = j_a / \rho_0. \tag{11.33}$$

If we now use Equation (11.29) as well as (11.28), k can be written as follows:

$$k = \frac{1}{\sum_1 \rho_1 / j_a} = \frac{1}{\sum_1 \rho_1 \alpha_1 / j_{pl}}. \tag{11.34}$$

It is clear from Equation (11.34) that the inverse of the growth rate can be written as the sum of a set of times, τ_1:

$$1/k = \sum_1 \tau_1, \tag{11.35}$$

where τ_1 is defined as

$$\tau_1 = \rho_1/j_a = \rho_1\alpha_1/j_{p1}. \qquad (11.36)$$

τ_1 is a suitable measure of the cost of a particular pathway. It is the ratio between the amount of polypeptide invested in that pathway and the total amount of peptide bonds made in the organism per unit time. When τ_1 is very large, the cost of that pathway reduces the growth rate significantly. When τ_1 is small, in contrast, the pathway has only a minor influence on the growth rate.

The parameter α_1 is the amount of product that is needed from the pathway (1) per peptide bond and the ratio j_{p1}/ρ_1 defines its efficiency, e_1:

$$e_1 = j_{p1}/\rho_1. \qquad (11.37)$$

A comparison between Equations (11.37) and (11.36) shows that an efficient pathway is also a low cost pathway which is favourable for fast growth. In order for a pathway to be efficient, it must be constituted by enzymes which are also efficient in the sense that they have as large a ratio as possible between their turnover rate (v_{1i}) and the number of amino acids (N_{1i}) invested in them. We define the efficiency (e_{1i}) of an enzyme (E_{1i}) from the relation

$$e_{1i} = \frac{v_{1i}}{N_{1i}} = \frac{1}{N_{1i}} \frac{[S_{1i}]k_{1i}}{[S_{1i}] + k_{1i}}. \qquad (11.38)$$

This definition implies, first, that the substitution of two enzymes which catalyse the same reaction and have the same efficiency according to Equation (11.38) does not change the efficiency or the kinetic cost, τ_1, of the pathway in which they operate. From the viewpoint of rapid growth, therefore, a slow enzyme can be as good as a fast one provided that its mass is correspondingly smaller. Since a high value of the efficiency of an enzyme reduces the costs associated with its pathway, it naïvely looks as if the bacterium's enzymes should always turn over near their maximal rates, k_{1i}.

According to Equation (11.38), e_{1i} increases with increasing values of $[S_{1i}]$ until it reaches its maximum value (k_{1i}/N_{1i}) at very high substrate concentrations. However, during exponential growth it is not only the proteins and nucleic acids of the cell but also all its intermediate metabolite pools that grow exponentially. Under these conditions, the rate of increase of a pool is proportional to its size. This implies that there exist optimal pool concentrations for all the cell's metabolites where the growth rate is as fast as possible.

The optimal concentration of a particular pool in a chemical pathway is determined by the balance between two opposing tendencies. When its steady state concentration increases, more enzymes are required in the preceding part of the chain to sustain its balanced growth. This implies that the density of the early parts of the chain increases and this will tend to increase the cost of the pathway. When the pool concentration decreases instead, the efficiency of the enzyme E_{1i} decreases according to Equation (11.38). In order to keep the same total flow rate, more enzymes are required and this will therefore also tend to increase the density and the kinetic cost of the pathway. When a mass increase in the early part of the chain, due to a small concentration increase in the pool concentration, exactly counterbalances the mass decrease in the later part of the chain due to the efficiency increase of the enzyme, then the pool concentration is optimal.

The best values of the pool concentrations will depend on the detailed kinetic properties of all the enzymes of the pathway, but the flow load of the reaction sequence is also relevant for optimal concentration choices (Ehrenberg and Kurland, 1984). In a pathway which has a large output rate of product molecules, the increase of substrate pools plays a relatively minor role in relation to the main flow from substrate to product. For a pathway with a small output rate, in contrast, the increase in the pool concentrations will constitute a much larger fraction of the main flow. This indicates that the choice of optimal substrate pool concentrations might be different in a high load pathway from that in a less frequently used, 'low load', pathway. A more detailed calculation shows that this is indeed the case (Ehrenberg and Kurland, 1984): the system is optimally adjusted when the low load pathway enzymes are less saturated with substrate than the high load pathway enzymes. These results are valid for general reaction pathways and they are also true for the special case of the translation apparatus itself. As pointed out by Maaløe (1979), it is particularly important for growing bacteria to use their very large ribosomal mass as efficiently as possible. The optimal use of the translation apparatus during exponential growth is the topic of our next section and we will end the present one with some concluding remarks on maximal growth rates.

From Equation (11.35), where the inverse of the growth rate is written as the sum of the kinetic costs of all the bacterium's pathways, it seems intuitively obvious that maximal growth is associated with minimal costs in all reaction sequences. This is justified with proper qualifications as follows (Ehrenberg and Kurland, 1984). Ascribe to each pathway (1) an initial substrate concentration $[S_{10}]$ and a final product concentration $[P_1]$. The maximum growth rate, k_{max}, is now obtained, given the concentrations of substrate and product, by minimizing all kinetic costs separately. However, to complete the calculation of k_{max} the analysis must be brought one step further and also include the best choices of initial substrate and final product concentrations. These pools couple the different pathways of the cell together and make all kinetic costs interdependent.

11.7 Low cost translation

Maaløe (1979) and Ingraham, Maaløe and Neidhardt (1983) have emphasized the importance for growing bacteria of using the translation apparatus as efficiently as possible. Those features of the translation system that make such an optimization of its performance critical are, first, that the mass of the ribosome is very large. There are almost one million daltons of protein and about 1.5 million daltons of rRNA in the complete 70S particle (Maaløe, 1979). Secondly, the substrates for protein synthesis are themselves macromolecular, ternary complexes. They are therefore expensive to produce and a proper adjustment of their concentrations is necessary. Thirdly, protein synthesis has a load which varies considerably with the growth conditions of the cell population (Maaløe, 1979; Ingraham, Maaløe and Neidhardt, 1983).

In poor growth media the total peptide bond formation rate per bacterial mass is small. Here, the density of the translation apparatus contributes only a minor fraction of the bacterium's total density. In rich media, in contrast, there is a much higher peptide bond formation rate per bacterial mass and the total density of the translation apparatus can be as much as 40% of the total density of the cell. Since the mass of the ribosome is large there will be, on the one hand, an advantage in making the elongation rate fast by having high concentrations of ternary complexes and other elongation factors. Indeed, an important idea in Koch's (1971) and Maaløe's (1979) analysis of bacterial growth is that the optimal elongation rate of translating ribosomes is always near its maximum and that the elongation rate is a constant when the growth medium changes. In order for the ribosomes to elongate very near their maximal rate, the concentration of ternary complexes must be very high. When the concentrations of different ternary complexes are far above their K_m values, a further concentration increase will only lead to an increase in the protein density of the translation apparatus and not to any significant increase in the peptide formation rate. This tends to reduce the efficiency ($e_a = j_a/\rho_a$) of translation and to decrease the growth rate. These arguments indicate that there exists a balance point where the ternary complex concentration is at its optimum with respect to rapid growth (Ehrenberg and Kurland, 1984; Kurland and Ehrenberg, 1984).

To see this we first note that Equation (11.35) can be simplified so that the inverse growth rate becomes the sum of two times only:

$$1/k = \tau_A + \tau_a. \tag{11.39}$$

The time τ_a is the kinetic cost of making peptide bonds:

$$\tau_a = \rho_a/j_a. \tag{11.40}$$

The other time, τ_A, is the kinetic cost of all other functions in the cell. They are performed by proteins at the density ρ_A and they supply the ribosomes with the required amino acids.

$$\tau_A = \rho_A/j_a. \qquad (11.41)$$

The sum of these partial densities equals the total protein density:

$$\rho_a + \rho_A = \rho_0 \qquad (11.42)$$

so that Equations (11.39)–(11.42) are consistent with (11.33).

Major biochemical pathways are active in the synthesis of amino acids in poor media (Ingraham, Maaløe and Neidhardt, 1983). The poorer the carbon source, the larger is the density of proteins that are invested in the production of amino acids. Under these conditions, the kinetic cost associated with amino acid synthesis is much larger than the kinetic cost of peptide bond formation.

In rich media, in contrast, the bacterium obtains the necessary flow of amino acids at a much lower cost. Here, the cost of peptide bond formation is roughly equal to the cost of amino acid acquisition.

A convenient approximation is now to put the protein density of the translation apparatus as the sum of ribosomes and ternary complexes only. This approximation is good since EF-Tu is the most abundant elongation factor (Neidhardt et al., 1977). It leads to the following expression:

$$\rho_a = N_R[R_0] + N_T[T_{30}] \qquad (11.43)$$

where $[R_0]$ and $[T_{30}]$ are the total concentrations of ribosomes and ternary complexes respectively, N_R ($\approx 10\,000$) is the number of amino acids in a ribosome and N_T (≈ 400) is the number of amino acids in a ternary complex. With support from in vitro experiments (Wagner et al., 1982; Ruusala et al., 1984; Bohman et al., 1984) we shall assume that in vivo translation follows ordinary Michaelis–Menten kinetics where the parameters k_{cat} and K_m are the same for all ternary complex – codon pairs (see Andersson, Buckingham and Kurland, 1983). Since there are sixty sense codons and also about sixty different ternary complexes, the flow into polypeptide can be written:

$$\frac{1}{j_a} = \frac{1}{[R_0]}\left(\frac{1}{k_{cat}} + \frac{K_m}{k_{cat}}\sum_{i=1}^{60}\frac{f_i}{[T_{3i}]}\right); \qquad (11.44)$$

where f_i is the normalized ($\Sigma f_i = 1$) frequency by which a particular codon (i) occurs in mRNA, $[T_{3i}]$ is the concentration of ternary complex cognate for a codon so that the sum of all the individual ternary complexes equals the total ternary complex concentration level ($\Sigma[T_{3i}] = [T_{30}]$). Equations (11.43) and

(11.44) together give the kinetic cost of a peptide bond according to our definition (Equation (11.40)):

$$\tau_a = \frac{N_R[R_0] + N_T[T_{30}]}{[R_0]}\left(\frac{1}{k_{cat}} + \frac{K_m}{k_{cat}}\sum_{i=1}^{60}\frac{f_i}{[T_{3i}]}\right). \tag{11.45}$$

One of the requirements of maximal growth rate is that the cost in Equation (11.45) is at its minimum. The most important features of minimal cost protein synthesis can be summarized as follows. The concentration of free ternary complex is optimally adjusted so that it is proportional to the square root of the codon frequency, f_i. The degree of saturation of the ribosomes varies with the growth conditions. In poor media, the density of the translation system is small and here the ribosomes are, relatively speaking, less saturated with ternary complex. Thus, the ribosomal elongation rate is expected to drop as the quality of the medium becomes poorer. This prediction has recently been verified for E. coli (Pedersen, 1984).

A minimized translation cost in very poor media, where ρ_a is much smaller than the total protein density, requires that the total amino acid mass ascribed to ternary complexes equals the mass ascribed to ribosomes. Under those conditions the ratio between the number of ternary complexes and the number of ribosomes in the cell has its highest value:

$$\frac{[T_{30}]}{[R_0]} = \frac{N_R}{N_T} = 25.$$

In very rich media, where ρ_a is large and comparable with ρ_A the degree to which ribosomes are saturated with ternary complex is expected to increase together with the rate of elongation. In a range of growth conditions the ratio between ternary complexes and ribosomes is expected to decrease with the inverse of the square root of the growth rate:

$$\frac{[T_{30}]}{[R_0]} \simeq \frac{1}{\sqrt{k}}$$

This prediction of the theory is consistent with experimental results (Neidhardt et al., 1977). The average kinetic cost of translating a codon depends in a characteristic way on the codon frequency distribution. In rich media, where the ribosomes are well saturated with ternary complex and elongate near their maximal rate, the peptide bond cost at optimal choices of ternary complex concentrations can be approximated as (Ehrenberg and Kurland, 1984):

$$\tau_a = \frac{N_R}{k_{cat}}\left[1 + \sqrt{\left(\frac{N_T K_m}{\rho_a}\right)u}\right]. \tag{11.46}$$

The kinetic cost under those conditions depends linearly on a parameter u which is defined as:

$$u = \sum_{i=1}^{60} \sqrt{f_i}. \tag{11.47}$$

When u increases, the cost of translation increases as well. A narrow codon usage, as has been observed for synonymous codons in high copy number proteins (Fiers *et al.*, 1971; Grantham *et al.*, 1981; Grosjean and Fiers, 1982), will tend to reduce u and to increase the efficiency of translation according to Equation (11.46). A broad codon usage, as for low copy number proteins, tends to increase u instead and thereby to make the translation process more costly. In rich media, where the translation apparatus plays such a dominant part, and where the elongation rate of ribosomes is so important for the growth rate, the bacteria take full advantage of a narrow codon usage. Here, it is the high copy number proteins, such as elongation factors and ribosomal proteins, that dominate the protein pools. Accordingly, the cost parameter u in Equation (11.47) is smaller in rich than in poor media where the high copy number proteins play a less prominent role and where the ribosomal elongation rate is not so critical. Further aspects of codon usage are discussed by Ehrenberg and Kurland (1984).

These considerations bring us naturally to another, fundamental question. This concerns the reason for the limitation of the elongation rate of ribosomes to values of between fifteen and twenty-two amino acids per second (Maaløe, 1979; Churchward, Bremer and Young, 1982). At least two alternative answers are conceivable. On the one hand it is possible that the maximal turnover rate is subject to a fundamental limitation. According to this interpretation the ribosomes are saturated with ternary complex in rich media as suggested by Gouy and Grantham (1980) and the translation rate is k_{cat}-limited. On the other hand, it is also possible that the K_m value for ribosomes, interacting with ternary complexes *in vivo*, is so high that minimum cost τ_a and maximal efficiency of translation e_a are obtained at elongation rates considerably below k_{cat}.

Our *in vitro* estimate of the K_m value for cognate ternary complex in poly(U) translation is between 3×10^{-7} M and 6×10^{-7} M (Bohman *et al.*, 1984; Ruusala *et al.*, 1984). These estimates agree well with those of Gouy and Grantham (1980) and if K_m is also the same *in vivo*, this indicates that here the ribosomes elongate the major protein species near k_{cat}, at least under rich growth conditions. It is, however, possible that our *in vitro* experiments underestimate the *in vivo* K_m values to a considerable extent since we know that k_{cat}, as measured *in vitro*, is between two- and four-fold smaller than the elongation rate *in vivo*. It is, therefore, impossible just now to say unambiguously what limits the elongation rate *in vivo*. It is of some importance to be able to do so for several reasons. One of these is that the answer to this question will influence a proper evaluation of the kinetic costs of accuracy in translation, the topic of the next section.

11.8 Optimal accuracy in translation

The *in vitro* experiments on translational accuracy described above are all compatible with a two-step selection on the ribosome: an initial selection followed by a proofreading step. A minimal scheme which is compatible with experimental data can be written as follows (see Kurland and Ehrenberg, 1984):

$$T_3^{c,w} + R_A \underset{k_{-1}}{\overset{k_1}{\rightleftharpoons}} R_1 \xrightarrow{k_2} R_2 \xrightarrow{k_3} R_3 \xdashrightarrow{k_4}$$

$$+ \, Tu \cdot GDP \qquad\qquad (11.48)$$

$$q_{-1} \Big\Updownarrow q_1$$

$$R_A + Tu \cdot GDP + AA\text{-}tRNA$$

A correct (T_3^c), or an incorrect (T_3^w) ternary complex binds to ribosomes with open A-site (R_A) and forms a complex (R_1). Subsequent hydrolysis of GTP leads to a new complex (R_2) which is a branch point. The steps which follow the release of $Tu \cdot GDP$ include peptidyl transfer and translocation and are summarized by the rate constant k_4. The R factor, defined in Equation (11.10), for correct ternary complex is given by:

$$R^c = k_{cat}^c / K_m^c = \frac{k_1}{(1 + a_1)(1 + a_2)}. \qquad (11.49)$$

The R factor for incorrect ternary complex is instead given by:

$$R^w = k_{cat}^w / K_m^w = \frac{k_1}{(1 + da_1)(1 + da_2)}. \qquad (11.50)$$

In Equations (11.49) and (11.50) we have introduced discard parameters, a_1, a_2, and an intrinsic selectivity, d, discussed above in more general terms, with the following definitions:

$$a_1 = k_{-1}^c / k_2^c \qquad da_1 = k_{-1}^w / k_2^w$$

$$\qquad\qquad (11.51)$$

$$a_2 = q_{-1}^c / k_3^c \qquad da_2 = q_{-1}^w / k_3^w$$

The normalized accuracy, A, introduced in Equation (11.22), can be interpreted for the special case in Reaction (11.48) as

$$A = IF = \frac{1 + da_1}{1 + a_1} \times \frac{1 + da_2}{1 + a_2}. \qquad (11.52)$$

In Equation (11.52) the accuracy increases with increasing cognate discard parameters, a_1, a_2. The parameter $1 + a_2$ is the number of EF-Tu cycles needed per correct peptide bond, f_c and $1 + da_2$ is the corresponding number for incorrect peptides, f_w, as introduced in Equation (11.23). The intrinsic selectivity is determined by the discrimination inherent in codon–anticodon base-pairing. We shall assume that d is constant, maximized and equal in the two selection steps (Kurland and Ehrenberg, 1984). The discard parameters can be varied simply by changing the standard free energy of interaction between ribosomes and ternary complexes. From Equation (11.49) we see that the correct R factor, R^c, decreases when the accuracy increases. The two selection steps appear symmetrical in R^c and they give identical contributions to the kinetic cost of translation in Equation (11.45). Although the initial selection costs can always be calculated from a reduction in the cognate R factor, R^c, the proofreading step has additional kinetic costs associated with it. First, it is inevitable that the proofreading discard parameter influences the maximal elongation rate for correct amino acid elongation (Ehrenberg and Kurland, 1984). This influence increases when the rate constant k_2 in Reaction (11.48) relative to the other rate constants of the elongation cycle decreases:

$$\frac{1}{k_{cat}^c} = \frac{1 + a_2}{k_2} + \frac{1}{k_3} + \frac{1}{k_4}. \qquad (11.53)$$

Therefore, kinetic proofreading influences the kinetic cost of translational accuracy in Equation (11.45), through both the R factor and the maximum turnover rate. Secondly, when a_2 increases, more GTPs become hydrolysed per peptide bond. This, in turn, gives a heavier load to the GTP-regenerating pathway of the bacterium so that more enzymes are needed here. This effect can be taken into account if we ascribe to the density, ρ_a, of the translation apparatus a partial density, ρ_R, which includes all amino acids in the proteins that are responsible for GTP regeneration. When the proofreading discard parameter increases, then ρ_R increases as well so that for a given peptide bond formation rate the translation system density must increase and the kinetic cost, which is the ratio ρ_a / j_a, must also increase.

Thirdly, when a_2 increases, more EF-Tu cycles per peptide bond are required, and in order to catalyse the release of GDP from EF-Tu, a higher concentration of EF-Ts is required (see Ruusala, Ehrenberg and Kurland,

1982b). This effect will also increase the density of the translation system at a fixed peptide bond formation rate and thus the kinetic cost in Equation (11.45) will increase as well.

In rich media, the density of the translation system is about equal to the rest of the protein density of the system. Under these conditions, it is particularly important to keep the kinetic cost at low values, since here it corresponds to a substantial contribution to the growth rate according to Equation (11.39). In very poor growth media, in contrast, the kinetic cost of translation is not such a critical parameter because here it corresponds to a much smaller contribution than the kinetic cost of amino acid synthesis. The translation cost will therefore influence the growth rate proportionally much less when the medium is poor than when it is rich according to Equation (11.39).

The analysis of the previous section has shown that the degree to which ribosomes are saturated with ternary complex is smaller in poor media than in rich media. The R factor term in the kinetic cost in Equation (11.45) will be more important under those conditions, while the k_{cat} term will tend to dominate more in rich media. From the analysis so far, it seems therefore that the accuracy costs under poor growth conditions will tend to be evenly distributed between initial selection and proofreading. In rich media, in contrast, proofreading becomes successively more costly than initial selection because of the increasing importance of the three extra costs associated with it. Since, finally, the translation cost contributes so little to the growth rate in poor media according to Equation (11.39), the negative effects of this kinetic cost at a high accuracy level seem to be less harmful than in rich media. If, therefore, the harmful effects of translation errors are the same in the two cases, it is beneficial for the organism to have a high translation accuracy in poor media but a low translation accuracy in rich media (Ehrenberg and Kurland, 1984).

We have argued previously that the rate constants of enzymes (k_{cat}, k_{cat}/K_m) have been maximized (Kurland and Ehrenberg, 1984). A perfectly transcribed and translated enzyme will, by this hypothesis, have its rate maximized. In other words, another enzyme, coded from the same gene but with amino acid substitutions in it, will tend to be slower than the perfect copy. This tendency has been summarized in a particularly simple way by a parameter β, which describes by what fraction k_{cat}/K_m or k_{cat} decreases as a result of each successive amino acid replacement in the canonical primary sequence. It could be shown that the efficiency of enzymes decreases from its canonical value, e_0, with increasing average levels, $\bar{p}$, of translation errors (Ehrenberg and Kurland, 1984) as follows:

$$e = e_0 \times \exp(N\bar{p}(\beta - 1)) \qquad (11.54)$$

where N is the number of amino acids in the enzyme. A finite error level must necessarily lead to a heterogeneous population of ribosomes. Therefore, there will not be a single translation error level but a distribution of errors around the level, p_0, of canonically transcribed and translated ribosomes. With a set of simplifying assumptions we could show that the average error rate, introduced in Equation (11.54), can be obtained from the canonical error as a root to the expression

$$\bar{p} = p_0 \times \exp(N_R \bar{p}(\lambda - 1)\beta). \tag{11.55}$$

To derive Equation (11.55) we assumed that the error changes on the average by a factor λ for each new amino acid substitution. It is clear from this equation that the translation system is stable against error propagation (see below) as long as

$$p_0 N_R (\lambda - 1)\beta < 1/\exp(1). \tag{11.56}$$

We have argued elsewhere (Kurland and Ehrenberg, 1984; Ehrenberg and Kurland, 1984) that the parameter λ in Equation (11.55) and in (11.56) is probably smaller rather than larger than one, so that the inequality (11.56) is automatically fulfilled for this reason. According to this hypothesis, amino acid substitutions in the ribosomes will tend to reduce the average translational error level rather than increase it. This suggestion can be motivated by the fact that translational accuracy cannot be maximized, since this would reduce the efficiency of the ribosomes and thereby inhibit bacterial growth. Instead, translational accuracy must be optimized for maximal growth rate and deviations in either direction from this optimum lead to an inferior behaviour as illustrated by the mutants described above. At such an optimum, the discard parameters are normally much smaller than one, so that amino acid substitutions in the ribosomes may increase as well as decrease them. The parameters increase when the affinity between ribosomes and ternary complex decreases. A tendency to decrease the affinity, rather than increase it, when amino acids are substituted at random in the ribosome appears more likely as there will probably be many more possible substitutions that lead to the weakening or destruction of a binding site than to its strengthening. If this is true, the tacit assumption in all previous studies on error propagation (Orgel, 1963; Orgel, 1970; Hoffman, 1974; Kirkwood and Holliday, 1975; Goel and Ycas, 1975; Kirkwood, 1977), that amino acid substitutions in the canonical sequence of ribosomes must lead to higher error levels, is false.

The problem of how to determine optimal error levels in growing bacteria takes a particularly simple and instructive form if we introduce a couple of rather drastic and probably unrealistic assumptions. These are that the parameter $N(\lambda - 1)$ in Equation (11.54) is similar for all enzymes in the cell and that the reduction in k_{cat}, due to amino acid replacements, is similar to the reduction in k_{cat}/K_m. With these simplifications, the error costs, as measured by their influence on the growth rate in Equation (11.39), can be introduced explicitly as follows:

$$k = \frac{\exp(N\bar{p}(\beta - 1))}{\tau_{A0} + \tau_{a0}} \qquad (11.57)$$

where τ_{a0} and τ_{A0} are the 'canonical' kinetic costs of translation and amino acid production so that the effects of errors on the efficiency of enzymes are now explicit in the numerator.

The optimal accuracy, which maximizes the growth rate in Equation (11.57), is obtained when a small increase in the translation cost τ_a, when the error level is reduced, is exactly counterbalanced by a corresponding increase in the kinetic efficiency of all the bacterium's enzymes.

11.9 Conclusions

We have shown that the maximal accuracy which can be attained in an enzymic selection with kinetic proofreading depends on (1) the intrinsic selectivity of the pathway, (2) the number of discard steps, and (3) the thermodynamic driving forces of the reaction. If the number of discard steps increases and the driving forces are raised, the accuracy of a selection can increase without ever reaching an upper limit, irrespective of the magnitude of the intrinsic selectivity of the pathway. It is, therefore, impossible to predict an upper limit to the accuracy of enzymic selections from calculations of standard free energies in equilibrium.

When the accuracy of an enzymic selection increases, the dissipative losses of free energy over this pathway necessarily increase as well. When these losses become too large, they become harmful for the growth of the cell. Thus, it is necessary that the accuracy in translation, as well as in other enzyme selections, be optimized rather than maximized.

This is illustrated by bacterial mutants which are more accurate than wild type bacteria. They are changed in their ribosomal proofreading activity and they have impaired ribosomal elongation rates, as well as growth rates. These mutated ribosomes operate with an increased GTP hydrolysis rate and with decreased Michaelis–Menten parameters (k_{cat}/K_m, k_{cat}), which are both associated with their selection of ternary complexes. The costs of accuracy are here a combination of dissipative losses and impaired kinetic properties.

Previous cost analyses based on dissipation only (Bennet, 1979; Savageau and Freter, 1979a, b; Freter and Savageau, 1980; Blomberg, Ehrenberg and Kurland, 1980) or only on rates (Thompson and Karim, 1982) cannot describe realistically these costs for growing bacteria. It is convenient to consider the efficiency of the translation apparatus itself, or of a biochemical pathway in general, as a ratio between the rate at which the reaction sequence works and the mass of the protein invested in it. In growing bacteria all costs associated with ribosomal accuracy tend to decrease the efficiency of the translation apparatus. These various, and seemingly unrelated, costs have one common measure which is the way that they influence the growth rate by reducing the efficiency of translation. The errors of translation tend to reduce the efficiency not only of the translation apparatus itself, but of all the pathways in the bacterium, since errors in enzymes reduce their kinetic parameters.

Hence, the translation accuracy is optimal when the cost increase associated with an improved translational accuracy is cancelled by the increased efficiency of all pathways in the cell, due to fewer amino acid replacements in their enzymes. It turns out that the costs of translational accuracy are less of a burden in poor media at slow growth rates than in rich media at fast growth rates. It would, therefore, pay for the bacterium to regulate its ribosomal accuracy so that it is higher in poor media than in rich ones. Accordingly, it may be that cells, in general, regulate their accuracy of gene expression to different levels under different physiological conditions.

References

Andersson, D. I., Bohman, K. T., Isaksson, L. A. and Kurland, C. G. (1982) Translation rates and misreading characteristics of *rpsD* mutants in *Escherichia coli*. *Mol. Gen. Genet.*, **187**, 467–472.

Andersson, D. I. and Kurland, C. G. (1983) Ram ribosomes are defective proofreaders. *Mol. Gen. Genet.*, **191**, 378–381.

Andersson, S. G. E., Buckingham, R. H. and Kurland, C. G. (1983) Does codon-composition influence ribosome function? *EMBO J.*, **3**, 91–94.

Bennet, Ch. H. (1979) Dissipation-error tradeoff in DNA replication. *Biosystems*, **11**, 85–91.

Blomberg, C. (1983a) Thermodynamic aspects on accuracy in the synthesis of biomolecules. *Int. J. Quant. Chem.*, **23**, 687–707.

Blomberg, C. (1983b) Free energy cost and accuracy in branched selection processes of biosynthesis. *Quart. Rev. Biophys.*, **16**, 415–519.

Blomberg, C. and Ehrenberg, M. (1981) Energy considerations for kinetic proofreading in biosynthesis. *J. Theor. Biol.*, **88**, 631–670

Blomberg, C., Ehrenberg, M. and Kurland, C. G. (1980) Free energy dissipation constraints on the accuracy of enzymatic selections. *Quart. Rev. Biophys.*, **13**, 231–254.

Bohman, K. T., Ruusala, T., Jelenc, P. C. and Kurland, C. G. (1984) Kinetic impairment of restrictive streptomycin resistant ribosomes. *Mol. Gen. Genet.*, **198**, 90–99.

Bouadloun, F., Donner, D. and Kurland, C. G. (1983) Codon-specific missense errors *in vivo*. *EMBO J.*, **2**, 1351–1356.

Churchward, G., Bremer, H. and Young, R. (1982). Macro-molecular composition of bacteria. *J. Theor. Biol.*, **94**, 651–670.

de Groot, S. R. and Mazur, P. (1969) *Non-equilibrium Thermodynamics*, North-Holland, Amsterdam and London.

Edelmann, P. and Gallant, J. (1977) Mistranslation in *E. coli*. *Cell*, **10**, 131–137.

Ehrenberg, M. and Blomberg, C. (1980) Thermodynamic constraints on kinetic proofreading in biosynthetic pathways. *Biophys. J.*, **31**, 333–358.

Ehrenberg, M. and Kurland, C. G. (1984) Costs of accuracy determined by a maximal growth rate constraint. *Quart. Rev. Biophys.*, **17**, 45–82.

Ehrenberg, M., Kurland, C. G. and Ruusala, T. (in press, 1985). Counting cycles of EF-Tu to measure proofreading in translation. *Biochimie*.

Fersht, A. (1977a) *Enzyme Structure and Mechanism*, W. H. Freeman and Company, San Francisco, p. 283.

Fersht, A. (1977b) Editing mechanisms in protein synthesis. Rejection of valine by the isoleucyl-tRNA synthetase. *Biochemistry*, **16**, 1025–1030.

Fersht, A. (1981) Enzymic editing mechanisms and the genetic code. *Proc. R. Soc. London B*, **212**, 351–379.

Fersht, A. and Dingwall, C. (1979) Establishing the misacylation/deacylation of the tRNA pathway for the editing mechanism of prokaryotic and eukaryotic valyl-tRNA synthetases. *Biochemistry*, **18**, 1238–1244.

Fersht, A. and Kaethner, M. (1976) Enzyme hyperspecificity. Rejection of threonine by the valyl-tRNA synthetase by misacylation and hydrolytic editing. *Biochemistry*, **15**, 3342–3346.

Fiers, W., Contreras, R., De Wachter, R., Haegeman, G., Merregaert, J., Min Jou, W. and Vanderberghe, A. (1971) Recent progress in the sequence determination of bacteriophage MS_2 RNA. *Biochimie*, **53**, 495–506.

Freter, R. R. and Savageau, M. A. (1980) Proofreading systems of multiple stages for improved accuracy of biological discrimination. *J. Theor. Biol.*, **85**, 99–123.

Galas, D. J. and Branscomb, E. W. (1976) *Nature, London*, **262**, 617–619.

Goel, N. S. and Ycas, M. (1975) The error catastrophe hypothesis with reference to ageing and the evolution of the protein synthesizing machinery. *J. Theor. Biol.*, **55**, 245–282.

Gorini, L. (1971) Ribosomal discrimination of tRNAs. *Nature New Biol.*, **234**, 261–264.

Gouy, M. and Grantham, R. (1980) Polypeptide elongation and tRNA cycling in *Escherichia coli*: A dynamic approach. *FEBS Lett.*, **115**, 151–155.

Grantham, R., Gautier, C., Gouy, M., Jacobzone, M. and Mercier, R. (1981). Codon catalog usage is a genome strategy modulated for gene expressivity. *Nucl. Acids Res.*, **9**, r 43–r 74.

Grosjean, H. and Fiers, W. (1982) Preferential codon usage in prokaryotic genes: The optimal codon–anticodon interaction energy and the selective codon usage in efficiently expressed genes. *Gene*, **18**, 199–209.

Hoffman, G. W. (1974) On the origin of the genetic code and the stability of the translation apparatus. *J. Mol. Biol.*, **86**, 349–362.

Hopfield, J. J. (1974) Kinetic proofreading: a new mechanism for reducing errors in biosynthetic processes requiring high specificity. *Proc. Natl Acad. Sci. USA*, **71**, 4135–4139.

Hopfield, J. J., Yamane, T., Yue, V. and Coutts, S. M. (1976) Direct experimental evidence for kinetic proofreading in aminoacylation of tRNA[Ile]. *Proc. Natl Acad.*

Sci. USA, **73**, 1164–1168.

Ingraham, J. L., Maaløe, O. and Neidhardt, F. C. (1983) *Growth of the Bacterial Cell*, Sinauer Associates Inc, Sunderland, Massachusetts.

Jelenc, P. C. and Kurland, C. G. (1979) Nucleoside triphosphate regeneration decreases the frequency of translation errors. *Proc. Natl Acad. Sci. USA*, **76**, 3174–3178.

Jelenc, P. C. and Kurland, C. G. (1984). Multiple effects of kanamycin on translational accuracy. *Mol. Gen. Genet.*, **194**, 195–199.

Kirkwood, T. B. L. (1977). Evolution of ageing. *Nature, London*, **270**, 301–304.

Kirkwood, T. B. L. and Holliday, R. (1975) The stability of the translation apparatus. *J. Mol. Biol.*, **97**, 257–265.

Koch, A. L. (1971) The adaptive responses of *Escherichia coli* to a feast and famine existence. *Adv. Microbiol.*, **6**, 147–217.

Kurland, C. G. (1978) The role of guanine nucleotides in protein biosynthesis. *Biophys. J.*, **22**, 373–388.

Kurland, C. G. and Ehrenberg, M. (1984) Optimization of translation accuracy. *Progr. Nucl. Acids Res. Mol. Biol.*, **31**, 191–219.

Loeb, L. A. and Kunkel, T. A. (1982) Fidelity of DNA synthesis. *Ann. Rev. Biochem.*, **51**, 429–457.

Loftfield, R. (1963) The frequency of errors in protein biosynthesis. *Biochem. J.*, **89**, 82–92.

Loftfield, R. and Vanderjagt, D. (1972) The frequency of errors in protein biosynthesis. *Biochem. J.*, **128**, 1353–1356.

Maaløe, O. (1979) Regulation of the protein-synthesizing machinery ribosomes, tRNA, factors and so on. In *Biological Regulation and Development* (ed. R. F. Goldberger), Plenum, New York, pp. 487–542.

Mulvey, R. and Fersht, A. (1977) Editing mechanisms in aminoacylation of tRNA: ATP consumption and the binding of aminoacyl-tRNA by elongation factor Tu. *Biochemistry*, **16**, 4731–4737.

Neidhardt, F. C., Bloch, P. L., Pedersen, S. and Reeh, S. (1977) Chemical measurement of steady-state levels of ten aminoacyl-transfer ribonucleic acid synthetases in *Escherichia coli*. *J. Bacteriol.*, **129**, 378–387.

Ninio, J. (1975) Kinetic amplification of enzyme discrimination. *Biochimie*, **57**, 587–595.

Ninio, J. (1982) *Molecular Evolution*, Pitman, London.

Orgel, L. E. (1963) The maintenance of the accuracy of protein synthesis and its relevance to ageing. *Proc. Natl Acad. Sci. USA*, **49**, 517–521.

Orgel, L. E. (1970) The maintenance of the accuracy of protein synthesis and its relevance to ageing: A correction. *Proc. Natl Acad. Sci. USA*, **67**, 1476–1480.

Pauling, L. (1957) The probability of errors in the process of synthesis of protein molecules. In *Festschrift Arthur Stoll*, Birkhauser, A.G. Basel, pp. 597–602.

Pedersen, S. (1984) *Escherichia coli* ribosomes translate *in vivo* with variable rate. *EMBO J.*, **3**, 2895–2898.

Pettersson, I. and Kurland, C. G. (1980) Protein L7/L12 is required for optimal translation. *Proc. Natl Acad. Sci. USA*, **77**, 4007–4010.

Ruusala, T. and Kurland, C. G. (1984) Streptomycin perturbs preferentially ribosomal proofreading. *Mol. Gen. Genet.*, **198**, 100–104.

Ruusala, T., Andersson, D. I., Ehrenberg, M. and Kurland, C. G. (1984) Hyper accurate ribosomes inhibit growth. *EMBO J.*, **1**, 2575–2580.

Ruusala, T., Ehrenberg, M. and Kurland, C. G. (1982a) Is there proofreading during polypeptide synthesis? *EMBO J.*, **1**, 741–745.

Ruusala, T., Ehrenberg, M. and Kurland, C. G. (1982b) Catalytic effects of elongation factor Ts on polypeptide synthesis. *EMBO J.*, **1**, 75–78.

Savageau, M. A. and Freter, R. R. (1979a) On the evolution of accuracy and cost of tRNA proofreading. *Proc. Natl Acad. Sci. USA*, **76**, 1902–1912.

Savageau, M. A. and Freter, R. R. (1979b) Energy cost of proofreading to increase fidelity of transfer ribonucleic acid amino acylation. *Biochemistry*, **18**, 3486–3492.

Thompson, R. C. and Dix, D. B. (1982) Accuracy of protein biosynthesis. A kinetic study of the reaction of poly(U)-programmed ribosomes with a leucyl tRNA$_2$-elongation factor Tu-GTP complex. *J. Biol. Chem.*, **257**, 6677–6682.

Thompson, R. C. and Karim, A. M. (1982) The accuracy of protein biosynthesis is limited by its speed: High fidelity selection by ribosomes of aminoacyl-tRNA ternary complexes containing GTP (γS). *Proc. Natl Acad. Sci. USA*, **79**, 4922–4926.

Thompson, R. C. and Stone, P. (1977) Proofreading of the codon-anticodon interaction on ribosomes. *Proc. Natl Acad. Sci. USA*, **74**, 198–202.

Thompson, R. C., Dix, D. B. and Eccleston, J. F. (1980) Single turnover studies of guanosine triphosphate hydrolysis and peptide formation in the elongation factor Tu-dependent binding of aminoacyl-tRNA to *Escherichia coli* ribosomes. *J. Biol. Chem.*, **255**, 11 088–11 090.

Thompson, R. C., Dix, D. B., Gerson, R. B. and Karim, A. M. (1981) A GTPase reaction accompanying the rejection of Leu-tRNA$_2$ by UUU-programmed ribosomes. *J. Biol. Chem.*, **256**, 81–86.

von der Haar, F. and Cramer, F. (1976) Hydrolytic action of aminoacyl-tRNA synthetases from Baker's yeast: Chemical proofreading preventing acylation of tRNA with misactivated valine. *Biochemistry*, **15**, 4131–4138.

Wagner, E. G. H., Jelenc, P. C., Ehrenberg, M. and Kurland, C. G. (1982) Rate of elongation of polyphenylalanine *in vitro*. *Eur. J. Biochem.*, **122**, 193–197.

Yamane, T. and Hopfield, J. J. (1977). Experimental evidence for kinetic proofreading in the aminoacylation of tRNA by synthetase. *Proc. Natl Acad. Sci. USA*, **74**, 2246–2250.

Yates, J. L. (1979) Role of ribosomal protein S12 in discrimination of aminoacyl-tRNA. *J. Biol. Chem.*, **254**, 11 550–11 554.

Zengel, J. M., Young, R., Dennis, P. P. and Nomura, M. (1977) Role of ribosomal protein S12 in peptide chain elongation: analysis of pleiotropic, streptomycin-resistant mutants of *Escherichia coli*. *J. Bacteriol.*, **129**, 1320–1329.

12 Selection for optimal accuracy and the evolution of ageing

T. B. L. KIRKWOOD and R. HOLLIDAY

12.1 Introduction

The processing of genetic information is fundamental to every aspect of life. Thus it is obvious that natural selection acts as much upon the accuracy of macromolecular synthesis as on any other trait. The outcome of this selection is readily apparent in the range of mechanisms which co-operate to produce the high fidelity of synthesis of DNA, RNA and proteins that is found in present-day species. Primitive organisms presumably lacked these mechanisms and were, therefore, far less accurate. In this chapter, we discuss whether the accuracy of synthesis of macromolecules is, in some sense, optimal, or merely the best which has so far been achieved. We also consider selection for the stability of translation, as well as its accuracy (see Chapter 2). We conclude that there is strong support for the view that both accuracy and translational stability have been optimized, and we explore their relationship to the evolution of ageing in higher organisms (see also Kirkwood, 1977; Kirkwood and Holliday, 1979).

In linking accuracy with ageing, we make the assumption, to be justified later, that the degenerative changes which are seen during senescence are the outward manifestation of the organism's failure to maintain indefinitely the viability of its organs, tissues and cells. More specifically, we describe an evolutionary basis for the hypothesis that ageing is due to an increasing number of cell deaths, caused either by a breakdown in the fidelity of macromolecular synthesis (Orgel, 1963, 1973; Kirkwood, Holliday and Rosenberger, 1984), or more generally, by a failure to maintain the integrity of macromolecules or replace defective with normal ones.

12.2 Evolution of accuracy in primitive organisms

Theoretical analysis of the stability of the translation apparatus, which was outlined in Chapter 2, suggests that a replicating cell can be either in a stable condition with a constant protein synthetic error level from one generation to the next, or in an unstable one when errors increase steadily until, after several cell generations, a lethal error catastrophe occurs. This poses a problem for the evolution of living organisms in the first place, since the primitive mechanisms for the synthesis of macromolecules, either nucleic acids or proteins, would undoubtedly have been very inaccurate. An organism containing a high load of errors would, on replicating itself, produce more errors and the eventual result would be an irreversible decline to a non-viable state. To survive and evolve further, such organisms have to achieve what is seemingly impossible, namely, an inherent stability in macromolecule synthesis, starting from an intrinsically unstable condition.

Only natural selection can provide an answer to this problem. For a population of unicellular organisms to multiply and survive each cell must, on average, produce more than one viable offspring; if an average of less than one viable offspring is produced, then the population will eventually die out. This can be illustrated with an example taken from Orgel (1973). Imagine a chemostat containing a population of growing bacteria, which is close to a source of radiation. If on average the number of induced lethal mutations (or any kind of lethal event) is less than one per generation, then the population will survive. On the other hand, if the number of lethal mutations or hits is more than one per cell generation, then the population will start to decline and may die out. However, in any such population there is genetic heterogeneity and a rare variant, which is for one reason or another resistant to the lethal effects of radiation, will survive and tend to take over the population. Even if the intensity of radiation is gradually increased the population will, at least for a long period, be expected to adapt itself. Since these cells are capable of dealing with damage in macromolecules, they have, in effect, become more accurate. The same argument can be applied to primitive organisms. In such a population we would expect extensive molecular heterogeneity and considerable genetic variability. Individual organisms incapable of maintaining themselves would be continually eliminated, and any device or mechanism which tended to maintain accuracy would be selected for. In these early stages of evolution, we might therefore expect strong selection for a progressive decline in mutability and a progressive increase in the fidelity of RNA and protein synthesis. This would be associated with a continuing rise in overall viability, or fitness.

12.3 Evolution of translational accuracy

In the evolution of enzymes from a primitive state to the structures we see today, natural selection would favour at least three separate characteristics: catalytic activity, substrate specificity and conformational stability. These properties are to a considerable extent independent of each other; for instance there could be a highly active, but relatively non-specific, enzyme, or a very specific enzyme with low catalytic activity. Greater conformational stability may increase, or possibly reduce, the other two properties.

Genetic variability will provide a wide range of alternative enzyme structures, and it is easy to envisage that in the first place there would be, in general, selection for an increase in catalytic activity, in substrate specificity and in stability. But this situation would not persist indefinitely, because the selective advantage for a continuing increase in each characteristic would not necessarily be the same. Once an enzyme is so active that the reaction it carries out is no longer rate limiting for growth, then there will be no further selection for catalytic activity. On the other hand, there may still be a need for increasing accuracy or specificity, if errors are likely to have deleterious consequences for the organism or its descendants. An even greater degree of accuracy might result in a slower reaction, which could now be rate limiting for growth. Thus, natural selection will result in a balance between specificity and activity.

In this chapter we are concerned mainly with the evolution of accuracy and in that group of enzymes or other proteins which are themselves necessary for protein synthesis, as well as for RNA and DNA synthesis. Errors in RNA or protein synthesis will have two consequences, one of which is general and one of which is more specific. The general consequence is a lowering of metabolic efficiency: there is little point in using energy and resources to synthesize an RNA or protein molecule which cannot perform its normal function. For this reason alone, we can expect the majority of such molecules to be error free. The specific consequence relates only to the proteins required for information transfer between macromolecules: it is the possibility of error propagation which could lead ultimately to cell death. Error propagation would not only increase the proportion of defective protein molecules, but it would also affect the fidelity of DNA replication itself. Therefore, to protect its DNA, the organism must necessarily achieve a considerable degree of accuracy in transcription and translation.

We do not yet know how much of the translation apparatus is related to the need to make proteins accurately, but it is likely that the ribosome itself is an essential component in preventing the misincorporation of amino acids. The cost of accuracy in protein synthesis may not be therefore just one or two proofreading steps (see below), which may use ATP, but also the synthesis of ribosomes and all the associated factors which control and co-ordinate the whole process (see Chapters 4, 5 and 6). In addition, it would be advantageous to get rid of any error-containing proteins which may be synthesized, so it is not surprising that protease scavenging mechanisms have also evolved (Goldberg and Dice, 1974; Goldberg and St John, 1976). The removal of such defective molecules is clearly another component in the overall machinery which prevents protein molecules with incorrect amino acid sequences accumulating in cells. However, we know that not all altered polypeptide chains are degraded. If this was so it would not be possible to purify and study mutant proteins. It is likely that only those molecules which are fairly seriously deranged in structure are rapidly degraded, others may persist in cells, but generally with a shorter half life than normal molecules.

Unicellular organisms are subject to fluctuations in the chemical and physical environment which may well affect the stability of their translation apparatus, or the fidelity of DNA replication. Adaptation to many possible environments would probably require an increased overall stability. The final outcome would be what we see today, namely, that unicellular organisms such as bacteria are intrinsically stable and in a variety of environments produce very few non-viable offspring. Nevertheless, for any organism, there are limits to the set of environments within which it can survive. Again, a balance has to be reached between optimum growth in a 'common' environment and ability to tolerate perturbations or sudden changes in that environment. In this context, the example of the chemostat is again instructive. A population of bacteria grown continuously under uniform conditions will slowly increase its growth rate, because there will be selection for variants which are better adapted to this particular environment. Cells which have lost their ability to utilize particular carbohydrates or other substances which are not present, or which can no longer tolerate a range of temperatures, may well have saved energy and resources and can therefore grow faster. However, if the environment is altered, such selected strains are likely to do less well than the starting population of 'wild type' cells. It can be concluded that the levels of accuracy, repair and maintenance in information transfer will be related directly or indirectly to the particular ecological niche which the organism inhabits.

12.4 The maintenance of the integrity of DNA

The evolution of accuracy in RNA and protein synthesis is governed by different selection pressures from those applying to DNA. Unlike DNA, RNA and proteins do not transmit information from generation to generation, but only ensure that the information in one cell generation will be expressed as the complete cell phenotype. A haploid organism has, in most cases, one copy of each gene and many of these are indispensable for cell growth. This gene can produce many messenger RNA molecules, which in turn produce many polypeptide chains. From the organism's point of view, the gene is the sensitive target, since once this is lost the cell will die. On the other hand, the loss of one or more messenger RNA molecules or of a proportion of a population of enzyme molecules will, in general, have no lasting effect on the cell. (There are some obvious exceptions to this, for example, a single altered repressor molecule may bind so strongly to DNA so that no further transcription of an essential gene is possible.)

Naturally occurring alkylating agents, ultraviolet light or other environmental hazards can damage DNA, RNA and protein, but it is the DNA damage which is by far the most dangerous to the cell. Cells must not only transmit their DNA intact during normal growth, but they must also protect it against induced damage or alteration of one kind or another. For this reason, organisms have evolved an elaborate series of devices to maintain their DNA (see Chapter 9). DNA replication is intrinsically very accurate, the mutation frequency per base pair being in the region of 10^{-7}–10^{-10} (Drake, 1969; Hubner and Alberts, 1980). This depends on fidelity of DNA polymerases, including proofreading, and the correction of mismatched bases at the replication fork (Radman *et al.*, 1978; Kornberg, 1980). DNA is also subject to spontaneous damage such as the loss of purine residues, or the deamination of cytosine to uracil. In addition, there is a variety of ways DNA can be damaged by external agents, such as alkylating agents or ultraviolet light. Most of this damage, whether spontaneous or induced, can be repaired by a variety of enzymic mechanisms which are reviewed in Chapter 9. In yeast, over fifty genes have been identified which in one way or another confer resistance to DNA damaging agents (Haynes *et al.*, 1982), and presumably at least as many exist in higher eukaryotes.

It is clear that organisms have invested considerable resources to maintain the structure of DNA molecules and it is therefore not surprising that the accuracy of DNA synthesis has been found to be several orders of magnitude higher than that of RNA and protein synthesis (Drake, 1969; Yarus, 1979). The DNA is, in effect, the germ line of every organism and the cost of preserving it is very high. Nevertheless, the law of diminishing return applies, since there would be no point in achieving complete accuracy in DNA replication. What is important is that the genome, with its array of several thousand indispensable genes, is transmitted with minimal or no damage from generation to generation. Complete accuracy would in any case be counter-productive, since it would eliminate the possibility of an organism adapting, by gene mutation and selection, to any new environment. For the same reason, perhaps, cells do not in general protect themselves by having multiple copies of the genome since this would also largely block their capacity to adapt.

12.5 Balancing the costs and benefits of accuracy

In this section we will examine in more detail the advantages and disadvantages of accuracy in macromolecular synthesis. The disadvantages are less obvious than the advantages. One direct disadvantage, which we have already mentioned, is that the evolution of greater accuracy in DNA synthesis and more successful maintenance of DNA structure by repair could, in theory, lead to extinction. In reality, changes in the environment will tend to select organisms which are capable of changing DNA sequences. However, the arguments concerning the optimal mutation rate for an organism are complex (Maynard Smith, 1978, pp. 188–194; see also Chapter 9) and we need not elaborate them here.

A much more general disadvantage of accuracy-promoting devices is their overall metabolic cost. That such mechanisms can be expensive, either directly in energy or equally important in the time taken for a synthesis step to be completed, is fairly self-evident. However, to make clear the precise sources of these costs, we review briefly some examples (see also Chapters 4 and 11). In kinetic proofreading (Hopfield, 1974), accuracy of synthesis is increased by means of an intermediate reaction step which drives the enzyme–substrate complex, through coupling to an additional energy-consuming reaction, into a high energy transitional state. From this transitional state, there is an increased probability of essentially irreversible spontaneous dissociation of noncognate bonds. In double sieve editing (Fersht and Dingwall, 1979; see also Chapter 4) accuracy is controlled by synthetases which have both an excision, as well as an insertion, activity. Each synthesis step involves discarding noncognate units (amino or nucleic acids), which consumes both energy and time. In Ninio's (1975) delayed reaction scheme, each step of synthesis is subjected to delay so as to allow the discrimination between cognate and noncognate bonds through their differential spontaneous dissociation rates to be amplified. The delay also adds to the likelihood of dissociating cognate bonds, thus requiring reinitiation of the whole synthesis step, which further increases the average step time.

In scavenging erroneous proteins by hydrolytic degradation, elimination of abnormal molecules is likely to be enhanced only at the expense of greater turnover of normal protein as well. Experimental evidence shows that the ability of a cell to hydrolyse abnormal protein does indeed require energy (Goldberg, 1972; Bukhari and Zipser, 1973; Waxman and Goldberg, 1982). In DNA proofreading or editing, the removal and replacement of damaged or incorrect bases involves the conversion of a proportion of nucleoside triphosphates into nucleoside monophosphates, with a consequent expenditure of energy. Finally, it should be emphasized that most of these mechanisms also involve the additional cost of synthesizing enzymes or other cellular components.

The above general considerations of the costs of accuracy are confirmed by a number of theoretical studies (Bennett, 1979; Savageau and Freter, 1979; Blomberg, Ehrenberg and Kurland, 1980; Ehrenberg and Kurland, 1984; see also Chapter 11). Savageau and Freter (1979) estimated the energy cost of proofreading the aminoacylation of transfer RNA and concluded that this activity alone may account for approximately 2% of the energy required to synthesize a bacterial cell. It is likely also that the evolution of ribosomes, which may account for up to 50% of the RNA and 30% of the protein synthesized in a rapidly growing bacterium (Maaløe, 1979), was directly related to the evolution of accurate translation. Although a full account of the total metabolic investment in accuracy is not yet possible, it seems very probable that it is a significant proportion of the full energy budget of a cell.

The conclusion that accuracy is selectively determined by balancing costs and benefits is further supported by experimental studies with bacterial mutants which have greater accuracy of protein synthesis than the wild type (Galas and Branscomb, 1976; Bohman, Ruusala, Jelenc and Kurland, 1984). Galas and Branscomb (1976) showed originally that the *str* A (*rps* L) ribosome mutant of *Escherichia coli*, which has increased accuracy of protein synthesis, also has a reduced rate of polypeptide elongation. Bohman *et al.* (1984) extended the analysis of this phenomenon and showed that the costs of elevated accuracy, revealed as a net reduction in cell growth rate, arise from a combination of (1) dissipative energy losses associated with changed ribosomal proofreading activity and (2) impaired kinetic efficiency of the translation apparatus (see also Chapter 11).

When these principles are extended also to eukaryotes and, in particular, to the cells of multicelled higher animals, it is clear that the same general relationship between accuracy and cost will apply: namely, increased accuracy may be attained only at significant metabolic cost. We should expect, therefore, that the accuracy of macromolecular synthesis will be limited in each type of cell to little more than the minimum level which is necessary.

12.6 Optimal accuracy of translation in reproductive and somatic cells

Following the distinction introduced by Weismann at the end of the 19th century, cells may be classed into two primary groups: reproductive cells and somatic cells. Reproductive cells constitute essential links in the chain of inheritance of genetic information and include unicellular micro-organisms, germ cells of higher animals, and a broad range of cells among plants and lower animals which reproduce asexually. In each of these, a high degree of translational stability is needed, since progressive deterioration in fidelity would be disastrous. As we have mentioned, this may be especially true for micro-organisms which inhabit unpredictable environments, where sudden exposure to heat or toxic chemicals could cause dramatic rises in error levels. By contrast, among somatic cells of higher organisms the future lineage of any single cell is limited since it is certain that, even in the absence of ageing, the body of which it forms a part will eventually die from disease, accident or starvation. In somatic cells there is therefore not the same need for long-term stability.

The main features of the relationship between accuracy and stability can be derived from a simplified model of the translation apparatus (Kirkwood and Holliday, 1975a; Kirkwood, Holliday and Rosenberger, 1984; see also Chapter 2). In general, there is a positive association between the two factors, though it is possible, within the terms of the model, to raise the stability of translation without significantly altering its basic accuracy. This can be done, for example, by scavenging erroneous proteins more efficiently.

For any cell, the minimum required level of accuracy is that which both (1) provides for efficient translation and (2) gives the necessary degree of stability. This raises a number of possibilities. The first is that the accuracy of (1) so greatly exceeds the accuracy of (2) that there is no additional selection for stability. This would imply that error propagation does not pose a significant threat to the stability of any present-day cell. In this case, all cells would be expected to have roughly similar accuracy. However, while this possibility cannot yet be excluded, there is evidence to suggest it is unlikely (see Chapter 2). The alternative is that selection *does* act on translational stability, in which case differences between somatic and reproductive cells may result either in different levels of accuracy, or in different amounts of protein degradation, or both. In any of these latter cases, we predict a lower investment of metabolic resources to maintain accuracy in somatic cells than in reproductive cells, and it is on this basis that a theory of the evolution of ageing can be developed (Kirkwood, 1977).

12.7 Evolution of ageing and longevity

The view that ageing in higher animals is due to an accumulation of defects is an attractive one, especially since the outward signs of senescence do seem to resemble the process of 'wear and tear' of inanimate objects. However, there is no fundamental biological reason why any organism could not be arranged in such a way as to maintain itself without deterioration. Examples can be found amongst the simpler animals, such as sea anemones, sponges or flatworms (see Comfort, 1979). These appear to be able to maintain themselves more or less indefinitely by regeneration or replacement of defective cells and tissues. If organisms can, in principle, maintain themselves in this way, then an explanation of the near universality of ageing among higher animals must be sought within the context of natural selection.

Logically, and historically, the first attempt to explain evolution of ageing has been in terms of a direct advantage to possessing only a finite lifespan. This approach was first adopted by Wallace (see Weismann, 1891, pp. 23–24), and the idea (often independently restated) has remained popular among gerontologists. The advantages which have been proposed can be summarized as follows. If evolution is to occur at all, then generation must follow generation. Thus, ageing is beneficial in eliminating old individuals from the population, thereby increasing the resources available to their progeny and allowing natural selection to take place more easily. However, there are two major objections to this view. Firstly, in most natural populations it is rare for an individual to survive long enough to show obvious signs of ageing (Medawar, 1952; see also, for example, Lack, 1954). Therefore, there is neither need, nor obvious opportunity, for the ageing process to have evolved. Secondly, the argument depends on group selection and suffers from all the difficulties associated with this view (Maynard Smith, 1976). For example, in a hypothetical population of animals that did age for the above reason, a mutant with increased longevity would, if other things were equal, have a selective advantage. Therefore, the process of ageing should not be stably maintained in the population.

In recognition of the difficulties confronting an adaptive explanation of ageing, Medawar (1952) proposed that ageing was the more or less accidental by-product of the process of natural selection. By noting that, even in the absence of ageing, the effect of environmental mortality would reduce the fraction of individuals surviving into progressively later portions of the lifespan, and thereby diminish the ability of natural selection to act against late-acting gene defects, Medawar suggested that ageing was due to an accumulation of harmful age-specific genes whose time of expression lay beyond the greater part of the natural range of lifespan. This concept was developed by Williams (1957) to place special emphasis on pleiotropic genes which have good effects early in life, but have bad effects later, and later authors have refined the concept still further (Hamilton, 1966; Charlesworth, 1980).

The significance of the work of Medawar and Williams is their clear elucidation of the effect of age on the force of natural selection, and hence the conclusion that selective control over the later portions of the lifespan must progressively tail away. However, there are aspects of the theory, especially Medawar's version, which are incomplete (see Kirkwood and Holliday, 1979). Furthermore, the theory is not at all specific about the possible molecular basis of ageing. These difficulties can be resolved by taking into account the *cost* of maintaining an organism's viability (Kirkwood, 1977; Kirkwood and Holliday, 1979).

At any time, the orderly regulation of an organism's metabolism may be disrupted by random errors and damage. Mechanisms to protect against or correct these defects require the consumption of metabolic resources which

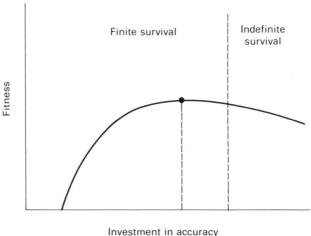

Figure 12.1 Relationship between the investment of metabolic resources in accuracy and evolutionary fitness. The disposable soma theory of ageing (see text) predicts that the optimum level of investment in somatic accuracy, i.e. that which maximizes fitness, will always be less than the minimum required for indefinite survival.

would otherwise be available for other activities, in particular, reproduction. The upshot is that while a long life is, in general, an advantage, the benefits of somatic longevity must be carefully weighed, through the optimizing influence of natural selection, against the benefits of greater investment in progeny.

Consider, for example, a species which experiences constant mortality through life of 10% per year. The probability of any one individual surviving to the age of 100 years is approximately 1 in 40 000. For practical purposes this probability is negligible, and there is therefore no value in investing more resources in somatic maintenance than are necessary for individuals to survive to 100 years, or perhaps somewhat less. Not only is there no value in doing this, but it is actually *disadvantageous* to waste these resources which could be better used to increase reproductive output in the earlier part of the lifespan. The optimum investment in somatic maintenance will therefore always be less than that required for indefinite survival (Fig. 12.1). In this way ageing is readily explained as the evolutionarily stable consequence of natural selection working to optimize an organism's use of its resources (Kirkwood, 1977, 1981; Kirkwood and Holliday, 1979, 1985). By obvious analogy with the manufacture of disposable goods, which have short expected lives and are therefore made as cheaply as possible, this is termed the 'disposable soma' theory.

Evolution of longevity is also readily explained by the disposable soma theory (Kirkwood and Holliday, 1979, 1985). For example, a species with a high risk of environmental mortality, such as the mouse (*Mus musculus*), can expect to survive only a short time. Therefore, it ought not to invest too heavily in the maintenance of each individual soma, which will consequently age early, but instead should concentrate resources on a high rate of reproduction. Conversely, a species which occupies a more secure ecological niche may profit by doing the reverse. The latter is especially likely to be true of the larger, more intelligent mammals, which require an extended period of infancy and parental care for education of the young.

So far, we have not considered which processes of somatic maintenance are the most critical for determining longevity. Studies of the physiology of ageing (see, for example, Comfort, 1979) indicate that there is a very general deterioration of function at each level of organization. This suggests that a key factor in senescence is the loss of viability of individual cells, whether dividing or nondividing (Martin, 1977a, b). This conclusion is supported in the case of dividing cells by the finite replicative lifespan in culture of somatic cells such as fibroblasts (Hayflick, 1965), and by the apparent negative correlation between this finite lifespan and the age of the cell donor (Martin, Sprague and Epstein, 1970). A number of hypotheses which implicate damage of one kind or another have been advanced to explain cellular ageing (see Strehler, 1977; Kirkwood, 1981, 1984; Holliday, 1984). Since all cellular functions depend ultimately on the integrity of protein synthesis and replication of DNA, it is natural to turn first to the possibility of a breakdown in macromolecular synthesis as the root cause of ageing. In this particular case, the disposable soma theory proposes that the accuracy of information transfer and the maintenance of macromolecules is reduced at some early stage in development and this places the cells in a metastable state. In this situation there is an appreciable chance that given cells will eventually become non-viable (Fig. 12.2). This interpretation is directly consistent with the commitment theory of cellular ageing (Kirkwood and Holliday, 1975b) which proposes that each somatic cell initially has the potential for unlimited division, but subsequently has a fixed probability in each cell generation of becoming committed to only a limited lifespan. The commitment theory explains the finite lifespan of fibroblasts in tissue culture and is supported by a range of experimental evidence (Holliday *et al.*, 1977; Holliday, Huschtscha and Kirkwood, 1981).

In summary, we predict that, during the course of the lifetime of a higher animal, the proportion of somatic cells which have become non-viable will gradually increase. This could occur by a breakdown in the fidelity of macromolecular synthesis, by the failure to maintain the structure of important macromolecules in the face of damage from free radicals or other toxic cellular products (see, for example, Ames, 1983), or even by the cells' inability to maintain the normal epigenetic controls of gene activity (Holliday, 1985). Senescence is seen as the end result of increasing numbers of cells

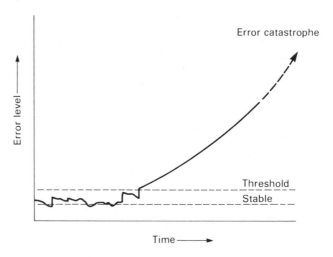

Figure 12.2 The level of errors in macromolecular synthesis is normally maintained close to the stable value. Random fluctuations (caused, for example, by a burst of protein errors following a transcription error, or by a disturbance in the cell's environment) will periodically cause the error level to rise. Provided this does not take the cell across the threshold error level above which irreversible error propagation occurs, the cell will revert to the stable level. Occasionally, however, a rise in error level may be so severe, or two or more rises may occur in such close succession, that the threshold is crossed and the error level progressively increases, leading ultimately to a lethal error catastrophe.

functioning abnormally or dying off at later ages. By the arguments outlined earlier, natural selection will determine the level of accuracy/stability to be such that the rate of accumulation of macromolecular defects is low enough not to interfere seriously with an organism's viability during its expected period of survival in a background of normal environmental hazard, but there will actually be selection *against* any further reduction in the error level (Kirkwood and Holliday, 1985).

12.8 Predictions and conclusions

The disposable soma theory makes several testable predictions. First of all, we would expect the accuracy of information transfer between macromolecules to be greater in germ cells than in somatic cells. No comparison of the accuracy of protein synthesis in germ cells, such as spermatogonia, and somatic cells, such as fibroblasts, has yet been made. However, it is also possible that the integrity of the germ line is preserved by eliminating defective, or error-committed, cells through cellular selection. This, too, can be investigated experimentally.

If differences in accuracy and maintenance do exist between somatic and germ cells, it does not necessarily follow from the theory that there would be the same differences in the accuracy of DNA synthesis as in the accuracy of protein synthesis. This is because the natures of the selective forces governing accuracy of these two processes are not entirely the same. For protein synthesis, natural selection will act, on the one hand, to optimize the efficiency of genetic translation and, on the other hand, to ensure sufficient stability that the germ line can be immortal. For DNA synthesis, selection pressure to maximize the fidelity of gene copying and maintenance may be balanced by counter-selection to preserve a degree of genetic variability and adaptability. The precise balance between this latter pair of selection pressures remains unclear (Maynard Smith, 1978), but it is likely that any selection for mutability *per se* will be more significant in organisms inhabiting unpredictably variable environments than in organisms whose environment is comparatively stable. In general, it may well be that there would not be any important difference in mutation frequency in young somatic cells and germ cells. On the other hand, senescent somatic cells should lose fidelity in DNA replication as enzymes, such as DNA polymerases, become more error prone, and such cells should have an increased mutation frequency. Experimental evidence bears this out (Fulder and Holliday, 1975; Morley, Cox and Holliday, 1982; Linn, Kairis and Holliday, 1976; Murray and Holliday, 1981).

Another prediction of the disposable soma theory is that there should be a higher intrinsic error level in the somatic cells of short-lived organisms, than in the same types of cell from long-lived ones. With regard to the fidelity of transcription and translation, this prediction has not yet been tested. Again, the expectation with regard to the accuracy of DNA synthesis is equivocal and the evidence so far suggests that mutation frequencies in cultured human or rodent cells are fairly similar (see Holliday, 1985). With regard to repair mechanisms, Hart and Setlow (1974) reported that the efficiency of excision repair of pyrimidine dimers induced by ultraviolet light was directly related to the lifespan of the species examined. In several subsequent studies on excision repair, one failed to confirm this relationship (Kato *et al.*, 1980), but three others confirmed it (Francis, Lee and Regan, 1981; Treton and Courtois, 1982; Hall *et al.*, 1984). The results of Treton and Courtois (1982) are particularly significant, since the cells used were in short-term cultures of eye lens epithelium, which in a natural environment is subject to damage from ultraviolet light. It is known that primary cultures from mouse embryonic tissue are capable of excision repair, but this ability is lost after several subcultures. Neither established lines of mouse cells, nor adult skin cells *in vitro* are able to excise more than a small fraction of pyrimidine dimers (Peleg, Raz and Ben Ishai, 1977; Ley *et al.*, 1977). These results provide evidence that some mechanisms for maintaining the structure of DNA in adult somatic cells are dispensed with in short-lived animals. Human cells survive in culture much longer than do mouse cells (barring transformation), and they also retain much greater excision repair capacity.

If our interpretation of commitment to senescence in cultured fibroblasts is correct, then we would predict that the probability of initiation of the unstable condition would be greater in cells from short-lived animals than in those from long-lived ones. We might also expect that the incubation period, during which errors are accumulating, would vary between species. Experimental investigations of these parameters in different species have not yet been carried out.

In conclusion, our theoretical considerations of the nature and probable outcome of selective forces on the accuracy of macromolecular information transfer suggest that a comprehensive study of all aspects of fidelity of synthesis of macromolecules would be particularly valuable. Of special significance will be comparative studies of accuracy in species of different longevities and investigations into the extent of error feedback and the level of translational stability in different types of cell. The data bearing on these points which are already available are sufficiently contradictory and controversial that it is clear that more careful and extensive studies are needed.

References

Ames, B. N. (1983) Dietary carcinogens and anticarcinogens. *Science*, **221**, 1256–1264.

Bennett, C. H. (1979) Dissipation-error tradeoff in proofreading. *Bio-Systems*, **11**, 85–91.

Blomberg, C., Ehrenberg, M. and Kurland, C. G. (1980) Free energy dissipation constraints on the accuracy of enzymatic selections. *Quart. Rev. Biophys.*, **13**, 231–254.

Bohman, K. T., Ruusala, T., Jelenc, P. C. and Kurland, C. G. (1984) Kinetic impairment of restrictive streptomycin resistant ribosomes. *Mol. Gen. Genet.*, **198**, 90–99.

Bukhari, A. I. and Zipser, D. (1973) Mutants of *Escherichia coli* with a defect in the degradation of nonsense fragments. *Nature New Biol.*, **243**, 238–241.

Charlesworth, B. (1980) *Evolution in Age-Structured Populations*. Cambridge University Press, Cambridge.

Comfort, A. (1979) *The Biology of Senescence*, 3rd edn, Churchill Livingstone, Edinburgh.

Drake, J. W. (1969) Comparative spontaneous mutation rates. *Nature*, **221**, 1128–1132.

Ehrenberg, M. and Kurland, C. G. (1984) Costs of accuracy determined by a maximal growth constraint. *Quart. Rev. Biophys.*, **17**, 45–82.

Fersht, A. R. and Dingwall, C. (1979) Evidence for the double-sieve editing mechanism in protein synthesis. Steric exclusion of isoleucine by valyl-tRNA synthetases. *Biochemistry*, **18**, 2627–2631.

Francis, A. A., Lee, W. H. and Regan, J. D. (1981) The relationship of repair of ultraviolet-induced lesions to the maximum life span of mai *Ageing Dev.*, **16**, 181–189.

Fulder, S. J. and Holliday, R. (1975) A rapid rise in cell variants during the of populations of human fibroblasts. *Cell*, **6**, 67–73.

Galas, D. J. and Branscomb, E. W. (1976) Ribosomes slowed by mutation to streptomycin resistance. *Nature*, **262**, 617–619.

Goldberg, A. L. (1972) Degradation of abnormal proteins in *Escherichia coli*. *Proc. Natl Acad. Sci. USA*, **69**, 422–426.

Goldberg, A. L. and Dice, J. F. (1974) Intracellular protein degradation in mammalian and bacterial cells, Part 1. *Ann. Rev. Biochem.*, **43**, 835–869.

Goldberg, A. L. and St John, A. C. (1976) Intracellular protein degradation in mammalian and bacterial cells, Part 2. *Ann. Rev. Biochem.*, **45**, 747–803.

Hall, K. Y., Hart, R. W., Bernischke, A. K. and Walford, R. L. (1984) Correlation between ultraviolet-induced DNA repair in primary lymphocytes and fibroblasts and species maximum achievable lifespan. *Mech. Ageing Dev.*, **24**, 163–173.

Hamilton, W. D. (1966) The moulding of senescence by natural selection. *J. Theor. Biol.*, **12**, 12–45.

Hart, R. W. and Setlow, R. B. (1974) Correlation between deoxyribonucleic acid excision repair and lifespan in a number of mammalian species. *Proc. Natl Acad. Sci. USA*, **71**, 2169–2173.

Hayflick, L. (1965) The limited *in vitro* lifetime of human diploid cell strains. *Expl Cell Res.*, **37**, 614–636.

Haynes, R. H., Barclay, B. J., Eckardt, F., Landman, O., Kunz, B. and Little, J. G. (1982) Genetic control of DNA repair in yeast. In *Proc. XIVth Int. Congr. Genetics* (ed. D. K. Beliayev), Nank Publishing House, Moscow.

Holliday, R. (1984) The unsolved problem of cellular ageing. *Monogr. dev. Biol.*, **17**, 60–77.

Holliday, R. (1985) The significance of DNA methylation in cellular ageing. In *The Molecular Biology of Aging* (eds A. D. Woodhead, A. D. Blackett and A. Hollaender), Plenum, New York, pp. 269–283.

Holliday, R., Huschtscha, L. I. and Kirkwood, T. B. L. (1981) Cellular ageing: further evidence for the commitment theory. *Science*, **213**, 1505–1508.

Holliday, R., Huschtscha, L. I., Tarrant, G. M. and Kirkwood, T. B. L. (1977) Testing the commitment theory of cellular ageing. *Science*, **198**, 366–372.

Hopfield, J. J. (1974) Kinetic proofreading: a new mechanism for reducing errors in biosynthetic processes requiring high specificity. *Proc. Natl Acad. Sci. USA*, **71**, 4135–4139.

Hubner, U. and Alberts, B. M. (1980) Fidelity of DNA replication catalyzed *in vitro* on a natural DNA template by the 4 bacteriophage multienzyme complex. *Nature*, **285**, 300–305.

Kato, H., Harada, M., Tsuchiya, K. and Moriwaki, K. (1980) Absence of correlation between DNA repair in ultraviolet irradiated mammalian cells and lifespan of the donor species. *Japan J. Genetics*, **55**, 99–108.

Kirkwood, T. B. L. (1977) Evolution of ageing. *Nature*, **270**, 301–304.

Kirkwood, T. B. L. (1981) Repair and its evolution: survival versus reproduction. In

Physiological Ecology: An Evolutionary Approach to Resource Use (eds C. R. Townsend and P. Calow), Blackwell Scientific, Oxford, pp. 165–189.

Kirkwood, T. B. L. (1984) Towards a unified theory of cellular ageing. *Monogr. dev. Biol.*, **17**, 9–20.

Kirkwood, T. B. L. and Holliday, R. (1975a) The stability of the translation apparatus. *J. Mol. Biol.*, **97**, 257–265.

Kirkwood, T. B. L. and Holliday, R. (1975b) Commitment to senescence: a model for the finite and infinite growth of diploid and transformed human fibroblasts in culture. *J. Theor. Biol.*, **53**, 481–496.

Kirkwood, T. B. L. and Holliday, R. (1979) The evolution of ageing and longevity. *Proc. R. Soc. Lond. B*, **205**, 531–546.

Kirkwood, T. B. L. and Holliday, R. (1985) Ageing as a consequence of natural selection. In *The Biology of Human Ageing* (eds K. J. Collins and A. H. Bittles), Cambridge University Press, Cambridge.

Kirkwood, T. B. L., Holliday, R. and Rosenberger, R. F. (1984) Stability of the cellular translation process. *Int. Rev. Cytol.*, **92**, 93–132.

Kornberg, A. (1980) *DNA Replication*, Freeman, San Francisco.

Lack, D. (1954) *The Natural Regulation of Animal Numbers*, Clarendon Press, Oxford.

Ley, R. D., Sedita, A., Grube, D. D. and Fry, R. J. M. (1977) Induction and persistence of pyrimidine dimers in the epidermal DNA of two strains of hairless mice. *Cancer Res.*, **37**, 3243–3248.

Linn, S., Kairis, M. and Holliday, R. (1976) Decreased fidelity of DNA polymerase activity isolated from ageing human fibroblasts. *Proc. Natl Acad. Sci. USA*, **73**, 2818–2822.

Maaløe, O. (1979) Regulation of the protein synthesising machinery – ribosomes, tRNA, factors and so on. In *Biological Regulation and Development*, Vol. 1 (ed. R. F. Goldberger), Plenum, New York, pp. 487–537.

Martin, G. M. (1977a) Cellular ageing – clonal senescence. *Am. J. Pathol.*, **89**, 484–511.

Martin, G. M. (1977b) Cellular ageing – postreplicative cells. *Am. J. Pathol.*, **89**, 513–530.

Martin, G. M., Sprague, C. A. and Epstein, C. J. (1970) Replicative lifespan of cultivated animal cells: effects of donor's age, tissue and genotype. *Lab. Invest.*, **23**, 86–92.

Maynard Smith, J. (1976) Group selection. *Q. Rev. Biol.*, **51**, 277–283.

Maynard Smith, J. (1978) *The Evolution of Sex*, Cambridge University Press, Cambridge.

Medawar, P. B. (1952) *An Unsolved Problem in Biology*, H. K. Lewis, London.

Morley, A., Cox, S. and Holliday, R. (1982) Human lymphocytes resistant to 6-thioguanine increase with age. *Mech. Ageing Dev.*, **19**, 21–36.

Murray, V. and Holliday, R. (1981) Increased error frequency of DNA polymerases from senescent human fibroblasts. *J. Mol. Biol.*, **146**, 55–76.

Ninio, J. (1975) Kinetic amplification of enzyme discrimination. *Biochimie*, **57**, 587–595.

Orgel, L. E. (1963) The maintenance of the accuracy of protein synthesis and its relevance to ageing. *Proc. Natl Acad. Sci. USA*, **49**, 517–521.

Orgel, L. E. (1973) Ageing of clones of mammalian cells. *Nature*, **243**, 441–445.

Peleg, L., Raz, E. and Ben Ishai, R. (1977) Changing capacity for DNA excision repair in mouse embryonic cells *in vitro*. *Exp. Cell Res.*, **104**, 301–307.

Radman, M., Villani, G., Boiteux, S., Kinsella, A. R., Glickman, B. W. and Spadari, S. (1978) Replication fidelity: mechanisms of mutation avoidance and mutation fixation. *Cold Spr. Harb. Symp. Quant. Biol.*, **43**, 937–946.

Savageau, M. A. and Freter, R. R. (1979) On the evolution of accuracy and cost of proofreading tRNA aminoacylation. *Proc. Natl Acad. Sci. USA*, **76**, 4507–4510.

Strehler, B. L. (1977) *Time, Cells and Aging*, 2nd edn, Academic Press, New York.

Treton, J. A. and Courtois, Y. (1982) Correlation between DNA excision repair and mammalian lifespan in lens epithelial cells. *Cell Biol. Int. Rep.*, **6**, 253–260.

Waxman, L. and Goldberg, A. L. (1982) Protease La from *Escherichia coli* hydrolyzes ATP and proteins in a linked fashion. *Proc. Natl Acad. Sci. USA*, **79**, 4883–4887.

Weismann, A. (1891) *Essays upon Heredity and Kindred Biological Problems*, Vol. 1, 2nd edn, Clarendon Press, Oxford.

Williams, G. C. (1957) Pleiotropy, natural selection and the evolution of senescence. *Evolution*, **11**, 398–411.

Yarus, M. (1979) The accuracy of translation. *Prog. Nucl. Acid Mol. Biol.*, **23**, 195–225.

13 Diversity and accuracy in molecular evolution: sketches past, present and future

J. NINIO

13.1 Sketch I

Species are eternal. To be sure, any individual deviates somewhat from the original type; usually, for any given character within a species (body size, hairiness, visual acuity, etc.) there is an observable diversity. Biometric studies show that individual properties and performances follow more or less Gaussian distributions in a population as a result of various environmental perturbations. External influences, by slightly perturbing at one point or another the development of the organism, will endow the individual with special traits. The child does not resemble his parents exactly.

All the observed diversity is simply fluctuation around an immutable prototype. Breeders claim that they do produce varieties with distinctive properties that are stable enough to be transmitted from one generation to the next. Actually, they are not changing the organisms in a fundamental way. They are artificially pulling on a spring and maintaining it in an unstable state. When domesticated plants or animals are allowed to return to the wild, they quickly recover all the attributes of the original species.

13.2 Sketch II

The metaphor of the spring may be fallacious, however. Under normal circumstances there is a rigid, reproducible pattern of organization. But it is possible to disturb it and make it more malleable. This occurs when the organism has to face external conditions to which it is not adapted. As noted by an eminent specialist in barnacle taxonomy, 'such changes of external conditions would, from their acting on the reproductive system, probably cause the organization of those beings which were most affected to become, as under domestication, plastic' (Darwin, 1859).

This suggests the possibility of a different kind of diversity. Normally, two individuals of the same species are but two members of a single Gaussian distribution, corresponding to the same prototype. However, an individual may represent a slightly altered prototype, with a new distribution of characters somewhat displaced from the original. A new range of possibilities may be presented by the new prototype. The changes may be favourable to the individual, conferring an advantage over other individuals of the same species. Then, however small the change, the new prototype will spread through the species, displacing the old prototype. A British naturalist wrote: 'Now the scale on which nature works is so vast – the numbers of individuals and periods of time with which she deals approach so near to infinity, that any cause, however slight, and however liable to be veiled and counteracted by accidental circumstances, must in the end produce its full legitimate results' (Wallace, 1859).

Such a view of organismic changes parallels Lyell's conception of geological changes. The spectacular events of the past (the raising of mountains, the formation of seas) can be accounted for by the slow and steady action of the same forces that we observe today, cumulating their effects over huge periods of time (see Gould, 1980).

13.3 Sketch III

The changes, however, are not gradual. The individuals are governed by their genes, which define their characteristics: the colour of the kernels in maize, the functionality of the wings in the fruit-fly, the number of fingers in humans. Mutations are rather rare, yet there is an important polymorphism in most animal populations. Actually, the main source of variation is combinatorial: sexual reproduction generates an enormous diversity in terms of gene combinations. Since 'most characters require for proper development a nicely adjusted train of processes . . . any change in the genes – no matter whether loss, gain, substitution or rearrangement – is more likely to throw the developmental mechanism out of gear, and give a "weaker" result, than to intensify it' (Muller, 1922).

It is possible to provoke inheritable changes using external agents like X-rays. Most of the provoked mutations are, of course, detrimental to the organism, leading to aberrations in development or to the loss of an important function. Therefore, viable mutations are relatively rare. There is an even stronger reason for their rarity. The gene reproduces itself by some kind of autocatalysis. A mutated gene can propagate only if it also has this extraordinary capacity to reproduce itself (Muller, 1922).

13.4 Sketch IV

Chromosomal aberrations, such as inversions, breaks, and transpositions, that lead to unequal segregations and chromosome losses at meiosis, form a rather limited source of mutations. Actually, genetic changes can be much less drastic than that: occasional deletions or insertions of one or a few nucleotides within the gene sequence or, more usually, replacement of a single nucleotide by another. The nucleotide bases can change spontaneously, due to fluctuations of their chemical structure, or they may be damaged by ionizing radiation or chemicals.

The replacement of one base by another in the gene will often provoke the replacement of one amino acid by another in the corresponding protein sequence. If the other amino acids are not found 'with considerable frequency at the relevant position of the polypeptide, their presence must be considered to be on the whole detrimental under present or recently existing natural conditions. Otherwise they would have arisen and not have been kept rare by elimination of the recurrent mutations.' (Sonneborn, 1965.)

Mutations may occur all along the gene's sequence. Two distant mutations in the same gene, both leading individually to the loss of the enzyme's activity, may have compensatory effects, and complement each other (Helinski and Yanofsky, 1963). Obviously, the functioning of an enzyme depends on its precise three-dimensional architecture. An amino acid change will often prevent the protein from folding correctly, unless its effects are more or less compensated by one or several changes elsewhere. Thus, 'the replacement in a population of a given gene by a mutant gene may often require two successive mutations, except when the population is very small (close inbreeding)' (Zuckerkandl and Pauling, 1962). Since most single-step variants of a protein are non-functional, it is important for the mechanism of protein synthesis to be very accurate. Although some preliminary data suggested some heterogeneity in the translation products of the genes, careful sequence analyses have always shown the proteins to be well-defined entities and not mixtures. For example, one compelling argument for the high accuracy of the translation apparatus lies in considering species differences in protein sequences. There is only one amino acid difference between human and gorilla haemoglobin beta chains. Undetected variants of the human protein 'must be rare, unless a human is more often like a gorilla than like a human' (Zuckerkandl and Pauling, 1962). The immune system would be quite intolerant of a translation system with low accuracy.

In any event, life cannot cope with a high level of translation errors. When errors occur at an appreciable level, some of the faulty proteins must generate more errors. The resulting accumulation of errors should lead ultimately to cellular death. Senescence might, in fact, be due to an error catastrophe (Orgel, 1963) (see Chapters 2 and 12). Order leads easily to chaos. On the other hand, there is at least one instance in which the reverse must have been true. 'The initial emergence from chaos of the master pattern of living material may possibly be of such transcendant improbability that it has occurred rarely – perhaps once in the history of the world.' (Hinshelwood, 1946, quoted by Pringle, 1953.)

13.5 Sketch V

The first protocells emerged in a context of chaotic chemical processes. Prebiotic syntheses produced racemic mixtures of amino acids. Every reaction that led to a 'biological' compound also produced many 'undesirable' chemicals along the way. Life was then compatible with chemical disorder, otherwise it could not have arisen. The protocells must have manufactured low-specificity enzymes through inaccurate translation. 'Thus, every gene must have been translated into a group of proteins, called a statistical protein, in which no two members had exactly the same primary structure, but all members bore some relationship to one another, and the group as a whole characterized the nucleic acid whence it originated.' (Woese, 1967.) Mathematical modelling shows, by the way, that a relatively inaccurate translation system can sustain itself at one or many steady-state levels of error production (Hoffman, 1974; Kirkwood and Holliday, 1975; see also Chapter 2). The requirements for catalytic activity are not too stringent. Single amino acids or peptides can act as catalysts. RNA molecules on their own can act upon RNA molecules and produce covalent changes (Kruger et al., 1982). Random 'proteinoids' produced by thermal condensation of amino acids can display a number of biologically interesting catalytic activities. Some synthetic organic polymers are even better catalysts than certain evolved enzymes. It is conceivable that many catalysts of the future will be obtained by screening of random peptide mixtures (Ninio, 1979).

Modern bacteria can cope with high levels of translation errors. They tolerate nonsense and missense suppressors that generate appreciable misreading of the genetic code. When a gene mutates and produces an inactive protein, one way of restoring its function is to grow the bacteria in the presence of streptomycin which induces translation errors, and thus allows the synthesis of variants of the mutant protein, some of which may be active. Bacteria with ambiguous ribosomes (*ram* mutants) grow slowly but are nevertheless viable (Rosset and Gorini, 1969). The level of translation errors in some of the high-ambiguity bacteria is such that out of a hundred molecules of β-galactosidase, there are not even two molecules with identical sequences. Yet, the overall level of β-galactosidase activity in such strains is about 30% of that found in wild type. Clearly, many close variants of a functional protein are very similar functional proteins.

Actually, whenever we look for viable variants of mutant proteins, we find them. Natural populations display a high degree of polymorphism for most of the genes in which polymorphism has been studied. Furthermore, the potential rate of evolutionary change in a gene is enormous. In mammals, 'on average, one nucleotide pair has been substituted in the population roughly every 2 yr' (Kimura, 1968). Selecting positively for each variant would require an enormous juvenile mortality; this is not observed. Thus, evolution is in large part a matter of random drift among populations (Kimura, 1968). Drifts may also occur at a lower level within families of closely related genes, in the genome of an organism. Thus, some multi-copy gene families (e.g., the trypanosome antigens or the MHC complex) will display a strange pattern of variation: substantial between species yet very small within and between individuals (Dover, 1982).

One can force bacteria to grow on novel substrates. Old enzymes can be recruited to new jobs. A few mutations can be sufficient to evolve an entirely novel metabolic pathway, built upon an existing set of enzymes (Clarke, 1978). Repetitive, non-coding DNA sequences can easily evolve into genes for fully active enzymes (Okada et al., 1983). Within a few hours or a few days, the immune system of a mammal will produce the appropriate antibody to recognize a completely new antigen. Somatic mutations are rarely harmful, and they can be solicited to produce rapidly a response to challenging external conditions. In lung or liver tissues which have sustained chemical injury, a few mutant cells are generally able to grow and regenerate the organ (Weill and Reynaud, 1980). Perhaps an organism keeps down its mutation rate when all is going well, and permits mutations to occur at an elevated rate when it is challenged. The overall rate of mutation may indeed be subject to control, but the direction cannot be specified. For it is 'axiomatic to modern biology that mutation is a wholly random process in so far as the nature of the resulting change in genetic "information" is concerned' (Burnet, 1974).

13.6 Sketch VI

A species may wander a bit around its central genotype, but this does not constitute evolution (Grassé, 1978). For example, the accuracy of replication in phage Qβ is rather low, so that 'each viable phage genome in a multiply passaged population differs in one or two positions from the "average" sequence of the parental population' (Domingo et al., 1978). Despite this high mutation rate, the central genotype remains stable and reproduces itself indefinitely, with its distribution of errors. It appears as if the distribution is being selected. Histones show remarkably stable sequences, yet their genes have abundantly mutated. In fact, genetic homeostasis is the rule (Lerner, 1954).

The neutrality of mutations can be both real and illusory. Consider two very different gene sequences that correspond to proteins that are equally important to the cell. Assume that the first is in a very sharp adaptive peak: all close variants of the gene produce virtually non-functional proteins. Assume that the second gene is 'neutral' – most of its variants give equally useful proteins. Obviously there will be many step-wise evolutionary pathways leading to the second gene, and perhaps none leading to the first. Thus, evolution should strongly favour genes with neutral neighbourhoods (Ninio, 1979). Perhaps, in some early critical stages of evolution, selection acted, not upon the gene alone, but upon the gene plus its variants. A possible strategy in molecular evolution would be for an organism to go through a stage of low accuracy of translation, thus exploring at once the whole neighbourhood of a sequence (see Bachinsky, 1980). Evidence in favour of this view comes from the following considerations.

When the genes that are responsible for the accuracy of translation or replication mutate, they can change in either direction, towards lower or higher fidelity. Considering that it is much easier to design error-prone enzymes than faithful ones, one expects two things. Mutations to lower accuracy should be much more frequent than mutations to higher accuracy. Low-fidelity ribosomes should produce erroneous DNA polymerases with, as a result, a significantly increased rate of mutations. This is not observed. The key genes for accuracy thus seem to be endowed with the property of being within easy mutational reach of high accuracy variants. In a sense, then, the stored genetic information of the cell extends beyond the gene's sequence to include the average properties of the gene's close variants.

Since the genes determining the accuracy of a process may easily mutate towards higher or lower accuracy, the observed fidelity levels are probably optimal. 'There must be gene mutations, but an excessive rate gives an array of freaks, not evolution; there must be selection but too severe a process destroys the field of variability and thus the basis for further advance' (Wright, 1932, quoted in Wright, 1955). Optimum fidelity will depend upon the circumstances. Some cells have evolved ways of regulating their error levels.

When bacteria are threatened by a severe imbalance in their amino acid pools they switch to a regime of higher accuracy of translation (Gallant, 1979). Translational ambiguity varies during cell differentiation in fungi (Picard-Bennoun, 1982). An increase in fidelity will block sporulation (Dequard-Chablat and Coppin-Raynal, 1984). Conversely, malignant cells that escape from senescence may have a less accurate translation apparatus than normal cells, as found in the case of SV-40 transformed fibroblasts (Harley et al., 1980). In the case of an oncogenic protein, it is possible that a lower specificity of translation may result from its diminished GTPase activity (Gibbs et al., 1984).

Conceivably, many 'unexpected' environmental perturbations will provoke, in the long term, a general decrease in translational accuracy, making the organism more malleable. A device for detecting errors in translation (for instance, a gene split by a nonsense codon) may inform the cell that something is going wrong, and cause it to switch to a different accuracy regime (Ninio, 1977) – perhaps a regime with a lower fidelity of replication and a higher fidelity of translation. Such a sophisticated mechanism does not seem to operate in bacteria, but may perhaps exist in higher cells.

13.7 Sketch VII

Consider a well-defined protein and one of its close variants and assume that the two differ markedly in their functional characteristics. If selection acts on the genes plus their neighbourhoods, the selective advantage of one gene over the other will be greatly diminished because of the averaging effects and the overlap between the neighbourhoods. A high level of translation errors (codon misreading or translational frameshifts) that mimic genetic changes (base substitutions, point insertions or deletions) makes such genetic changes less severe than they would have been in the context of error-free translation. Obviously, the evolution of many sequences, like histones, tRNAs, or 5S RNAs in mammals, has reached a virtual standstill. The sequences are fixed in their informational neighbourhoods. Major reorganizations of the genetic material (like gene fusions, tandem gene duplications, or massive frameshifts) would appear to be necessary to protein evolution. 'Without major steps, many proteins would reach an evolutionary stagnation point in which every conceivable amino acid substitution was slightly harmful' (McLachlan, 1980).

There are ways of modifying proteins or RNA beyond the range of variation obtained through translation or transcription errors. It may well be that at a certain point in tRNA evolution, no significant improvement could be reached through base substitutions so that chemical modifications became the preferred way of perfecting the tRNAs (Ninio, 1979). Many proteins, especially in eukaryotes, have chemical modifications (phosphorylations, acetylations, glycosylations, etc.) that are essential for their activities. Accuracy in translation may be less important in some cases than accuracy of post-translational modification. In Rous Sarcoma virus, as in several other cases now known, the gene responsible for malignant transformation codes for a protein kinase (Collett and Erikson, 1978).

Genes wear a pattern of methyl groups by which they are tagged. Errors in the replication of the methylation pattern provoke subtle transmissible variations. Thanks to these and other submutational mechanisms, the organism can, with an unchanged DNA sequence, adapt its gene expression to a range of environmental variations.

Gene expression in eukaryotes involves an elaborate network of screens and counter-screens, involving proteins that mask or unmask sections of DNA and RNA (in its various stages of processing) and all the proteins that modify the enzymes or modify the masking proteins. Perhaps the cell can in some way monitor the usage of each of its gene products (as suggested by Garel, 1974, in the case of tRNA). If an old enzyme has, by chance, some affinity for a novel substrate, and hence becomes less available for its normal function, the cell might respond by amplifying the gene for the old enzyme, thus providing genetic material for evolving an enzyme acting on the novel substrate (Glogoczowski, 1981). What is true of the immunological system as a whole could be partially true for genes at the cellular level.

13.8 Sketch VIII

The sources of diversity in natural populations (spontaneous mutations, lesions, recombinations, transpositions, intra- and inter-species gene transfer) have such different consequences, and the resulting population dynamics are so complex, that all explanations finish by cancelling one another out. Analytical evolution must be abandoned and replaced by constructive evolution. Mitsuyama's proposal (quoted in Ninio, 1983) to 'put some order back into the inextricable tangle of varieties and species' by favouring, within natural populations, the acquisition of genetic mechanisms (like duplicative transpositions) that drive them towards homogeneity, ought to be taken seriously.

Another approach to constructive evolution is to test hypotheses about evolution on new forms of life made in the laboratory, and which obey simple genetic laws. The first goal is to create cells that would be mirror images of natural cells. Success in the non-enzymatic replication of mirror image DNA (Joyce et al., 1984) encouraged chemists to synthesize the mirror-image of a DNA polymerase (Zhu and Thang, 1985). Unfortunately, the synthesized protein does not adopt the predicted conformation and behaves strangely: it replicates natural DNA much faster than the natural polymerase, and is totally inhibited by mirror image DNA. Another strategy, developed simultaneously in several laboratories, takes advantage of the properties of an analogue of DNA in which the sugar-phosphate backbone is replaced by a polyester chain. When this plastic DNA is introduced in various cells it is initially replicated with the help of a small oligopeptide catalyst which supplements the nutrient medium. It has been observed that plastic DNA, despite its much lower fidelity of replication, gradually takes control of its environment. Within a few years it becomes independent of the external oligopeptide and starts to control the host DNA (Margulis and Crick, 1989). Thus, natural DNA is not the most efficient of all possible DNAs. The same result is obtained with most of the newly created genetic materials, whatever the precise chemistry of their backbones or the choice of the complementary units. Artificial DNA ends up by dominating natural DNA.

In the converse situation, i.e., when plastic-feeding strains of bacteria whose chromosomes are entirely made of plastic DNA are injected with natural DNA and supplemented with appropriate replication catalysts, natural DNA ultimately takes over.

In the first experiments, artificial DNA was new to the natural organisms and it won. In the last experiment, natural DNA was new to plastic bacteria, and it won. This paradoxical result clarifies the whole field of evolution. The rule is that, given enough time, novelty always wins. New variants replace old variants by lassitude. When moths started to bear black wings, at the turn of last century in industrialized areas of Great Britain, the trees turned black just to keep up, and if dinosaurs finally pulled out of terrestrial existence this is just because they went out of fashion. Although these theoretical ideas fit nicely with observations made on planet Earth, we must await further experimental evidence to see whether or not they apply equally well to other inhabited planets.

References

Bachinsky, A. G. (1980) The qualitative estimation of the influence of template processes ambiguity on the population fitness. In *Mathematical Models of Evolutionary Genetics* (in Russian) (ed. V. A. Ratner), Academy of Sciences Press, Novosibirsk, pp. 49–65.

Burnet, F. M. (1974) *Intrinsic Mutagenesis: A Genetic Approach to Ageing*, Medical and Technical Publishing Co., Lancaster.

Clarke, P. H. (1978) Experiments in microbial evolution. In *The Bacteria. A Treatise on Structure and Function. Vol. VI: Bacterial Diversity* (eds. L. N. Ornston and J. R. Sokatch), Academic Press, New York, pp. 137–218.

Collett, M. S. and Erikson, R. L. (1978) Protein kinase activity associated with the avian sarcoma virus *src* gene product. *Proc. Natl Acad. Sci. USA*, **75**, 2021–2024.

Darwin, C. (1859) Extract from an unpublished work on species by C. Darwin, Esq., *J. Linn. Soc. Lond. (Zool.)*, **3**, 46–50. Reprinted in *Evolutionary Genetics* (1977) (ed. D. L. Jameson), Dowden, Hutchinson and Ross, Stroudsburg, Pennsylvania, pp. 22–31.

Dequard-Chablat, M. and Coppin-Raynal, E. (1984) Increase of translational fidelity blocks sporulation in the fungus *Podospora anserina*. *Mol. Gen. Genet.*, **195**, 294–299.

Domingo, E., Sabo, D., Taniguchi, T. and Weissman, C. (1978) Nucleotide sequence heterogeneity of an RNA phage population. *Cell*, **13**, 735–744.

Dover, G. (1982) Molecular drive: a cohesive mode of species evolution. *Nature*, **299**, 111–117.

Gallant, J. A. (1979) Stringent control in *E. coli*. *Ann. Rev. Genet.*, **13**, 393–415.

Garel, J.-P. (1974) Functional adaptation of tRNA population. *J. Theoret. Biol.*, **43**, 211–225.

Gibbs, J. B., Sigal, I. S., Poe, M. and Scolnick, E. M. (1984) Intrinsic GTPase activity distinguishes normal and oncogenic *ras* p21 molecules. *Proc. Natl Acad. Sci. USA*, **81**, 5704–5708.

Glogoczowski, M. (1981) Open letter to biologists. *Fundamenta Scienciae*, **2**, 233–254.

Gould, S. J. (1980) *Ever Since Darwin*, Penguin Books, Harmondsworth and New York.

Grassé, P.-P. (1978) *Biologie Moléculaire, Mutagenèse et Evolution*, Masson, Paris.

Harley, C. B., Pollard, J. W., Chamberlain, J. W., Stanners, C. P. and Goldstein, S. (1980) Protein synthetic errors do not increase during aging of cultured human fibroblasts. *Proc. Natl Acad. Sci. USA*, **77**, 1885–1889.

Helinski, D. R. and Yanofsky, C. (1963) A genetic and biochemical analysis of second site reversion. *J. Biol. Chem.*, **238**, 1043–1048.

Hinshelwood, C. N. (1946) *The Chemical Kinetics of the Bacterial Cell*, Clarendon Press, Oxford.

Hoffman, G. W. (1974) On the origin of the genetic code and the stability of the translation apparatus. *J. Mol. Biol.*, **36**, 349–362.

Joyce, G. F., Visser, G. M., van Boeckel, C. A. A., van Boom, J. H., Orgel, L. E. and van Westrenen, J. (1984). Chiral selection in poly(C)-directed synthesis of oligo (G). *Nature*, **310**, 602–604.

Kimura, M. (1968) Evolutionary rate at the molecular level. *Nature*, **217**, 624–626.

Kirkwood, T. B. L. and Holliday, R. (1975) The stability of the translation apparatus. *J. Mol. Biol.*, **97**, 257–265.

Kruger, K., Grabowski, P. J., Zaug, A. J., Sands, J., Gottschling, D. E. and Cech, T. R. (1982) Self-splicing RNA: autoexcision and autocyclisation of the ribosomal RNA intervening sequence of *Tetrahymena*. *Cell*, **31**, 147–157.

Lerner, I. M. (1954) *Genetic Homeostasis*. John Wiley and Sons, New York.

Margulis, L. and Crick, F. H. C. (1989) Plastic DNA: the ultimate endosymbiont. *J. Artificial Biol.*, **3**, 26–34.

McLachlan, A. D. (1980) Pseudo-symmetric structural elements and the folding of domains. In *Protein Folding* (ed. R. Jaenicke), Elsevier/North-Holland, Amsterdam, pp. 79–96.

Muller, H. J. (1922) Variation due to change in the individual gene. *Am. Nat.*, **56**, 32–50. Reprinted in *Classic Papers in Genetics* (1961) 3rd edn (ed. J. A. Peters), Prentice-Hall, Englewood Cliffs, New Jersey, pp. 104–116.

Ninio, J. (1977) Ageing and the control of accuracy. *Trends in Biochem. Sci.*, **2** (8), N185–N186.

Ninio, J. (1979) *Approches Moléculaires de l'Evolution*, Masson, Paris. English translation. *Molecular Approaches to Evolution*, Pitman, London (1982) and Princeton University Press, Princeton (1983).

Ninio, J. (1983) L'explosion des séquences: les années folles 1980–1990. *Biochemical Systematics and Ecology*, **11**, 305–313.

Okada, H., Negoro, S., Kimura, H. and Nakamura, S. (1983) Evolutionary adaptation of plasmid-encoded enzymes for degrading nylon oligomers. *Nature*, **306**, 203–206.

Orgel, L. E. (1963) The maintenance of the accuracy of protein synthesis and its relevance to ageing. *Proc. Natl Acad. Sci. USA*, **49**, 517–521.

Picard-Bennoun, M. (1982) Does translational ambiguity increase during cell differentiation? *FEBS Lett.*, **149**, 167–170.

Pringle, J. W. S. (1953) The origin of life. In *Symposia of the Society for Experimental Biology. Number VII: Evolution*, The University Press, Cambridge, pp. 1–21.

Rosset, R. and Gorini, L. (1969) A ribosomal ambiguity mutation. *J. Mol. Biol.*, **39**, 95–112.

Sonneborn, T. M. (1965) Degeneracy of the genetic code: extent, nature, and genetic implications. In *Evolving Genes and Proteins* (eds V. Bryson and H. J. Vogel), Academic Press, New York, pp. 377–397.

Wallace, A. R. (1859) On the tendency of varieties to depart indefinitely from the original type. *J. Linn. Soc. Lond. (Zool.)*, **3**, 53–62. Reprinted in *Evolutionary Genetics* (1977) (ed. D. L. Jameson), Dowden, Hutchinson and Ross, Stroudsburg, Pennsylvania, pp. 21–31.

Weill, J.-C. and Reynaud, C. A. (1980) Somatic darwinism *in vivo*. *BioSystems*, **12**, 23–25.

Woese, C. R. (1967) *The Genetic Code*, Harper, New York.

Wright, S. (1955) Classification of the factors of evolution. *Cold Spring Harbor Symp. Quant. Biol.*, **20**, 16–24.

Zhu, X. H. and Thang, M. D. (1985) Studies on the polymerization of D-aminoacids. 48: complete synthesis of the D form of *E. coli* DNA polymerase 1. *Acta Chimica Sinica Shangai*, **22**, 1–327.

Zuckerkandl, E. and Pauling, L. (1962) Molecular disease, evolution, and genetic heterogeneity. In *Horizons in Biochemistry* (eds M. Kasha and B. Pullman), Academic Press, New York, pp. 189–225.

Index